PROGRAMMABLE CONTROLLERS
Operation and Application

PROGRAMMABLE CONTROLLERS
Operation and Application

Ian G. Warnock

Microelectronics Educational Development Centre (MEDC)
Paisley College of Technology, Scotland

PRENTICE HALL

New York   London   Toronto   Sydney   Tokyo   Singapore

First published in 1988 by
Prentice Hall Europe
Campus 400, Maylands Avenue, Hemel Hempstead
Hertfordshire, HP2 7EZ
A division of
Simon & Schuster International Group

Printed and bound in Great Britain by
Redwood Books, Trowbridge, Wiltshire

Library of Congress Cataloging-in-Publication Data

Warnock, Ian G., 1955–
  Programmable controllers.

  Bibliography: p.
  Includes index.
  1. Programmable controllers. I. Title.
TJ223.P76W37 1988   629.8′95   87–7299
ISBN 0–13–730037–9

British Library Cataloguing in Publication Data

Warnock, Ian G.
  Programmable controllers: operation and application.
  1. Programmable controllers
  I. Title
  629.8′95   TJ223.P76

  ISBN 0–13–730037–9

---

10   97 96

ISBN 0-13-730037-9

To Ruth, Laura and Geoffrey

Contents

2 Programmable controllers (PLCs) 46

3 Programming of PLC systems 82

4 Ladder program development 128

Preface

In today's competitive world, a company must be efficient, cost-effective and flexible if it wishes to survive. In the manufacturing and process industries, this has resulted in a greatly increased demand for industrial control systems in order to streamline operations in terms of speed, reliability, versatility and material throughput.

Established control media, including relay, logic and computer systems, can and do provide effective control of industrial processes and plant. However, each of the above control media has limitations or disadvantages that may often be overcome through the use of a *programmable controller* (PC or PLC).

A programmable controller is defined by Capiel (1982) as:

A digitally operating electronic system designed for use in an industrial environment, which uses a programmable memory for the internal storage of instructions for implementing specific functions such as logic, sequencing, timing, counting and arithmetic to control through analog or digital input/output modules, various types of machines or processes.

BACKGROUND

PLCs were first developed during the early 1970s and were used mainly in the motor industry where they replaced large relay panels. They were able to provide the required control while taking up much less space than the relay equivalent and were also much more reliable in operation over long periods. Most importantly, they were extremely flexible in terms of modifying the control sequence if this became necessary. It was now possible to alter a control system without having to disconnect a single wire: it was only a case of changing the resident program, using a small keypad or VDU terminal attached to the programmable controller.

In the last ten years the UK market for programmable controllers has grown from almost nil to over $40 million annually, involving over 30 UK suppliers.

For example, a selection in alphabetical order:

Allan Bradley Co.	Mitsubishi Electric Ltd
General Electric Fanuc	MTE
GEC	Saab
Gould Ltd	SattControl
Klockner Moeller	Siemens Ltd

Square D   Toshiba
Texas Instruments   Westinghouse
Telemechanique

The increased rate of application of programmable controllers within the process industry has encouraged manufacturers to develop whole families of microprocessor-based systems having various levels of performance. The range of available PLCs now extends from small self-contained units with 20 digital input/output points and 500 program steps, up to sophisticated modular systems with a range of add-on function modules for tasks such as *analog input/output and communications*. This modular approach allows the expansion or upgrading of a control system with minimum costs and disturbance.

AIMS OF THIS BOOK

The engineer and manager in industry, and the engineering student, require comprehensive information on the operation and application of all elements of automated manufacturing and plant control systems. This book is intended to meet this need in the area of programmable controllers. Most modern computer-controlled devices may be operated by staff with relatively little knowledge of the underlying technologies that make up the system, and programmable controllers are at the forefront of 'user-friendliness' in this respect. However, it requires a greater insight than this to select the best controller or control system for a particular application and then to design successfully and produce the necessary control programs. The engineer requires an understanding of the operation and capabilities of the different programmable controllers available, together with an appreciation of common programming facilities and their use in various applications.

CONTENT AND STRUCTURE

The content of this text was the outcome of the factors mentioned above. In order to allow the less experienced reader to realistically compare PLCs with other means of control, Chapter 1 gives a general introduction to control systems, what they are used for and the different forms of control required. It then looks at the operation and use of different control technologies – relays, analog and digital electronics, and computers – highlighting the advantages and disadvantages of each in turn. Readers having reasonable knowledge or experience of these topics may wish to move on to Chapter 2, using Chapter 1 for reference as necessary. Chapter 2 describes the development of programmable controllers, leading on to the internal design and operation of the different types. This is followed in Chapter 3 by an examination of fundamental PLC software facilities, including ladder programming and functions such as timers, internal auxiliary relays and data handling. Chapter 4 examines ladder program development for combinational and sequential control tasks, introducing the use of function charts and flowcharts as design aids. This continues in Chapter 5, where specific sequential control facilities are investigated. These include shift registers, stepladder functions and sequencers. Several detailed programming examples are included in Chapters 3–5. Turning to the hardware, Chapter 6 considers several advanced functions that can be obtained on modern programmable controllers, including analog input and output, three-term continuous control and remote I/O units. Another increasingly desirable facility –communications – is the subject of Chapter 7. The need for PLC communications is considered in the context of plant automation, looking at serial links, local area networks and distributed control systems. This includes an examination of GM MAP and its effects on industry. Chapter 8 contains several case-studies of PLC applications, and, finally, Chapter 9 considers the factors involved in the choice, installation and commissioning of a PLC system. Appendices contain typical PLC technical details and sample documentation.

THE USER

Because programmable controllers are by nature applications-oriented rather than for general-purpose development work, it is felt that this book will be of greatest benefit to undergraduate and higher certificate students on industrial-control or control-engineering options, and will be of value to those studying microcomputer control applications and industrial automation. Industrial engineers and managers will benefit from a greater knowledge of the capabilities and role of programmable controllers in advancing automation and distributed control systems. Equally, it can only be advantageous for them to keep abreast with the technology associated with all microprocessor-based equipment.

ACKNOWLEDGMENTS

The author wishes to acknowledge the generous cooperation of several manufacturers, in particular General Electric Fanuc (UK), GEC, Mitsubishi and SattControl. Many examples and illustrations in the text are based on information contained in user manuals and programming guides for their specific controllers (but any mistakes are completely the responsibility of the author, who checked them!). Thanks also go to the Microelectronics Educational Development Centre at Paisley College of Technology, which promoted the author's interest and use of PLCs, as well as providing the opportunity to become involved in various industrial applications and interesting projects.

I.G.W.

## ONE

Introduction to Control Systems

This chapter provides an introduction to control methods, strategies and systems. It examines the important developments and features of relay, electronic and computer control, introducing concepts and terminology to be used in the subsequent chapters on programmable controllers.

1.1 AUTOMATION AND CONTROL

In virtually all forms of industry, the route towards increased productivity is through increased automation of processes and machines. This automation may be required to directly increase output quantities, or to improve product quality and precision. In any form, automation involves replacing some or all human input and effort required to both carry out and control particular operations.

Many factories and plants place the workers in control of machines and equipment, instead of requiring them to physically carry out the task. This control requires the worker to know how a particular process operates, and what inputs are necessary to achieve and maintain the desired output.

To achieve process automation, the operator must be replaced by some form of automatic system that is able to control the process with little or no human intervention. This requires a system that has the ability to start, regulate and stop a process in response to monitored or measured variables within the process, in order to obtain the desired output. A system that possesses all these abilities is called a *control system*.

Automatic control

Any control system can be divided into three constituent sections: input, processing and output. The model in Fig. 1.1 can also be described in terms of *actions*, consisting of input measurements, control processing carried out on these inputs, and the resultant output actions produced. The task of the *processing section* or *control plan* is to produce predetermined responses (in the form of outputs) as a result of information provided by the input signal measurements. There are several different methods available for implementing the processing function, but they all use similar inputs and outputs.

This model also represents control by a human operator acting as the

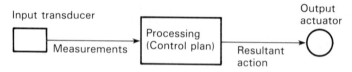

Fig. 1.1   Elements of a control system.

'processing section'. The operator knows what the desired process output is to be, visually monitoring the relevant variables − i.e. inputs (using gages, etc.). In response to these readings, the operator will alter the settings of appropriate controls (valves, heaters, etc.) to obtain the desired process output.

Inputs

Input signals are normally provided by various *transducers* that convert physical quantities into electrical signals. These transducers may be simple push-buttons, switches, thermostats or strain gages, etc. They all transmit information about the quantity that is being measured. Depending on the transducer used, this information may be *discontinuous on/off* (*binary*) or a *continuous* (*analog*) representation of the input quantity.

Table 1.1   Types of input transducer.

Transducer	Measured quantity	Output quantity
Switch	Movement/position	Binary voltage (on/off)
Limit switch	Movement/position	Binary voltage (on/off)
Thermostat	Temperature	Binary voltage
Thermocouple	Temperature	Varying voltage
Thermistor	Temperature	Varying resistance
Strain gage	Pressure/movement	Varying resistance
Photocell	Light	Varying voltage
Proximity cell	Presence of objects	Varying resistance

Outputs

The control system must be able to alter certain key elements or quantities within the process, if it is to exercise control over the way that process performs. This is achieved by using output devices such as pumps, motors, pistons, relays, etc. which convert signals from the control system into other necessary quantities. A motor, for example, converts electrical signals into rotary motion. In other words, output devices are also transducers but in the other direction. As with the input transducers,

Table 1.2   Types of output device.

Output device	Quantity produced	Input
Motor	Rotational motion	Electrical
Pump	Rotational motion plus displacement of product	Electrical
Piston	Linear motion/pressure	Hydraulic/pneumatic
Solenoid	Linear motion/pressure	Electrical
Heater	Heat	Electrical
Valve	Orifice variation	Electrical/hydraulic/pneumatic
Relay	Electrical switching/ Limited physical movement	Electrical

output devices can be simple on/off (binary) units or be continuously variable in operation between fully off and fully on (analog).

Processing section

This corresponds to the operator's knowledge of the operations that are required to keep a process 'in control'. The operator uses this knowledge in conjunction with information obtained from input readings, producing resultant output actions.

From the input information the automatic control system has to produce the necessary output signals in response to the *control plan* built in to the processing section. This control plan can be implemented in two different ways, using either *hard-wired control* or *programmable control*.

Hard-wired systems have the control function fixed permanently when the system elements are connected together (e.g. electrically), whereas in a programmable system the control function is *programmed* (and stored) within a *memory unit* and may be altered by reprogramming if this becomes necessary. Table 1.3 lists examples of hard-wired and programmable systems, together with the type of control they can perform – digital (switched) or analog (continuous).

Table 1.3   Control systems and the form of control provided; (digital or analog).

Hard-wired systems	Type	Programmable systems	Type
Relays	Digital	Computers	Digital/analog
Electronic logic	Digital	Microcomputer	Digital/analog
Pneumatic logic	Digital	P–C systems	Digital/analog
Hydraulic logic			
Analog electronics	Analog		

1.2 DIGITAL (BINARY) AND ANALOG SYSTEMS

In the real world, most naturally varying quantities such as temperature, speed, position etc., change gradually and continuously across an infinite range of values; they are *analog quantities*. Fig. 1.2(a) shows an analog waveform with varying amplitude and frequency. Many sensors generate analog signals that vary either in frequency or amplitude depending on the sensor and measured quantity. Typical signals have amplitude variation of 0–5 V; for sensors with varying frequency output, the range depends on 60th process and signal conditioning.

Many devices produce or respond to digital signals, however, where there are only two possible levels (see Fig. 1.2(b)). The two states may represent:

on	or	off
open	or	closed
yes	or	no
+ 5 V	or	0 V
1	or	0
true	or	false

These two-state signals can be represented using the binary (base 2) numbering system, where one level is assigned the value of 1 and the other the value 0. Thus, in

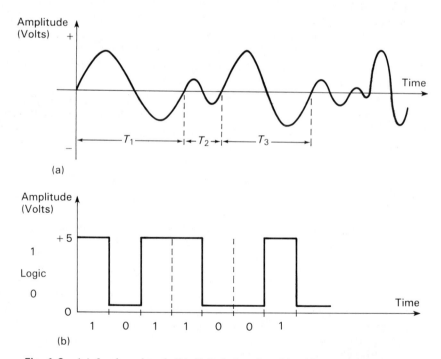

Fig. 1.2   (a) Analog signal; (b) digital signal and its binary representation.

digital (binary) systems, the actual signal level (e.g. voltage) is important only in terms of representing a logic 1 or 0, rather than for its exact (analog) quantity.

Analog (continuous) control

Continuous or *analog* control uses directly representative input signals from sensors, and drives associated output devices (actuators) such as valves, pumps, heaters, etc. These actuators may also be continuously variable, or simply be on/off switched units (Fig. 1.3).

The signal processing that takes place within the control system depends on the individual process involved, but typically involves signal amplification and some form of mathematical function, such as summation or integration, in order to bring about the desired change in an output device.

Continuous controllers include *analog electronic systems, computers* and *microcomputers.*

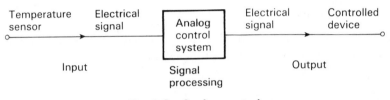

Fig. 1.3 Analog control.

Discontinuous (binary) control

Discontinuous or *on/off* control is commonplace in most industries, since many machines and processes consist of units (that can be in only one of two conditions) controlled by a larger number of simple operations or sequence steps. There are many instances where the input signal is commonly discrete in form, e.g., pulses from a switch, 'bits' of data from a keyboard, etc. In such cases we employ *binary switching techniques.* This should not be regarded as an inferior form of control

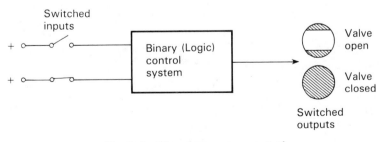

Fig. 1.4 Binary sequence control.

compared to the continuous method – each has its own area of application and is most efficient in that area. The levels of control plan sophistication found in binary systems can often match or surpass that found in continuous control systems.

Binary sequence controllers (Fig. 1.4) include relays, electronic/pneumatic/ hydraulic logic systems, computers and microcomputers, programmable controllers.

Both analog and digital control methods are dealt with in the remainder of this chapter.

1.3   TYPES OF INDUSTRIAL PROCESS

There is a vast number of different manufacturing processes being carried out in industry today. All processes can be grouped into three main categories in relation to the *type of operations* that take place within the process:

1.   continuous production;
2.   batch production;
3.   discrete item production.

Every process has individual characteristics and requirements that must be considered when a control system is to be designed.

Continuous processes

A *continuous process* (Fig. 1.5(a)) takes in raw materials at the input and runs continuously, producing finished material or product at the output. The process

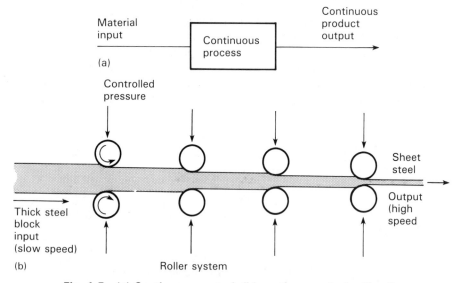

Fig. 1.5   (a) Continuous control; (b) continuous steel rolling line.

may run for long periods of time, typically minutes, hours or even weeks in certain cases.

The production of sheet steel (Fig. 1.5(b)) is an example of a continuous process. This involves large red-hot blocks of steel being fed through a series of rollers, which compress the blocks. This reduces the thickness of steel in stages, finally producing sheet steel at the end of the roller line (output). This process can take several minutes to complete, depending on the lengths of steel fed in.

As the rollers reduce the thickness of the steel plate, the speed of the sheet through the system increases to around 500 m/minute (1600 ft/minute). Control must be exercised over each set of rollers to maintain a constant output sheet thickness. Accurate and rapid control of the first set of rollers is particularly important to allow compensation for input steel blocks of varying thickness.

Batch processes

A batch process uses a set quantity of source (input) materials and performs process operations on this material, producing a specific quantity of finished product (output) or product that will undergo further stages of processing. Fig. 1.6 shows a typical batch processing system.

Here chemicals are pumped into tanks 1 and 2 from reservoirs of polymer and alkali. The polymer is heated to $60\,^{\circ}$C and is then transferred to tank 3 where it is mixed with the alkali for a set period of time. The mixture now passes through other stages in the process as a batch, rather than as a continuous process. This includes passing through a filter into tank 4 as a complete batch of product.

Several different batch processes make up the complete cycle: measuring quantities of materials, heating, stirring and filtering, etc. Each activity requires some form of control to ensure the correct formation of each batch. For example,

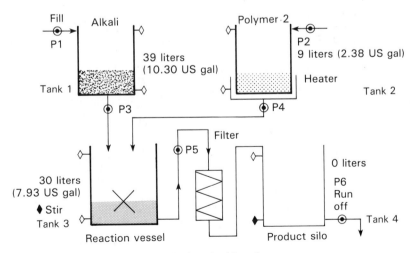

Fig. 1.6   A chemical batch process.

correct volumes of alkali and polymer, correct temperature of polymer when mixed with alkali, etc. The type of control used at each stage depends on the individual process, but typically would be on/off discontinuous control for most of these tasks.

Discrete parts production

In this type of process an individual item undergoes various operations before being produced in a final form. Alternatively, several components may be combined or assembled within the process to emerge as an individual item or unit.

An example of discrete parts production is illustrated in Fig. 1.7. Here metal blanks are loaded onto a drilling table by a robot arm, where they are securely clamped. The blanks are then drilled with a pattern of holes by different drill bits mounted in a rotating 'selector' head. The finished workpiece is then released from the table and placed on an adjacent conveyor belt by the robot arm, and so on.

This example involves several operating sequences: robot picks up blank and places on table; blank clamped to table; blank drilled with a preset pattern by drilling machine, etc. Most of these activities will be controlled on a binary on/off

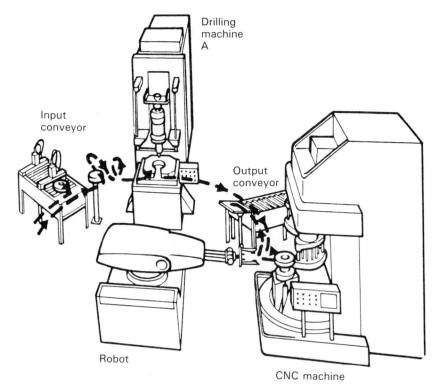

Fig. 1.7 Work cell producing discrete parts.

basis, including linking between each machine to pass information that the previous operation has finished, and the next may begin. This is commonly known as *interlocking*.

1.4   CONTROL STRATEGY

The simplest form of control is known as *open loop* (Fig. 1.8). The basic idea is to set up a system as accurately as possible, producing the desired output by adjusting the system output action. No information is fed back to the controller to determine whether the desired output continues to be achieved; therefore large errors can develop in this type of system. An example of open-loop control is where room A is heated by a radiator unit with no thermostat to open or close the radiator valve. The overall house temperature is set by a thermostat in room B, which switches the main boiler on and off to maintain the temperature in that room, irrespective of the temperature in the other rooms. Room A may be in a particularly cold part of the house, and will never warm up, because room B is already at the desired thermostat setting.

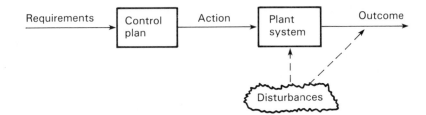

Fig. 1.8   Open-loop control.

Open-loop control

Open-loop control may provide acceptable performance in certain situations, particularly when it is not financially or physically possible to fit a more sophisticated system. For example, a lathe may be set up for a certain speed, feed rate and depth of cut, then left to perform the cutting. Provided there are no unusual disturbances the job will be satisfactory, although in practice the lathe should be monitored. If there is any unforeseen effect or *disturbance* that acts on the process, for example tool wear or nonstandard workpiece hardness, then there is no way of compensating for it — the system may go out of control.

However, if outside influences are known to affect the operation of a system, it is unwise to ignore them. It is often possible to *monitor* the disturbances in a system and use this information to take compensating control action. This strategy is termed *feedforward control* and is illustrated in Fig. 1.9.

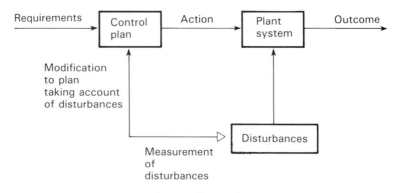

Fig. 1.9   Feedforward control.

Feedforward control

Feedforward is suitable when disturbances are few in number and can be measured accurately, but becomes harder and more expensive to implement when they are numerous. If this is the case it may be necessary to adopt a control strategy that deals with disturbances in a different way (see Fig. 1.10 below). Feedforward is often used where it is not possible to measure the output directly but where it is possible to measure disturbances. An example is a steel extrusion plant, where the very high temperatures involved (around $900°C$) make measurement of steel output virtually impossible. In this case the disturbances in process rollers, etc. are monitored (see section on continuous processes).

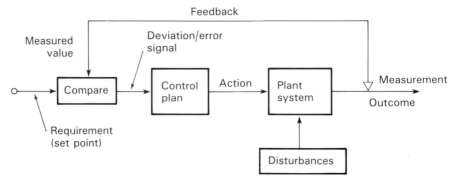

Fig. 1.10   Feedback control.

Closed-loop (feedback) control

Feedback or *closed-loop control* corrects for disturbances by measuring their effect on the system output and then calculating a *correcting action* that counteracts the disturbances and maintains the desired output. Feedback of a fraction of the output signal to the input occurs to allow a comparison between the *actual process condition* and the *desired condition*. The difference between these two signals (the *error*) is fed into the control 'plan' which uses this information to alter the output signal to the required value:

$$\text{i.e.: error signal} = \text{set point} - \text{measured value}$$
$$e = SP - MV$$

this is an extremely important area of continuous control, and many textbooks cover this subject alone. In this text we will deal with feedback closed-loop only as it applies to the various topics discussed.

Control precision

Ideally, any applied control would be exact and precise, with the process instantly responding to changes in the *target input value*. However, in practice, absolute precision is not possible and all processes display a certain amount of acceptable *deviation* between the desired output and the actual output. The smaller the permitted variation in output, then the more complex (and more expensive) the control system becomes.

1.5   CONTINUOUS CONTROL PLANS

In closed-loop feedback systems the control action performed on the error signal may be based on several different *actions* or *terms*.

Proportional action

Proportional action is the simplest form of continuous control, producing a controller output that is directly proportional (by an amplifying factor K) to the error input, for example $K = 5$ (five times greater).

$$\text{output} = K \times \text{error signal}$$

However, proportional action is rarely sufficient in itself because, as the system output approaches the desired set point (SP) the error reduces proportionally, and so reduces the control output. This results in a *steady-state error* or gap between the set point and the measured value (MV) at the output (Fig. 1.11). This steady-state error can be reduced by increasing controller gain, but this in turn can lead to *system instability* and *oscillation*.

To overcome these problems it is common to use proportional action in combination with *derivative* and *integral actions*.

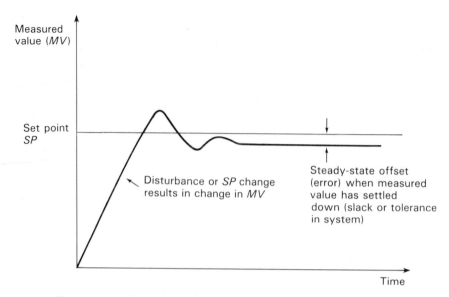

Fig. 1.11 Steady-state error, with the response of a proportional controller to a change in MV.

Derivative action

Derivative action provides an output signal proportional to the rate of change of the error. Thus, if a rapidly increasing error occurs, then a large correcting output will be produced. As the rate of error change slows, so the derivative output decreases. This improves system response to dynamic errors but does not improve steady-state error.

Integral action

Integral action generates an output signal proportional to the mathematical integral of the error, meaning the *summed history of the error*. This is used to overcome steady-state error, since the integral term provides a *matching* error output value; i.e. the output of the integrator will vary as long as the error input is nonzero. When the measured error input reaches zero (steady-state error now occurring – achieved by proportional action alone) then the integral output will *equal* the steady-state error, offsetting it and driving the system into alignment.

PID control

As stated, it is usual to have a combination of proportional, integral and derivative

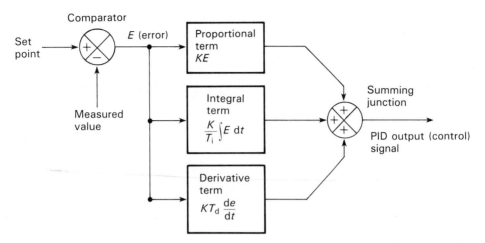

Fig. 1.12   Proportional, integral and derivative (PID) control actions.

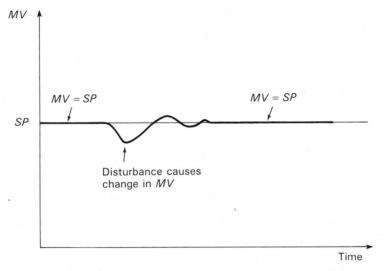

Fig. 1.13   Typical response of a PID controller to a change in measured value.

actions in a controller. This is then called *PID control* and is one of the most widely used control algorithms in industry today (Figs. 1.12, 1.13).

1.6 CONVENTIONAL CONTROL SYSTEMS

This section includes a brief account of the main types of conventional control systems, looking at the advantages and disadvantages of each in turn.

Relays

The relay is an important part of many modern control systems, being an electrical switch with a high current capacity that is indirectly operated by a relatively low control current.

Relays were first developed in the mid-nineteenth century, when they were used to increase the power of weak telegraph signals for transmission over long distances. In the early twentieth century they were rapidly incorporated into control tasks, which had been mechanically based up to that time. This allowed the construction of more sophisticated control systems, which were becoming desirable in many factories and workshops to provide increased output and efficiency.

Relays have since undergone considerable development in terms of operating speed and physical size, to the extent that many modern control panels and equipment still incorporate various types of relay to act as 'interfaces' between the low signal levels (5–12 V) from controllers to high-current devices.

However, all relays contain moving parts and/or electrical contacts (Fig. 1.14) and are therefore limited in terms of operating speed, lifespan and reliability. They are also physically bulky, requiring large mounting racks in most applications, and have a high unit cost per switching function. (A modern semiconductor device termed a 'solid-state relay' with no moving parts is often used for high-speed switching of high-current devices, providing much improved reliability and speed over the traditional relay (see Fig. 1.15). Semiconductor solid-state relays are available which can handle loads up to 400 V, 40 A d.c.).)

A *relay control system* is made up of several tens or hundreds of relays that are energized by the opening and closing of input contacts. The overall control function or plan is determined by the particular way the input and output contacts are connected during the design and assembly of the relay system (Fig. 1.16).

A typical relay system may consist of several hundred or thousand switching contacts, which presents the design engineer with a considerable task. Even for relatively simple control tasks, the number of relays required can result in a large

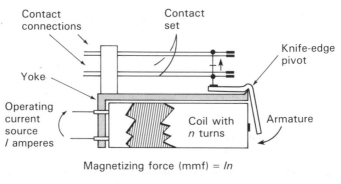

Magnetizing force (mmf) = In

Fig. 1.14 Relay construction.

Fig. 1.15   Dual-in-line (DIL) solid-state relay.

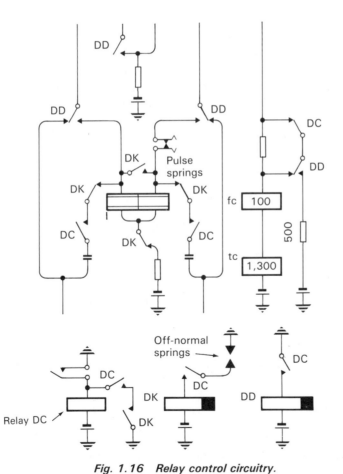

Fig. 1.16   Relay control circuitry.

control panel, since each relay can support only a small number of contacts (e.g. $\leqslant 10$). The function of a relay system is normally described (and designed) on a *relay diagram* which illustrates the interconnection of all electrical contacts and relay coils, together with details of the electrical and mechanical construction of the system. Fig. 1.16 shows contacts controlled by three relays: relay DC with 6 related (DC) contacts: relay DD with 4 contacts, etc.

Relay logic functions

By connecting the input and output contacts in series and/or parallel, any desired logic functions may be produced. For example, when contacts are connected in *series*, the AND function is produced.

The example given in Fig. 1.17(a) shows a simple relay circuit with two normally open contacts A and B which, on closure, will cause relay RA to operate. Relay RA has a contact RA_1 which will operate with the relay, energizing a second relay RB, provided contact C is not operated (opened). Note that (conceptual) *power flow* through the relays will only occur when the intervening switches are positioned to create a continuous circuit.

Combinations of various logic elements may be used to create fairly complex control plans, although for even a simple task the number of relays required can result in a large control panel. It is *also extremely difficult to change the control*

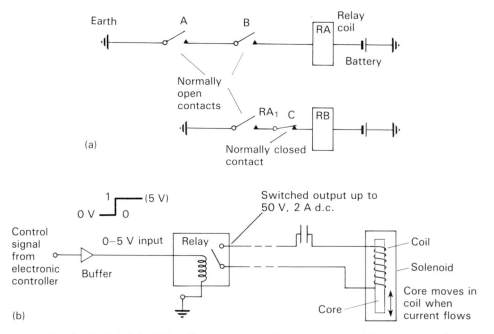

Fig. 1.17 *(a) Relay RA will operate only if contacts A and B are operated (closed); (b) relay used as an output device.*

function of a panel once it has been wired up, and is likely to involve a complete re-wiring of the system. Together with the other disadvantages of cost, speed and reliability described earlier in this section, the above points have led to the replacement of relay control systems by modern alternatives based on electronics and microprocessors.

Relays continue to be used extensively as *output devices* (*actuators*) on other types of control system, being ideal for the conversion of small control signals to higher-current/higher-voltage driving signals. An example of using a relay to drive a 50 V, 2 A solenoid is shown in Fig. 1.17(b).

1.7 ELECTRONIC SYSTEMS

Transistors

Germanium transistors began to be used as switching elements in the mid-1950s. Having a very small physical size and no moving parts, semiconductors offered greater reliability and faster switching at a lower cost than the relay equivalent, although their power-handling capability was very much less.

The control applications of transistors were not confined to logic switching (Fig. 1.18), with the two states 1 and 0 represented by the transistor being 'in saturation' (fully conducting) or 'cut off' (non-conducting). Also they possess signal-amplifying characteristics that proved ideal for *analog* (*continuous*) *control* (Fig. 1.19). Transistor circuits were designed to perform a variety of signal conditioning in order to generate a required controller output, including most mathematical operations.

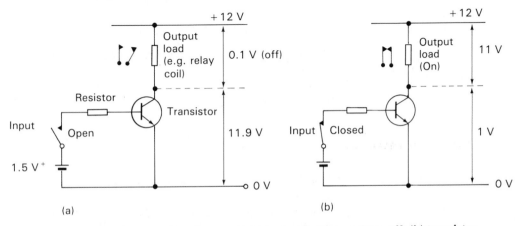

Fig. 1.18 *Basic transistor switching circuit: (a) transistor off; (b) transistor biased on.*

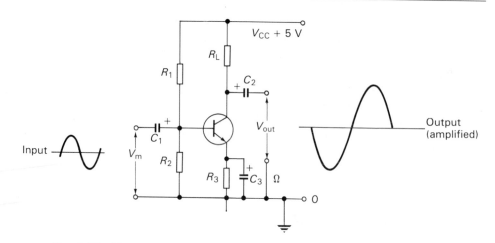

Fig. 1.19 Transistor amplifier circuit with biasing components that determine the operating range and provide protection against overdriving and instability.

Integrated circuits (ICs)

As new improved devices became available through developments in semiconductor technology, they were quickly adopted in control situations. Until the late 1960s all electronic circuits were constructed using *discrete* (*individual*) components. Integrated circuits were then introduced, having a large number of components 'integrated' onto a single chip of silicon. This caused a major revolution across the whole electronics field, since sophisticated and complex systems could now be implemented simply by the appropriate interconnection of suitable ICs – the designer no longer had to consider the *details* of every transistor stage in a circuit, but now dealt with *building blocks* that were selected from a range of functional chips.

Integrated circuits were developed in two distinct fields:

1. *Linear ICs*, which handle analog signals, carrying information in terms of amplitude and waveshape.
2. *Digital ICs*, which deal exclusively with binary signals, processing this information through various logic 'gates'.

Linear integrated circuits

One of the more important and useful linear ICs developed was the *operational amplifier* (op-amp - see Fig. 1.20).

The term 'op-amp' is now generally used to describe a *high-gain voltage amplifier* within a single chip. The name comes from the use of these amplifiers in analog computing operations, which involve the performance of mathematical operations such as integration, differentiation, etc.

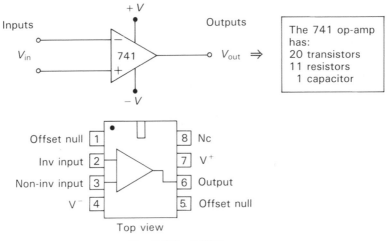

Fig. 1.20   A 741 op-amp.

Electronic control

Op-amps were quickly adopted into the field of continuous control and provided a much simplified solution to complex control functions compared with existing discrete electronic systems. Analog control is now heavily based on linear integrated circuits (Figs. 1.21, 1.22), and remains *the fastest form of control available.*

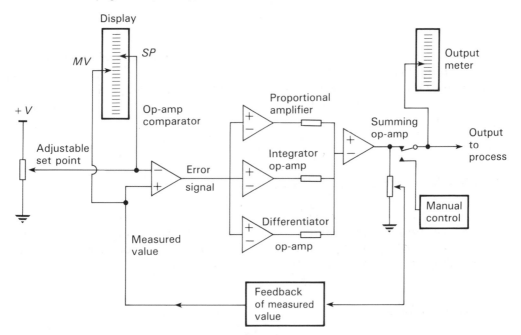

Fig. 1.21   Schematic of a linear control system using linear integrated circuits.

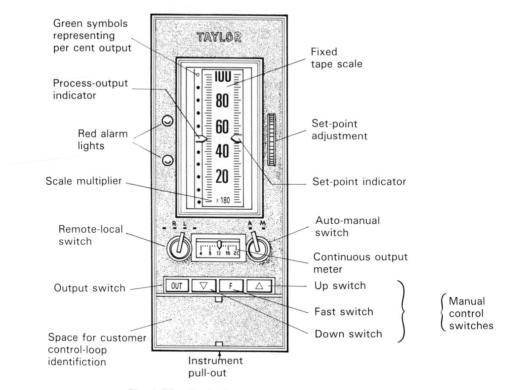

Fig. 1.22   Typical control panel for a linear controller.

However, the 'fine tuning' of feedback systems during design and commissioning remains a difficult task. This, coupled with the fixed nature of electronic circuit construction, results in a control medium that cannot easily have its function changed – the complete electronic system may have to be replaced if this proves necessary.

1.8   DIGITAL INTEGRATED CIRCUITS

The first digital ICs produced had only a few logic gates on each chip of silicon, with each gate consisting of several diodes or transistors plus associated resistors and capacitors. Logic gates operate at much higher speeds and consume considerably less power than an equivalent relay circuit.

Logic families

Several different families of logic circuits have been developed, each having different

Table 1.4 Logic families.

Logic family	Gate propagation time (nanoseconds)	Power per gate (milliwatts)
Reed Relay (for comparison)	2000	200
TTL transistor-transistor logic	10	10
ECL emitter coupled logic	3	60
pMOS p-type MOS (Metal Oxide Semiconductor)	100	1.5
CMOS complementary MOS	40	0.05 (50 nW)

The values given for TTL apply to standard TTL. There are several other forms of TTL with different characteristics.

operational characteristics and applications. It is worthwhile comparing their main features as listed in Table 1.4.

TTL (transistor-transistor logic) and CMOS (complementary MOS) are the most popular logic families currently in use, with several IC manufacturers offering an enormous range of functions in both families. Whilst TTL and CMOS families form the basis for the majority of digital designs, ECL (emitter-coupled logic) is used for applications requiring the highest possible operating speed, and MOS logic is used extensively for large scale integration (LSI).

In general, TTL is fast in operation but consumes quite large amounts of power (other forms of TTL are available that trade off some speed for lower power consumption). For low-power applications CMOS logic is invariably chosen. Logic circuit systems can be created by selecting digital ICs from the range available, using the ICs as building blocks to make up the required functions.

Logic symbols

When designing any system with ICs and logic gates, a logic diagram is made up using *logic symbols*, which indicate the particular function of each gate, rather than its detailed circuitry. Different sets of standard circuit symbols have been developed in both Britain and the US, those of US origin being the most widely used. The symbols show only the binary inputs and outputs, omitting power-supply connections for clarity. Once the diagram is complete, the designer determines the type and number of integrated circuits required to implement this design. Finally a circuit board is prepared with the necessary hardwiring between the IC sockets and additional components (Fig. 1.23).

Boolean algebra

This is an area of mathematics that is particularly suited to the analysis and design of

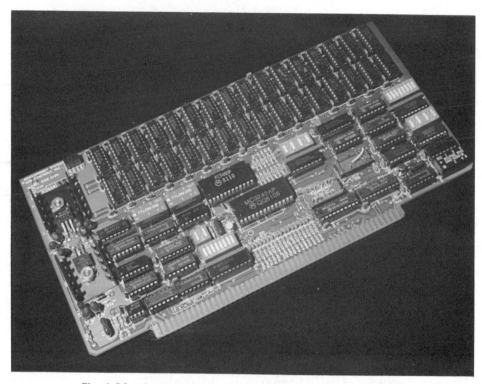

Fig. 1.23 A printed circuit board (PCB) and its components.

switching circuits. Its use enables the hardware engineer to design and interconnect logic circuits in the most efficient and economic manner. *Boolean algebra deals purely with binary variables*, i.e. two states only.

The logic operations given in Table 1.5 are Boolean descriptions of gate operation: for example,

$$X = A.B$$

For more complex circuits, Boolean algebra can be used to calculate the overall logic function, for example, of the circuits shown in Fig. 1.24. Appendix B contains further rules and examples of Boolean algebra and logic switching circuits.

Scales of integration – SSI, MSI, LSI, VSLI

In the early days of integration only a few logic gates could be fabricated onto one chip, using the various logic families listed above. As technology has improved so the number of gates which could be integrated onto a single chip greatly increased. This has resulted in approximate descriptive terms being used to categorize *levels of*

Table 1.5 **Basic gates and symbols.**

Gate	Symbol	Operation	Truth table		
			Inputs		Outputs
			A	B	X
NOT	$A \;\triangleright\!\!\!\circ\; \overline{A}$	OUTPUT $X = \text{NOT } A$ $= \overline{A}$	0	—	1
			1	—	0
AND	$\begin{matrix} A \\ B \end{matrix} \;\sqsupset\; A \cdot B$	$X = A \text{ AND } B$ $= A \cdot B$	0	0	0
			0	1	0
			1	0	0
			1	1	1
OR	$\begin{matrix} A \\ B \end{matrix} \;\supset\; A + B$	$X = A \text{ OR } B$ $= A + B$	0	0	0
			0	1	1
			1	0	1
			1	1	1
NAND	$\begin{matrix} A \\ B \end{matrix} \;\sqsupset\!\!\circ\; \overline{A \cdot B}$	$X = \text{NOT } (A \text{ AND } B)$ $= \overline{A \cdot B}$ $= A \text{ NAND } B$	0	0	1
			0	1	1
			1	0	1
			1	1	0
NOR	$\begin{matrix} A \\ B \end{matrix} \;\supset\!\!\circ\; \overline{A + B}$	$X = \text{NOT } (A \text{ OR } B)$ $= \overline{A + B}$ $= A \text{ NOR } B$	0	0	1
			0	1	0
			1	0	0
			1	1	0
EXCLUSIVE-OR	$\begin{matrix} A \\ B \end{matrix} \;\supset\supset\; A \otimes B$	$X = A \text{ AND NOT } B \text{ OR}$ $B \text{ AND NOT } A$ $= A\overline{B} + \overline{A}B$	0	0	0
			0	1	1
			1	0	1
			1	1	0

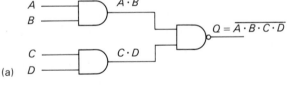

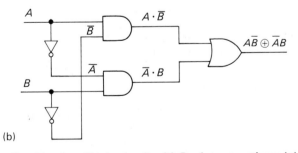

Fig. 1.24 **Combinational logic circuit with Boolean equations: (a) four-input NAND gate; (b) construction of an EXCLUSIVE-OR gate.**

integration, as follows:

SSI   small-scale integration   − less than 10 gates per IC
MSI   medium-scale integration   − 10 to 100 gates per IC
LSI   large-scale integration   − 100 to 1000 gates per IC
VLSI very large-scale integration − over 1000 gates per IC

Most logic families have SSI and MSI chips containing *standard functions* 'available off the shelf'. The LSI and VLSI chips, however, are more often specialist chips containing massive amounts of logic, structured to produce complex functions. Examples of each level are given below.

SSI   basic gates, e.g. quad two-input NOR, NAND, OR, etc.
MSI   bistables, registers, counters, etc.
LSI   arithmetic/logic units, memory chips, microprocessors
VLSI microprocessors, single-chip microcomputers, high-capacity memory
        devices, etc.

The following sections briefly examine several aspects of these integrated circuit functions, since *MSI and LSI circuits play a vital role in programmable control systems such as microcomputers and programmable controllers*. The topics of microcomputers and memory are discussed again in Section 1.9 on computer control.

Sequential systems − the basis of computer operation

The examples discussed so far have been based upon *combinational logic*, where the outputs at a given time are dependent *only* on the values of the inputs at that time.

Many digital systems have outputs which are dependent upon the *previous state of the system* − these are known as sequential systems. Such a system requires some form of storage or *memory element* in order to hold information from a previous operation, as in Fig. 1.25.

A basic storage element is required to hold the last output produced. This form

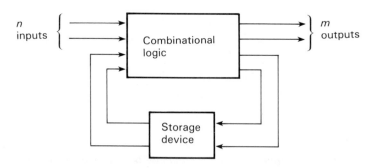

Fig. 1.25 A sequential circuit with memory.

of circuit is known as a *bistable* or *flip-flop*, and provides the basic building block in data storage circuitry.

Storage of data

This basic storage device – the bistable – can store only one *bit* (binary digit) *of data at a time*. Large digital systems and computers require to store information *in groups of bits or words*; therefore a number of bistables are required to store a single word. Such an arrangement of bistables is called a *register* and to store an 8-bit word the register must consist of 8 bistables.

Semiconductor memory

A digital system or computer may be required to store a large number of words in memory. Memory effectively consists of an *array of registers with individual addresses* (Fig. 1.26), allowing each word location to be exclusively accessed (i.e. read from or written to) by the system. The size or width of registers in the array should ideally be the same as the system *data bus* (8 or 16 bits wide), which transfers data words to and from the memory. An addressable register array forms the basis for the memory used in programmable control systems such as computers or microprocessors.

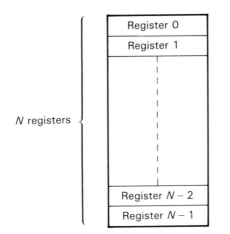

Fig. 1.26   Memory as an addressable array of locations.

RAM and ROM

Semiconductor memories are divided into two main groups:

(a)   Those which lose their contents when power is turned off: *volatile memories*, including RAM devices.

(b)   Those that retain their contents when power is removed: *non-volatile memories*, including ROM devices.

(a)   Volatile memory

Random access memory (RAM) is the name given to read/write memory, which allows individual signals or data words to be written in or read out when correct control signals are present (Fig. 1.27). RAM devices may have 'static' or 'dynamic' characteristics of data retention:

> *Static RAM cells* can be directly compared to the bistable or 'flip-flop' circuits examined earlier, in that when a cell or register group of cells is set to 1 or 0 it retains that status until it is reset, or the power is removed.
> *Dynamic RAM (DRAM)* effectively consists of capacitors which must be kept at a constant level of charge if they are to retain stored information. For this purpose we must supply *refresh facilities*, which will regularly 'top-up' the charge on each cell. DRAM cells are considerably smaller than the equivalent static RAM cell, so where large memory capacities are required the chip will normally be based on dynamic RAM.

Using RAM, every word location can be written to, read or erased as often as required, but because it is volatile the entire contents will be lost if power is turned off. *If it is desired to retain stored data*, a back-up battery must be fitted to the RAM – often CMOS RAM is chosen for its very low power consumption. It is also common to design the circuit to permit re-charging of a Ni–Cd secondary cell as the RAM battery back-up.

(b)   Non-volatile memory

This group does not lose data when power is removed. We can subdivide the group into erasable and non-erasable memories.

(i)   *Non-erasable memories* consist of:
Read only memory (ROM) which is permanently programmed at manufacture

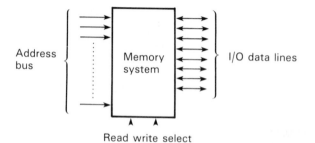

Address bus   Memory system   I/O data lines

Read write select

Fig. 1.27   *RAM circuit, showing data and address lines – bus connections.*

and cannot be subsequently altered. ROMs are used for the production of large batches of identically programmed devices.

Programmable ROMs (PROMs) which can be programmed by the user (via a PROM programmer). This involves the breaking of 'weak links' in the memory chip to form the desired circuit, and is not a reversible process. (Similar to a *programmable logic array (PLA)*.)

    (ii)   *Erasable memories* consist of:

Erasable PROMs (EPROMs) which are programmed in a similar way to PROMs, but can have their contents erased by exposure to ultraviolet light for approximately 30 minutes. (they have a transparent quartz window over the actual chip – see Fig. 1.29). EPROMs can then be re-programmed, and this procedure may be carried out several times.

Electrically Erasable PROMs (EEPROMs) are similar to EPROMs but can be erased electrically while connected in the circuit.

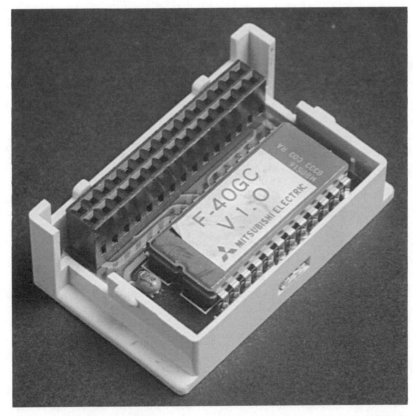

Fig. 1.28   Plug-in EPROM cartridge (courtesy Mitsubishi Electric UK Ltd).

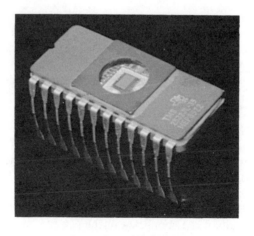

Fig. 1.29   EPROM ICs.

Memory devices may be found in both medium- and large-scale integration, depending on the amount of storage provided on the chip. For example,

MSI RAM: 4-bit by 16-word read/write memory
VLSI RAM: 8-bit by 64k (thousand) words.

(Memory systems are investigated further in a later section on computers.)

Progammable logic arrays (PLAs) are very useful LSI devices if there is a need for a large array of combinational logic elements. PLAs contain gates that can be interconnected in order to form the desired circuit function. PLAs are made which can be programmed at manufacture (and are then called mask-programmable) to a user's requirements, together with those which can be programmed in the field by fusing internal wire links which connect the logic gates internally. An example is the fusible-link 82S100 PLA which can produce eight outputs from a programmed logic combination of up to sixteen input lines.

Processing of data

When it is necessary to perform arithmetic operations on binary data, an *adder circuit* is used. This is an example of medium-scale integration (MSI) and is a frequently used functional block. The circuit can add two 4-bit numbers to generate a 4-bit sum plus a carry bit. It is a simple matter to expand the circuit to add larger numbers by employing additional adder chips.

A device known as an *arithmetic logic unit* (ALU) can also be used as an adder, but has the capability of performing additional functions such as subtraction, number comparison and shifting. Typically an ALU carries out these functions in tens of nanoseconds.

Microprocessors

The single most important development in LSI was the creation of the micro-processor – *the central processing unit of a computer on a single chip*. This signaled the start of the 'microcomputer era', with microprocessors and associated LSI *support chips* (such as memory and input/output devices) being connected together to create microcomputer systems. These devices are discussed in more detail in the next section.

Fig. 1.30   Microprocessor IC package.

Control applications of digital systems

The functions available in digital electronic logic families allow the design and construction of hardwired systems capable of very high-speed control in sequence and continuous control situations (through the use of analog-to-digital converters).

If, however, the control plan requires amending or renewing, the inflexible nature of a hardwired digital electronic solution will mean total replacement of the physical control hardware, together with associated time lags for redesign and construction of the 'new' solution.

1.9   COMPUTER CONTROL

Digital computers are electronic machines that process information in binary form. They are ideally suited to numeric calculations and storing large quantities of information.

Computer operation

The general structure (or architecture) of all computer systems is similar to that shown in Fig. 1.31, consisting of a central processing unit, memory units and input/output units, all connected together by 'bus lines' that carry information between the various devices.

The central processing unit (CPU) supervises and controls all operations within the computer, by carrying out instructions that are part of the computer program stored in the memory area. Memory is also used to store temporary data during the execution of a program. The computer system is linked to the outside world by means of the input/output units, to which transducers and actuators are connected for control applications. The input units are used to pass process data to the central processor, together with any operator data that may be required. This data is operated upon by the central processor, using the stored program of instructions which defines the operations to be carried out. This produces resulting data, which may be for storage in memory, or for direct output to devices connected to the output unit.

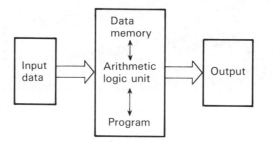

Fig. 1.31   Computer architecture.

Computers in control – history

Computers have been used in process control since the mid-1950s, when attempts were made to take advantage of post-war electronics development. Computers performed operations such as data logging and alarm recording in addition to various control strategies where the designers could implement complex functions that might (in theory) be quickly and easily altered.

Several of these applications were in the chemical industry, resulting in successful computer control of processes such as ammonia, vinyl chloride, and rubber production. (Skrokov, 1980)

At that time *computer hardware was very expensive*, *slow in speed* (because of the magnetic drum-type memory used), and *physically massive* (in comparison to later microelectronic systems) since all logic circuitry was based on thermionic valves (vacuum tubes).

Unfortunately, from then until the late 1960s many attempted applications of computer control were unsuccessful, for several reasons:

1. Computer company staff rarely had any understanding of the process to be controlled, and most process engineers had little or no computer experience.
2. Computer manufacturers tried to make the industrial application 'fit' the available hardware (usually based on a single mainframe computer located in a remote control room and requiring long cabling to connect up input/output devices).
3. The installation and maintenance of these systems was usually very difficult and time-consuming.

These factors often led to inadequate (or even incorrect) specification of the computer system required, with resulting deficiencies in the physical implementation and performance of the control system. These failures and the resulting publicity created further barriers against computer control. Most large computer manufacturers dropped out of the control market to concentrate on business data-processing applications instead.

As a result, there were very few applications of computer control until the mid-1970s, when developments in electronics resulted in smaller, faster and less expensive computers, making them far more attractive to industry. Also, computer languages more suitable for control applications were becoming available. The disadvantages of computers remained their high cost (although this rapidly dropped) and the need for highly trained pesonnel to program them, together with relatively slow program execution compared to hardwired logic or analog electronics.

Today, powerful low-cost micro- and minicomputers are available, and are often used in digital and analog control systems. However, although component costs have fallen, the need for highly trained programming staff remains.

Microprocessors

The development of microprocessors acted as a further catalyst in the growth of computer control, bringing down the size and cost of hardware dramatically, whilst providing the processing power of previous minicomputers.

As shown in Fig. 1.32, the microprocessor replaces the computer system CPU, containing the logic necessary to recognize and execute the program instructions, with ROM or EPROM memory containing the operating software and fixed programs. RAM integrated circuits provide read/write memory for temporary data storage, and standard input/output ICs form the connection to the outside world.

For completion, this system requires only a power supply, an oscillator (or clock) to synchronize internal operations, plus a few logic gates to interconnect the ICs via the bus lines.

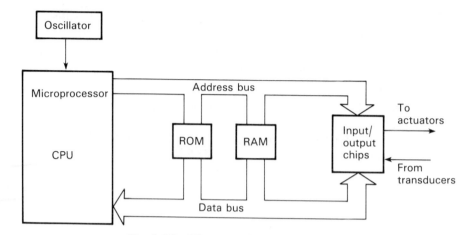

Fig. 1.32 Microcomputer architecture.

State of the art

Having moved through several stages of development that produced many 8-bit microprocessors (which handle data in 8-bit words), such as the Zilog Z80, we now have powerful VLSI 'number-crunchers' like the Motorola 68000 family and Intel's 80286/80386 – microprocessors having 32-bit internal operation (32-bit data words) and a 16-bit external data bus (or also having a 32-bit external bus on the 68020 and Intel 80386 microprocessors), plus the ability to address up to 8 megabytes (8 million bytes) or more of memory (Fig. 1.33.) The ability to address larger memory allows the storage of greater quantities of data, including much larger and more sophisticated programs, whilst longer data-word size allows the processing of larger numeric values, resulting in much improved data throughput. Many other advanced features are now supported, such as instruction look-ahead and hardware multiplication/division.

Also, there are VLSI *single-chip microcomputers* (SCMs), which have the CPU plus all the necessary memory and input/output functions implemented on a single chip, reducing the physical size of a microprocessor-based system to the minimum. Fig. 1.34 gives details of the Intel 8048 range of SCMPs. Single-chip microcomputers are aimed at dedicated stand-alone applications such as controllers of mass-produced items of plant and equipment. For example, domestic washing machines, $X-Y$ plotter tables, etc.

The Intel 8048 is a typical example of a single-chip micro, possessing 1 K of ROM and 64 bytes of RAM together with three input/output ports. There is a family of products in this range, each designed for a particular area of use. These include SCMPs that use only the internal memory, requiring it to be programmed by the manufacturer, and one device which can be fitted with 'piggy back' ROM or EPROM memory for development use, allowing the testing of applications programs, or for small-quantity applications which would not justify having several hundred ICs programmed by the manufacturer.

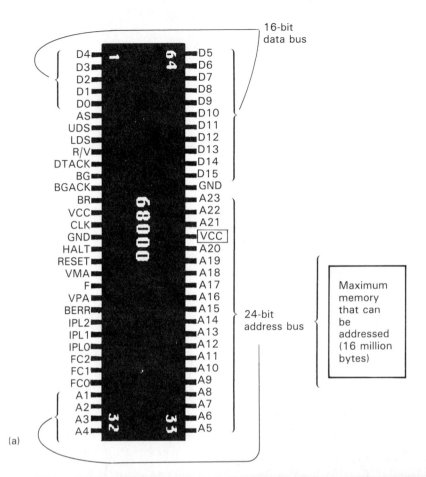

16-bit
data bus

D4	D5
D3	D6
D2	D7
D1	D8
D0	D9
AS	D10
UDS	D11
LDS	D12
R/V	D13
DTACK	D14
BG	D15
BGACK	GND
BR	A23
VCC	A22
CLK	A21
GND	VCC
HALT	A20
RESET	A19
VMA	A18
F	A17
VPA	A16
BERR	A15
IPL2	A14
IPL1	A13
IPL0	A12
FC2	A11
FC1	A10
FC0	A9
A1	A8
A2	A7
A3	A6
A4	A5

68000

1   64
32   33

24-bit
address bus

Maximum memory that can be addressed (16 million bytes)

(a)

(b)

Fig. 1.33 (a) 68000 microprocessor; (b) surface-mounted Intel 80286 package (centre).

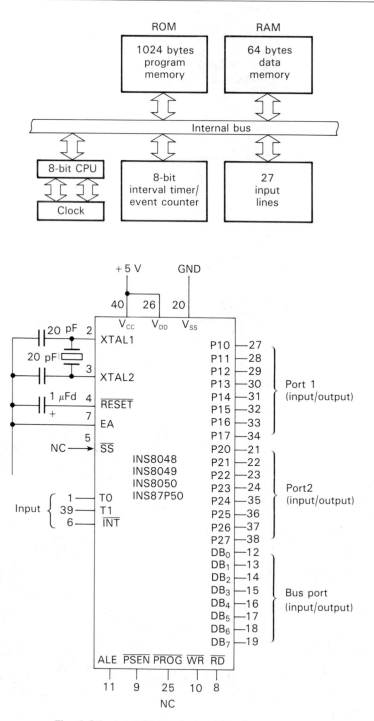

Fig. 1.34   Intel 8048 single-chip microcomputer.

Single-chip microcomputers are widely used in control applications, usually when a large number of similar units are required. Although programmable, they require similar program-development facilities to conventional microprocessors. This constitutes a significant disadvantage against their use in control situations where program alterations may be necessary.

Input/output

Data within most microprocessor systems is transferred from point to point in parallel form, i.e. using one line for each bit. Whilst this allows very fast data transfer, it is not always economical to transmit data in this form, particularly over long distances. This is because special line drivers and receivers have to be used, together with multicore cable.

Data may be converted from parallel to serial form by using appropriate communications ICs, allowing a single pair of wires to be used for transmission. The receiving computer must clock in the incoming data in such a way that the transmitted data is reconstructed (Fig. 1.35(a) and (b)).

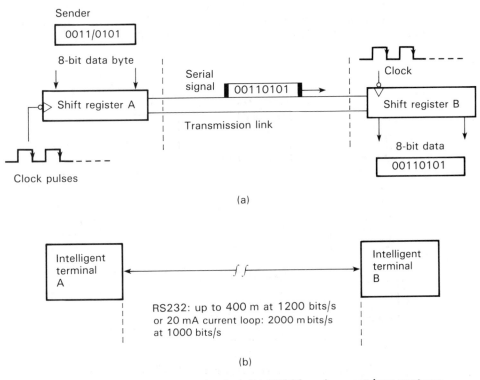

Fig. 1.35 (a) Serial communication: (b) RS232 and current-loop systems.

Communications

A topic related to both computers and their use in larger control systems is the communication of data between devices (Fig. 1.35). This may be simple *one-way transmission*, for example sending logged data to a printer or display screen (VDU) over a *standard serial link such as RS232*, where the binary words (bytes) are transmitted one bit after the other. RS232 is simply an agreed standard of physical and electrical connections plus data formats.

Long distance *two-way transmission* is often required between intelligent devices or computers, and may be provided using RS232 or a current loop. The 20 mA current loop is used in many industrial systems, employing a switched 20 mA circuit to generate the serial data.

When it becomes desirable or necessary to interconnect a large number of terminals, a *communications network* is used. A *local area network* (LAN – Fig. 1.36) provides a physical link between all devices plus providing overall data *exchange management* or protocol, ensuring that each terminal can 'talk to' other machines and understand data received from them.

In industry, with increasing automation of plants and processes to raise efficiency and output, the need for networked communications has been recognized. This is necessary to allow communications and control on a plant-wide basis, interconnecting microcomputers, minicomputers, robots, CNC machines and programmable controllers used to form the production plant.

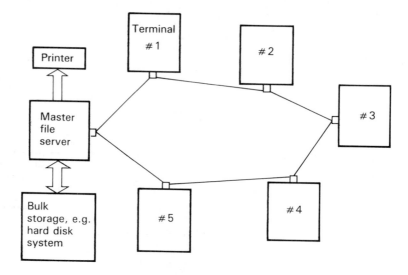

Fig. 1.36   Local area network – a ring system.

Programs

The function of any computer is totally dependent on the program stored in memory. This can be thought of as a long list of instructions that have to be carried out consecutively by the CPU. For control applications, it is likely the program will perform the following tasks:

1. read information from input devices;
2. carry out a logical or arithmetic operation on the input data;
3. output control signals as a result of (2).

Programming languages

Computers work in binary codes – combinations of 1s and 0s arranged in groups that form *words* or *bytes*. This is known as *machine code*, and is tedious for people to deal with, so other forms of representing data and instructions are desirable.

Assembly language permits a set of mnemonic codes to be used to represent computer instructions. Mnemonics are 'memory aiding' in that they help you to remember the operation they represent (Fig. 1.37)

Assembly language covers all the instructions which a given machine can execute, and any assembly program has to be converted into machine code before the computer can execute it – this is done by an *assembler*. When designing an assembly program to carry out control functions, a detailed knowledge of the computer and its operation is essential.

High-level languages are available which provide commands and statements very close to the actual functions that the programmer wants done, making the task of programming easier and faster. These languages are designed to remove the need

Address	Code	Labels	Instructions		Comments
	00006000	DATA	EQU	$6000	
	00004000	PROGRAM	EQU	$4000	
	00006000		ORG	DATA	
006000	00000002	VALUE1	DS.W	1	FIRST VALUE
006002	00000002	VALUE2	DS.W	1	SECOND VALUE
006004	00000002	RESULT	DS.W	1	16 BIT STORAGE FOR ADDITION RESULT
	00004000		ORG	PROGRAM	
004000	207C00006000		MOVEA.L	#VALUE1,A0	INITIALIZE A0 WITH ADDRESS OF VALUE
004006	3010		MOVE.W	(A0),D0	GET FIRST VALUE IN D0
004008	D1FC00000002		ADDA.L	#2,A0	INCREMENT ADDRESS REGISTER A0 BY 2
00400E	D050		ADD.W	(A0),D0	ADD SECOND VALUE TO FIRST VALUE
004010	D1FC00000002		ADDA.L	#2,A0	INCREMENT A0 BY 2 AGAIN
004016	3080		MOVE.W	D0,(A0)	STORE RESULT OF ADDITION
004018	4E75		RTS		
			END		

Fig. 1.37 *Assembly listing for a 68000 microprocessor.*

```
program add(input,output);

var sum, next :real;

begin
  sum := 0;

  read(next);
  repeat
    sum := sum + next;
    read(next)
  until next < 0;

  writeln('sum is ', sum)
end.
```

Fig. 1.38   Pascal example.

for users to have a detailed knowledge of machine code and computer operation. Often the language is *independent of the computer* on which it is going to run, meaning the program written in that language may be *transportable* to a different type of computer – a very desirable feature.

Many different languages exist for various applications, with BASIC, C, FORTH, FORTRAN and PASCAL (Fig. 1.38) being used for scientific purposes including control. Most high-level languages produce considerably more final object (machine) code than a hand-produced assembly equivalent, but this is less important today due to the low cost of large-capacity memory devices. For real-time control applications, the execution speed of a control program may be a critical factor, and this must be taken into consideration when initially selecting a computer and language. A rule of thumb guide to the C language is that a 'good' program can be written in a quarter the time taken to do the same job in assembler, and the C program will run at approximately half the speed, i.e. 50 per cent slower. This illustrates the production/execution tradeoff.

Efficient use of these languages still requires highly trained and experienced personnel.

Process-control languages

In order to reduce this *unfriendliness*, several manufacturers of process-control systems have developed high-level *process-control languages* based on a 'conversational style' of input. These are specifically designed to allow non-programming engineers to build the operating structure of their computer control system, using a question-and-answer form of presentation. Here the user can input the required strategies as answers to questions on the type of control desired, together with relevant parameters. The user may also build up *process mimic diagrams* on a VDU screen for use by operators.

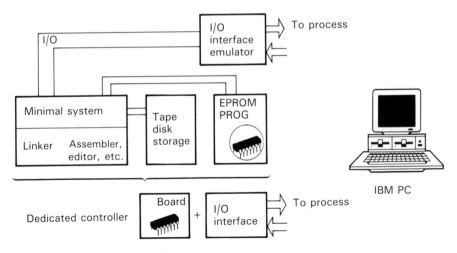

Fig. 1.39   MDS and IBM PC.

Examples of process-control languages (PCLs) include: PROTEUS (West-inghouse); BICEPS (Honeywell); PCL (Hitachi).

Unfortunately most PCLs are implemented only on mainframes and minicom-puters and are not available for microprocessor systems.

Microcomputer software development is normally carried out using a develop-ment system (MDS) consisting of a powerful micro-/minicomputer, disk system, PROM programing facility and associated software (usually editor, assembler or compiler, etc.). An in-circuit emulator may also be employed to drive the target micro directly from the MDS (Fig. 1.39). This helps prove software operation before actual installation in the microcontroller. Such systems are normally very expensive.

This is all necessary for the production of object code, to be blown into EPROM for running on the target microcontroller. (Software for real-time control is unlikely to be disk-based due to the access time delays and vulnerable nature of disks.) The use of this equipment requires highly trained staff, adding to the cost. It should be noted that if software modifications are necessary after installation, the development system must again be used to alter and test the program.

An alternative chosen by several firms uses a personal computer (with suitable software) as a low-cost development system.

Computers in control

Today, powerful low-cost micro- and mini-computers are available, and are often used in both sequence and continuous control systems. Microprocessor-based control panels are small enough to locate at (or near) the point of final control, simplifying connection requirements.

In large processes it is now common for several microcontrollers to be used

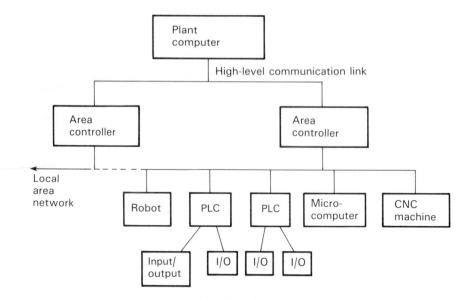

Fig. 1.40   Distributed control structure.

instead of a single large mainframe control computer, with resulting benefits in performance, cost and reliability. Each micro can provide optimal local control, as well as being able to send or receive control data via other microcontrollers or a *host supervisory computer* (mini or micro). This is termed *distributed control* (Fig. 1.40) and allows far greater sophistication of control than was possible with a centralized strategy using a single large computer, since the control function is divided between several dedicated processors.

A distributed control hierarchy need not consist exclusively of mini- and microcomputers, but can include other intelligent devices such as CNC machines, robots and programmable controllers.

Sequence control

Microcomputers can deal easily then with switched input/output requirements of sequence control, providing suitable signal levels are available. The application of any microcomputer to a 'physical task' involves a certain amount of *interfacing*. This may be minimal, requiring say, the use of buffer/driver ICs between the micro I/O device and the target process (see Fig. 1.41). Normally however, the interfacing requirement is much greater, involving:

Signal level changing – e.g.: 5–24 V d.c. or 240 V a.c.
Electrical isolation between controller and process – up to 2000 V a.c. is often required.
Data conversion – serial/parallel, etc.

The resultant costs contribute to the total price for hardware. In addition, hardware for the target microcontroller and associated interfacing may have to be designed and built, unless an off-the-shelf product is available.

Often the large basic investment required for this equipment means that for many small production lines and small companies, custom-designed microcomputer control is not financially viable.

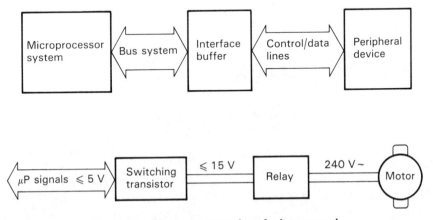

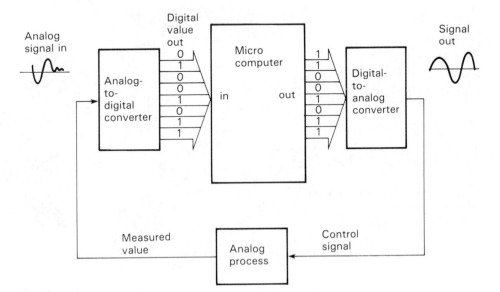

Fig. 1.41 Microprocessor interfacing example.

Fig. 1.42 DAC/ADC layout.

Continuous control

For digital microcomputers to process analog (continuous) quantities they must be equipped with *analog-to-digital converters* (ADCs) at the input ports, together with *digital-to-analog converters* (DACs) for any continuous output devices (Fig. 1.42). The ADCs will produce a binary approximation of the offered input voltage in a certain time. For example,

Analog input of 0–10 V
8-bit ADC has a range of 0–255 steps in binary

(a)

Fig. 1.43   (a) Fitting an interface card to a personal computer; (b) available expansion slots on a PC.

If (e.g.) 5 V offered to the input, the ADC will output 127, $= 0111\ 1111$ in binary ($\frac{5}{10} \times 255$).

The binary data can now be processed within the microcomputer using programmed functions held in memory, producing output data that may be converted back into analog form (by a DAC) to drive output actuators.

Direct digital control (DDC)

If a continuous-control algorithm is implemented using a digital computer it is

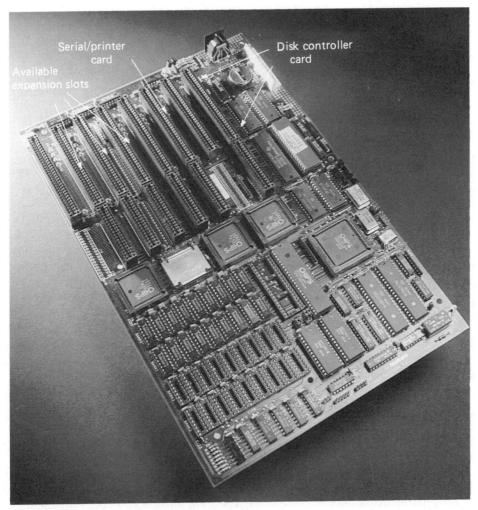

(b)

termed direct digital control. The control software must be designed to perform the mathematical functions required, such as integrals and differentials for a PID algorithm. In practice all measured values are sampled at a rate dependent on the computer clock speed (1–12 MHz) and also the conversion time of intermediate hardware units such as ADCs (ms). In the above example this means the integral and derivative terms can only be approximated, but this is adequate for most industrial applications.

The nature of both high- and low-level computer programs allows any mathematical control algorithm to be implemented, including those previously too complex or cumbersome for analog circuitry, since parameters within the algorithm can be altered to 'fine tune' the control system if necessary. However, efficient programming of computers for control applications (using high- or low-level languages) requires experienced staff, unlikely to be found in the user's company. Thus the main advantage of computers, the ability to modify or add to programs, can rarely be exploited by the actual user.

For large production runs, where size is at a premium, single-chip microcomputers are likely to be employed. They are even more remote from the user, normally being programmed at manufacture.

Current trends in computer control

The availability of powerful, low-cost personal and industrial microcomputers based on the IBM PC standard has led to the development of a wide range of add-on interface boards that equip the PC to act as an effective industrial controller. This has resulted in increased use of such systems for many control applications.

The range of interfaces includes multipoint digital I/O cards, analog I/O and communications cards. These are fitted to the chosen personal computer or to a 'ruggedized' industrial version, both machines having several expansion slots that are use to hold up to six interface cards (see Fig. 1.43).

Appropriate software applications packages are then loaded and run on the system. This software handles the configuring of the I/O cards, together with data processing required to carry out the control plan. There is frequently a very 'friendly' operator interface via a dynamic graphics display. This presents process data to the operator, including system status and exception conditions. It may allow limited operator intervention in certain circumstances.

1.10 SUMMARY

The main features of several control media have been examined in this chapter, allowing the reader to compare these alternatives with the main subject of this text – programmable controllers (PLCs). These are introduced in the next chapter.

A table comparing PLCs with conventional control media appears in Chapter 2 (Table 2.2).

1.11 EXERCISES

1.1 Draw a diagram to show the main elements of a control system.

1.2 Add to the diagram in Exercise 1.1 the elements necessary to form closed-loop control, and name them.

1.3 Name and describe the function of four different transducers and actuators.

1.4 Draw a relay logic circuit and a logic gate circuit to perform the boolean function: $X = A.B + A.C.D$

1.5 Describe the differences between continuous and discontinuous control methods.

1.6 List the advantages and disadvantages of each control system described in this chapter, i.e. relays, digital logic, electronics, and computers.

1.7 Explain the differences between open- and closed-loop control systems.

1.8 Describe why the control plans for closed-loop continuous-control systems tend to be fairly complex.

1.9 What is the meaning of the terms *measured value*, and *set point* in relation to control variables?

1.10 Draw a block diagram of a microprocessor-based controller, and describe the function of each block.

1.11 List the different types of semiconductor memory available and their characteristics.

1.12 Give the advantages of distributed control over centralized control, with a diagram showing typical elements and their configuration.

## TWO

Programmable Controllers (PLCs)

2.1 INTRODUCTION

The need for low-cost, versatile and easily commissioned controllers has resulted in the development of *programmable-control* systems – standard units based on a hardware CPU and memory for the control of machines or processes. Originally designed as a replacement for the hard-wired relay and timer logic to be found in traditional control panels, PLCs provide ease and flexibility of control based on programming and executing simple logic instructions (often in ladder diagram form). PLCs have internal functions such as timers, counters and shift registers, making sophisticated control possible using even the smallest PLC.

A programmable controller operates by examining the input signals from a process and carrying out logic instructions (which have been programmed into its memory) on these input signals, producing output signals to drive process equipment or machinery. Standard interfaces built in to PLCs allow them to be *directly connected* to process actuators and transducers (e.g. pumps and valves) without the need for intermediate circuitry or relays.

Through using PLCs it became possible to modify a control system without having to disconnect or re-route a single wire; it was necessary to change only the

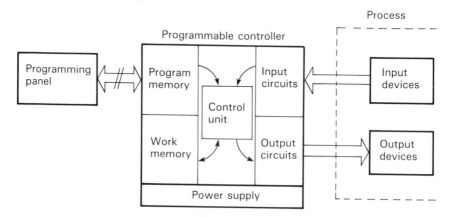

Fig. 2.1 Programmable controller structure.

control program using a keypad or VDU terminal. Programmable controllers also require shorter installation and commissioning times than do hardwired systems. Although PLCs are similar to 'conventional' computers in terms of hardware technology, they have specific features suited to industrial control:

- rugged, noise immune equipment;
- modular plug-in construction, allowing easy replacement/addition of units (e.g. input/output);
- standard input/output connections and signal levels;
- easily understood programming language (e.g. ladder diagram or function chart);
- ease of programming and reprogramming in-plant.

These features make programmable controllers highly desirable in a wide variety of industrial-plant and process-control situations.

2.2 BACKGROUND

The programmable controller was initially conceived by a group of engineers from General Motors in 1968, where an initial specification was provided: the controller must be:

1. Easily programmed and reprogrammed, preferably in-plant, to alter its sequence of operations.
2. Easily maintained and repaired – preferably using plug-in modules.
3. (a) More reliable in a plant environment.
   (b) Smaller than its relay equivalent.
4. Cost-competitive, with solid-state and relay panels then in use.

This provoked a keen interest from engineers of all disciplines in how the PLC could be used for industrial control. With this came demands for additional PLC capabilities and facilities, which were rapidly implemented as the technology became available. The instruction sets quickly moved from simple logic instructions to include counters, timers and shift registers, then onto more advanced mathematical functions on the larger machines. Developments in hardware were also occurring, with larger memory and greater numbers of input/output points featuring on new models. In 1976 it became possible to control *remote I/O racks*, where large numbers of distant I/O points were monitored and updated via a communications link, often several hundred meters from the main PLC. A *microprocessor-based PLC* was introduced in 1977 by the Allan–Bradley Corporation in America. It was based on an 8080 microprocessor but used an extra processor to handle bit logic instructions at high speed.

The increased rate of application of programmable controllers within industry has encouraged manufacturers to develop whole families of microprocessor-based systems having various levels of performance. The range of available PLCs now

Table 2.1   Chart of Programmable controller developments.

Year	Nature of developments
1968	Programmable controller concept developed
1969	Hardware CPU controller, with logic instructions, 1 K of memory and 128 I/O points
1974	Use of several (multi) processors within a PLC – timers and counters; arithmetic operations; 12 K of memory and 1024 I/O points
1976	Remote input/output systems introduced
1977	Microprocessor-based PLC introduced
1980	Intelligent I/O modules developed
	Enhanced communications facilities
	Enhanced software features (e.g. documentation)
	Use of personal microcomputers as programming aids
1983	Low-cost small PLCs introduced
1985 on	Networking of all levels of PLC, computer and Machine under standard GM MAP specification.
	Distributed, hierarchical control of industrial plants.

extends from small self-contained units with 20 digital I/O points and 500 program steps (see Fig. 2.2), up to modular systems with add-on function modules for, e.g:

- analog I/O;
- PID control (proportional, integral and derivative terms);
- communications;
- graphics display;
- additional I/O;
- additional memory.

This modular approach allows the expansion or upgrading of a control system with minimum costs and disturbance.

Programmable controllers are developing at virtually the same pace as microcomputers, with particular emphasis on small controllers, positioning/numeric control and communication networks. The market for small controllers has grown rapidly since the early 1980s when a number of Japanese companies introduced very small, low-cost units that were much cheaper than others available at that time. This brought programmable controllers within the budget of many potential users in the manufacturing and process industries, and this trend continues with PLCs offering ever-increasing performance at ever-decreasing cost.

The Mitsubishi F40 PLC shown in Fig. 2.2(a) is a typical example of a modern small PLC, providing 40 I/O points, 16 timers and counters, plus other functions. The controller uses a microprocessor and has 890 RAM locations for user programs. The 24-input channels of the F40 operate at 24 V d.c. Whilst the 16 outputs may be 24 V d.c. or 240 V a.c. to provide easy interfacing to industrial equipment. The programming panel is also shown in the figure (a).

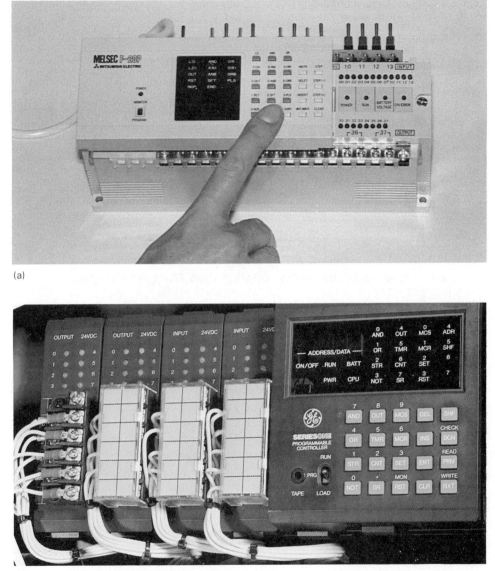

(a)

(b)

Fig. 2.2 Small PLCs: (a) Mitsubishi F series (courtesy Mitsubishi Electric UK Ltd); (b) GE series 1 (courtesy General Electric).

2.3 TERMINOLOGY – PC OR PLC?

There are several different terms used to describe programmable controllers, most referring to the functional operation of the machine in question:

PC   programmable controller (UK origin)
PLC   programmable logic controller (American origin);
PBS   programmable binary system (Swedish origin).

By their nature these terms tend to describe controllers that normally work in a binary (on/off) environment. Since all but the smallest programmable controllers can now be equipped to process analog inputs and outputs these 'labels' are not representative of their capabilities. For this reason (and others) the overall term *programmable controller* (abbreviated to PC) has been widely adopted to describe the family of freely programmable controllers. However, to avoid confusion with the personal computer 'PC', this text uses the abbreviation PLC for programmable (logic) controller.

2.4 COMPARISON WITH OTHER CONTROL SYSTEMS

Table 2.2 provides a comparison between various control media. This is only an approximate guide to their capabilities, and further technical information can be obtained from the manufacturers' data sheets on each specific system.

Programmable controllers emerge from the comparison as the best overall choice for a control system, unless the ultimate in operating speed or resistance to electrical noise is required, in which case hardwired digital logic and relays are chosen respectively. For handling complex functions a conventional computer is still marginally superior to a large PLC equipped with relevant function cards, but only in terms of creating the functions, not using them. Here the PLC is more efficient through passing values to the special function module, which then handles the control function independently of the main processor – a multiprocessor system.

Programmable controllers have both hardware and software features that make them attractive as controllers of a wide range of industrial equipment. We shall now examine these features in some detail.

2.5 PLCs – HARDWARE DESIGN

Programmable controllers are purpose-built computers consisting of three functional areas: *processing*, *memory* and *input/output*. Input conditions to the PLC are sensed and then stored in memory, where the PLC performs the programmed logic instructions on these input states. Output conditions are then generated to drive associated equipment. The action taken depends totally on the control program held in memory.

Table 2.2   Comparison of control systems.

Characteristic	Relay systems	Digital logic	Computers	PLC systems
Price per function	Fairly low	Low	High	Low
Physical size	Bulky	Very compact	Fairly compact	Very compact
Operating speed	Slow	Very fast	Fairly fast	Fast
Electrical noise immunity	Excellent	Good	Quite good	Good
Installation	Time-consuming to design and install	Design time-consuming	Programming extremely time-consuming	Simple to program and install
Capable of complicated operations	No	Yes	Yes	Yes
Ease of changing function	Very difficult	Difficult	Quite simple	Very simple
Ease of maintenance	Poor — large number of contacts	Poor if ICs soldered	Poor — several custom boards	Good — few standard cards

In smaller PLCs these functions are performed by individual printed circuit cards within a single compact unit, whilst larger PLCs are constructed on a modular basis with function modules slotted into the *backplane connectors* of the mounting rack. This allows simple expansion of the system when necessary. In both these cases the individual circuit boards are easily removed and replaced, facilitating rapid repair of the system should faults develop.

In addition a *programming unit* is necessary to *download* control programs to the PLC memory.

Central processing unit (CPU)

The CPU controls and supervises all operations within the PLC, carrying out programmed instructions stored in the memory. An internal communications highway, or *bus system*, carries information to and from the CPU, memory and I/O units, under control of the CPU. The CPU is supplied with a clock frequency by an external quartz crystal or RC oscillator, typically between 1 and 8 megahertz depending on the microprocessor used and the area of application. The clock determines the operating speed of the PLC and provides timing/synchronization for all elements in the system (see Fig. 2.3(a)).

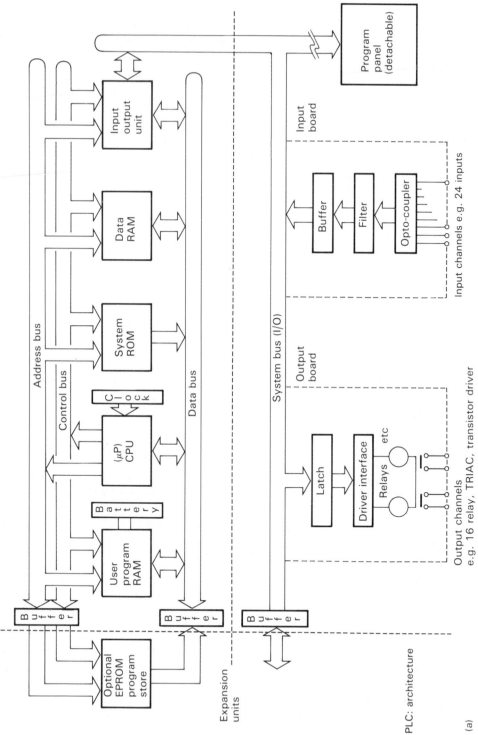

PLC: architecture

(a)

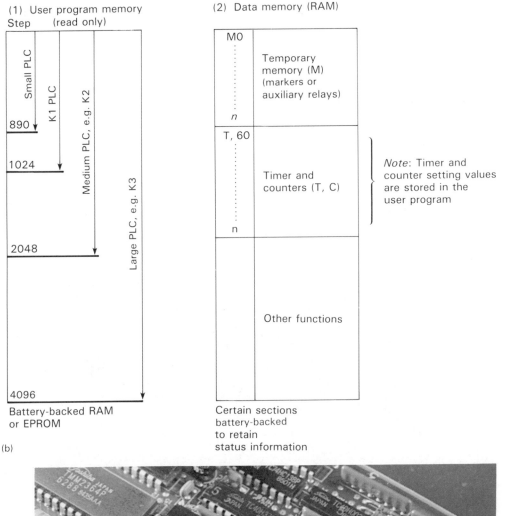

(1) User program memory
Step (read only)

Small PLC

K1 PLC

Medium PLC, e.g. K2

Large PLC, e.g. K3

890

1024

2048

4096

Battery-backed RAM
or EPROM

(b)

(2) Data memory (RAM)

M0
....
....
....
....
n

Temporary
memory (M)
(markers or
auxiliary relays)

T, 60
....
....
....
....
n

Timer and
counters (T, C)

Other functions

Note: Timer and
counter setting values
are stored in the
user program

Certain sections
battery-backed
to retain
status information

(c)

Fig. 2.3 (a) Block diagram of PLC internal architecture; (b) memory allocation for a range of PLCs (courtesy Mitsubishi Electric UK Ltd); (c) INTEL 8051 SCMP on a small PLC board.

Virtually all modern programmable controllers are *microprocessor-based*, using a 'micro' as the system CPU (see Fig. 2.3(c)). Some larger PLCs also employ additional microprocessors to control complex, time-consuming functions such as mathematical processing, three-term PID control, etc.

Memory

(a) For *program storage* all modern programmable controllers use semiconductor memory devices such as RAM read/write memory, or a programmable read-only memory of the EPROM or EEPROM families (see Fig. 1.29).

In virtually all cases RAM is used for initial program development and testing, as it allows changes to be easily made in the program. The current trend

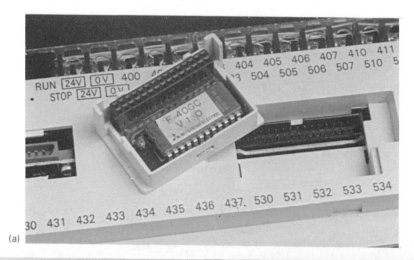

(a)

(b)

Fig. 2.4 (a) EPROM facility on Mitsubishi PLC; (b) battery-backed unit on a small PLC (courtesy Mitsubishi Electric UK Ltd).

is to provide CMOS RAM because of its very low power consumption, and to provide battery back-up to this RAM in order to maintain the contents when power is removed from the PLC system. (RAM is volatile in nature – see Chapter 1.) This battery has a lifespan of at least one year before replacement is necessary, or alternatively a rechargeable type may be supplied with the system, being recharged whenever the main PLC power supply is on.

This feature makes programs stored in RAM virtually permanent. Many users operate their PLC systems on this basis alone, since it permits future program alterations if and when necessary.

After a program is fully developed and tested it may be loaded (blown) into a PROM or EPROM memory chip, which are normally cheaper than RAM devices. PROM programming is usually carried out with a special-purpose programming unit, although many programmable controllers now have this facility built-in, allowing programs in the PLC RAM to be *downloaded* into a PROM IC placed in a socket provided on the PLC itself (Fig. 2.4.)

(b) In addition to program storage, a programmable controller may require memory for other functions:

   1. Temporary buffer store for input/output channel status—I/O RAM (see input/output copying below.
   2. Temporary storage for status of internal functions, e.g. timers, counters, marker relays, etc.

Since these consist of *changing data* (e.g. an input point changing state) they require RAM read/write memory, which may be battery-backed in sections (see Fig. 2.3(b)).

Memory size

Smaller programmable controllers normally have a fixed memory size, due in part to the physical dimensions of the unit. This varies in capacity between 300 and 1000 instructions depending on the manufacturer. This capacity may not appear large enough to be very useful, but it has been estimated that 90% of all binary control

Table 2.3 Hardware features of typical PLCs.

	Small PLC	Medium PLC
Model	Mitsubishi F2 40	Allan–Bradley Mini PLC 2
CPU	Intel 8031	Zilog Z80A
Scan time	7 m/s per 1K memory	20 m/s per 1K of memory
Memory	CMOS RAM 1K	CMOS RAM 1K
Max I/O	40 points	256 points
Analog I/O	No	128 points max plus PID
Language	Relay ladder/list	Relay ladder

(See Appendix A for details of other controllers.)

tasks can be solved using less than 1000 instructions, so there is sufficient space to meet most users' needs.

Larger PLCs utilize memory modules of between 1K and 64K in size, allowing the system to be expanded by fitting additional RAM or PROM memory cards to the PLC rack.

As integrated circuit memory costs continue to fall, the PLC manufacturers are providing larger program memories on all products.

Input/output units

Most PLCs operate internally at between 5 and 15 V d.c. (common TTL and CMOS voltages), whilst *process signals* can be much greater, typically 24 V d.c. to 240 V a.c. at several amperes!

The I/O units form the *interface* between the microelectronics of the programmable controller and the real world outside, and must therefore provide all necessary signal conditioning and isolation functions. This often allows a PLC to be *directly connected* to process actuators and transducers (e.g. pumps and valves) without the need for intermediate circuitry or relays (Fig. 2.5).

To provide this signal conversion programmable controllers are available with a choice of input/output units to suit different requirements. For example,

Inputs	(choice of):	5 V (TTL level) switched I/P
		24 V switched I/P
		110 V switched I/P
		240 V switched I/P
Outputs	(choice of):	24 V 100 mA switched O/P
		110 V 1 amp
		240 V 1 A a.c. (triac)
		240 V 2 A a.c. (relay)

It is standard practice for all I/O channels to be electrically isolated from the controlled process, using opto-isolator circuits (Fig. 2.6) on the I/O modules. An opto-isolator circuit consists of a light-emitting diode and a photo-transistor, forming an opto-coupled pair that allows small signals to pass through, but will clamp any high-voltage spikes or surges down to the same small level. This provides protection against switching transients and power-supply surges, normally up to 1500 V.

In small self-contained PLCs in which all I/O points are physically located on the one casing, all inputs will be of one type (e.g. 24 V) and the same for outputs (e.g. all 240 V triac). This is because manufacturers supply only standard function boards for economic reasons. Modular PLCs have greater flexibility of I/O, however, since the user can select from several different types and combinations of input and output modules.

In all cases the input/output units are designed with the aim of simplifying the connection of process transducers and actuators to the programmable controller.

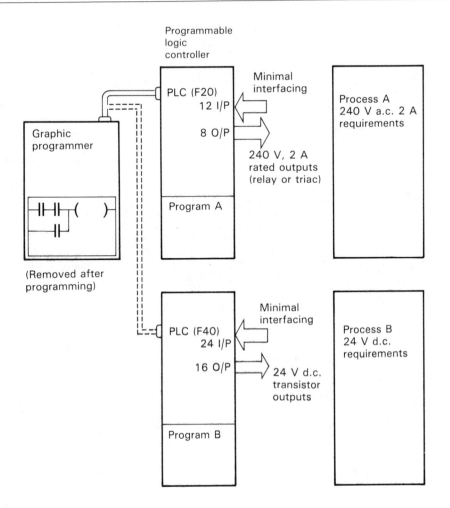

Programmable
logic
controller

Graphic
programmer

(Removed after
programming)

PLC (F20)
12 I/P

8 O/P

Minimal
interfacing

240 V, 2 A
rated outputs
(relay or triac)

Program A

Process A
240 V a.c. 2 A
requirements

PLC (F40)
24 I/P

16 O/P

Minimal
interfacing

24 V d.c.
transistor
outputs

Program B

Process B
24 V d.c.
requirements

PLC Control.
Easily programmed/altered by the USER.
Used for switched input/output.

Fig. 2.5 PLC input/output connected to plant equipment.

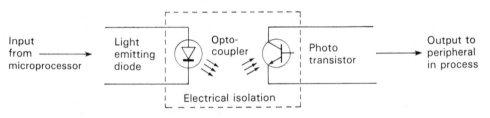

Input
from
microprocessor

Light
emitting
diode

Opto-
coupler

Photo
transistor

Output to
peripheral
in process

Electrical isolation

Fig. 2.6 Opto-isolator circuit.

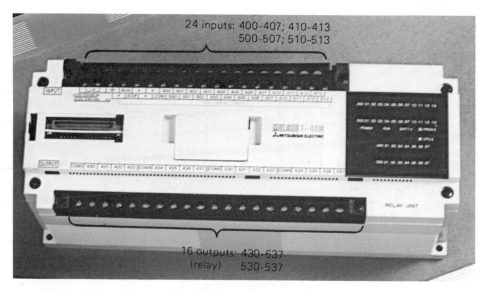

Fig. 2.7 *Input/output numbering for an F40 PLC (courtesy Mitsubishi Electric UK Ltd).*

For this purpose all PLCs are equipped with standard screw terminals or plugs on every I/O point, allowing the rapid and simple removal and replacement of a faulty I/O card.

Every input/output point has a unique address or channel number which is used during program development to specify the monitoring of an input or the activating of a particular output within the program. Indication of the status of input/output channels is provided by light-emitting diodes (LEDs) on the PLC or I/O unit, making it simple to check the operation of process inputs and outputs from the PLC itself (see Fig. 2.7).

Programming units

All but the most simple programming panels contain enough RAM to enable semipermanent storage of a program under development or modification. If the programming panel is a portable unit, its RAM is normally CMOS type with battery backup, allowing the unit to retain programs whilst being carried around a plant or factory floor (Fig. 2.8). Only when a program is ready for use/testing will it be transferred to the PLC. Once the installed program has been fully tested and debugged, the programming panel is removed and is free to be used on other controllers.

The terminal may have a 'monitoring' and forcing facility, allowing real-time observation of switches, gates and functions during program execution – this can be valuable for troubleshooting, especially when the target process is remote or inaccessible.

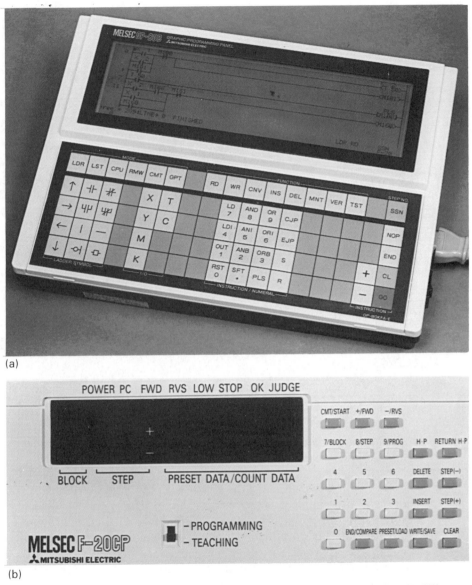

(a)

(b)

Fig. 2.8   Graphic programming panels (courtesy Mitsubishi Electric UK Ltd).

Larger PLCs are often programmed using a visual display unit (VDU) with a full keyboard and screen display, connected to the controller via a serial link (normally RS232). VDUs provide improved programming facilities such as screen graphics and the inclusion of text comments that assist in the readability of a program. Recently personal computers have been configured as program development workstations by most PC manufacturers. The high-speed operation and screen

(a)

Fig. 2.9   Personal computer workstations: (a) IBM PC-compatible; (b) enhanced IBM type; (c) portable version.

graphics facilities of machines like the IBM PC and COMPAQ (Fig. 2.9) are ideal for graphics programming of ladder circuits, etc. (Davie, 1985).

Also, the large memory available on modern 16-bit microcomputers is ideal for storage of several PLC programs complete with comments and documentation. Selecting a personal computer as a programmable controller workstation also provides the user with access to other useful software facilities for project management, such as databases, spreadsheets, word processing and financial planning packages.

2.6   PROGRAMMING PLCs

The main requirement from any PLC programming language is that it may be easily understood and used in a control situation. This implies the need for a *high-level*

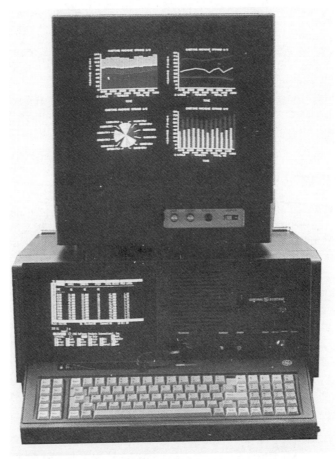

(b)

(c)

language to provide commands very close to the functions required by a control engineer, but without the complexity and learning time associated with most high-level computer languages.

Ladder diagrams have been the most common method of describing relay logic circuits, so it was only natural to base PLC programming on them in order to create a familiar environment for the user and designer of small logic control systems.

Explanation of ladder diagrams

To show the relationship between a physical circuit and a ladder representation, consider the electric motor circuit in Fig. 2.10(a). The motor is connected to a power source via three switches in series plus an overload switch, (a)–(d). The motor will

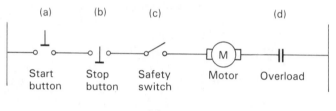

(a)

The ladder diagram could be as represented below:

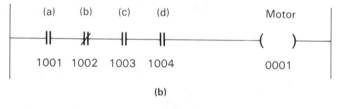

(b)

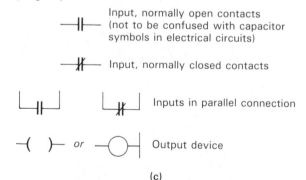

(c)

Fig. 2.10 *(a) Motor circuit electrical symbols/ladder symbols; (b) Equivalent relay ladder diagram; (c) ladder symbols.*

turn on if all switches are made (closed). (b) shows the equivalent ladder diagram. The ladder diagram uses standard symbols to represent the circuit elements and functions found in a control system.

The ladder diagram consists of two vertical lines representing the power rails, plus circuit symbols that make up a rung of the ladder. Here the symbols represent three normally open switch contacts, one normally closed contact and one output device – the motor coil.

Examples of simple logic circuits are given in Fig. 2.11.

Ladder symbols are used to construct any form of switched logic control system and the diagrams produced can be as complex as necessary for a particular application. An essential part of any ladder design is the *documentation* of the system and its operation, to allow any user to understand the ladder solution quickly.

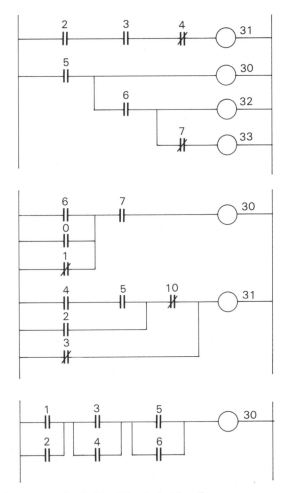

Fig. 2.11 Simple ladder diagrams.

Logic instruction set

The most common technique used for programming small PLCs is to draw a ladder diagram of the logic to be used, then convert this into mnemonic instructions which will be keyed into a programming panel attached to the programmable controller. These instructions are similar in appearance to assembly-type codes, but refer to physical inputs, outputs and functions within the PLC itself.

The instruction set consists of logic instructions (mnemonics) that represent the actions that may be performed within a given programmable controller. Instruction sets vary between PLCs from different manufacturers, but are similar in terms of the control actions performed. An example is shown in Table 2.4.

Because the PLC logic instruction set tends to be small, it can be quickly mastered and used by control technicians and engineers.

Each program instruction is made up of two parts: a *mnemonic operation component* or *opcode*, and an *address* or *operand component* that identifies particular elements (e.g. outputs) within the PLC. For example,

opcode	*operand* (unique address)
OUT	Y430
device symbol	identifier

Here the instruction refers to output (Y) number 430.

Table 2.4 Typical instruction sets.

Texas Instruments Mnemonic action		Mitsubishi F series Mnemonic action	
STR	store (start a new rung of a ladder diagram)	LD	start rung with an open contact
OUT	output	OUT	output
AND	series components	AND	series elements
OR	parallel components	OR	parallel elements
NOT	inverse action (used in conjunction with other instructions to invert their function, e.g. ANDNOT = NAND function.	..I	as for NOT, e.g.; ORI meaning NOR function

Input/output numbering

These instructions are used to program logic control circuits that have been designed in ladder diagram form, by assigning all physical inputs and outputs with an operand (address) suitable to the PLC being used. The numbering systems used differ between manufacturers, but certain common terms exist. For example, Texas

Instruments and Mitsubishi use the symbol X to represent inputs, and Y to label outputs.

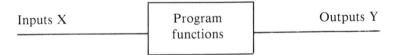

Programmable controller

Inputs X   Program functions   Outputs Y

A range of addresses will be allocated to particular elements: for example

Mitsubishi F40 PLC: (note the octal (base 8) numbering)
    24 Inputs: X400–407, 410–413
                       X500–507, 510–513
    16 Outputs: Y430–437
                       Y530–537

Inspection of these number ranges will reveal that there is no overlap of addresses between functions; that is, 400 must be an input, 533 must be an output. Thus for these programmable controllers the symbol *X* or *Y* is redundant, being used purely for the benefit of the user, who is unlikely to remember what element 533 represents. (Other number ranges are assigned to internal functions of the PLC, and are examined in Chapter 3). However, for many PLCs both parts of the address are essential, since the I/O number ranges are identical. For example, the Klockner–Moeller range of controllers:

Sucos PS 21 PLC: 8 Inputs: I 0 to 7, etc.
                          8 Outputs: Q 0 to 7, etc.

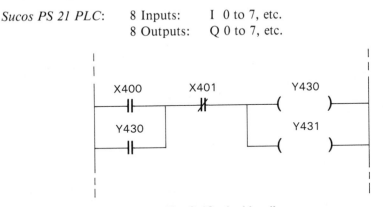

Fig. 2.12 Ladder diagram.

To program the ladder diagram given in Fig. 2.12, the following code would be written, then programmed into a keypad or terminal. (Mitsubishi PLC code)

1	LD X400	start a rung with a normally open contact
2	OR Y430	connect a normally open contact (associated with output Y430) in parallel
3	ANI X401	connect a normally closed contact in series

4	OUT Y430	drive an output channel
5	OUT Y431	drive another channel
6	END	end of program – return to start

Notice the contact Y430 that forms a latch across X400. The Y contact is not a physical element, but is simulated within the programmable controller and will operate in unison with the output point Y430. The programmer may create as many contacts associated with an output as necessary. This is a fundamental feature of programmable controllers and will be expanded in Chapter 3.

2.7 PLCs – INTERNAL OPERATION AND SIGNAL PROCESSING

When a program is loaded into a PLC, each instruction is placed in an individual memory location (address).

The CPU contains a *program counter register* which points to the next instruction to be fetched from memory. When an instruction is received by the CPU it is placed in the *instruction register* for decoding into the internal operations (microinstructions) required by that particular instruction. For example, this may result in further instructions being read from memory, or in a physical device being driven by the CPU.

When the programmable controller is initially set to run, the program counter will point to address 0000 – the location of the first instruction as shown in Table 2.5 and Fig. 2.13. The CPU then fetches, decodes and executes this program instruction, here LD X400. Thus the CPU finds it has to examine the first element of a logic circuit – a normally open contact (LD) associated with input channel (X) number 400. Input status is held in the I/O RAM, so the CPU scans the RAM location allocated to X400 and reads its status into a temporary memory location or internal register. The CPU is now expecting further instructions to complete the logic circuit. The program counter automatically increments to point to the next address, where the second instruction – AND X401 – is fetched and executed. This causes the CPU to scan the I/O RAM location of X401 and then perform a logical AND function with the values of X400 and X401, temporarily storing the result. The program

Table 2.5 PLC Instruction list for Fig. 2.13.

Memory address	Opcode	Operand
0000	LD X400	
0001	AND X401	rung 1
0002	OUT Y430	
0003	LDI Y430	
0004	OUT Y431	rung 2
0005	etc	

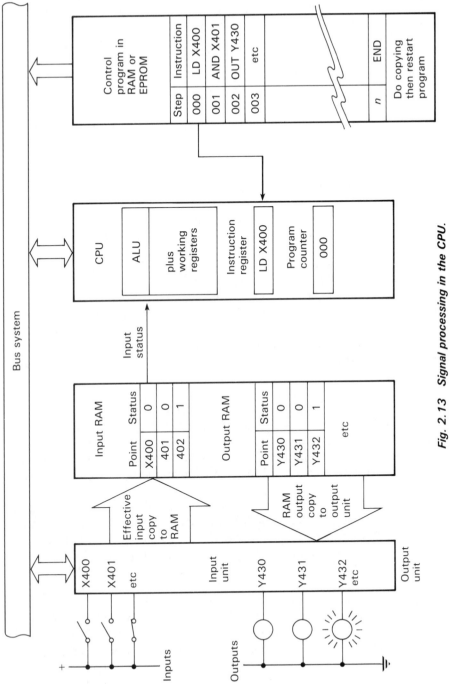

Fig. 2.13 Signal processing in the CPU.

counter steps on to address 0003 where the instruction OUT Y430 is fetched and decoded, instructing the CPU to pass on the result of the previous logical operation (X400 AND X401) to the location in I/O RAM corresponding to output Y430.

The sequence then continues until either the last memory location is reached by the program counter, or an END instruction is encountered. In both cases the program counter is reset to 0000, causing the program to start at the beginning again – i.e: it will cycle continuously.

The program counter can be programmed to jump over one or more addresses if required, so that the jumped instructions are not being processed. This is caused by the use of a conditional jump instruction in a logic program (see Chapter 3).

Input/output processing

There are two different methods used for I/O processing in programmable controllers: (a) continuous updating, and (b) mass input/output copying.

(a) Continuous updating

This involves the CPU in scanning input channels as they occur in program instructions, with a built-in delay to ensure that only valid input signals are read into the processor. (The delay – typically 3 ms – prevents contact-bounce pulses and other noise from entering the PLC.) Output channels are driven (directly) when OUT instructions are executed following a logical operation. Outputs are latched in the I/O unit so they retain their status until the next updating (Fig. 2.14(a)).

(b) Mass I/O copying

In larger PLCs there may be several hundred input/output points. Since the CPU can deal with only one instruction at a time during program execution, the status of each input point must be examined individually to determine its effect on the program. Since we require a 3 ms delay on each input (see above), the total cycle time for a continuously sampled system becomes progressively longer as the number of inputs rises.

To allow rapid program execution, input and output updating may be carried out at one particular point in the program. Here a specific RAM area within the PLC is used as a buffer store between the control logic and the I/O unit. Each input and output has a cell in this I/O RAM. During I/O copying, the CPU scans all the inputs in the I/O unit and copies their status into the I/O RAM cells. This happens at the start/end of each program cycle (Fig. 2.14(b)).

As the program is executed, the stored input data is read one location at a time from the I/O RAM. Logic operations are performed on the input data, and the resulting output signals are stored in the output section of the I/O RAM. Then, at the end of each program cycle the I/O copying routine transfers all output signals from the I/O RAM to the corresponding output channel, driving the output stages of the I/O unit. These output stages are latched, that is they retain their status until they are updated by the next I/O routine.

(a)

Start

Fetch, decode and execute first instruction	Scan necessary contacts	Next instruction	Scan or actuate devices	Next instruction	Scan I/O or actuate O/P	etc
Typical time delay ≈ 5 µs	3 ms delay	5 µs	3 ms delay	5 µs	3 ms delay	

(b)

Start   I/O copy   Program execution   End I/O copy   Prog

Copy all inputs into RAM	Fetch, decode and execute all instructions in sequence	Copy all outputs from O/P RAM to output unit, and inputs into RAM	

Time depends on length of total program
e.g. 1K program ≈ 5 ms

Fixed length delay
e.g. 5 ms

Fig. 2.14   Input scanning and reaction times: (a) continuous updating; (b) mass I/O copying.

This task is carried out automatically by the CPU as a subroutine to the normal program. (A subroutine is a small program designed to perform a specific function, that may be called by the main program. In this case the I/O subroutine is automatically accessed by the underlying CPU control program—the monitor.) I/O copying takes place between the end of one program cycle and the start of the next.

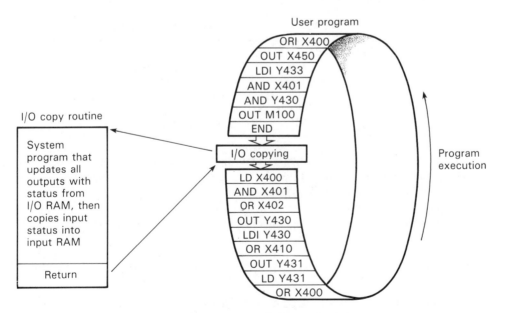

Fig. 2.15   Input/output copying.

Timing considerations

Note that by virtue of the cyclic nature of the program I/O copy, the status of inputs and outputs *cannot be changed* within the same program cycle. If an input signal changes state after the copy routine, it will not be recognized until the next copy occurs.

The time to update all inputs and outputs depends on the total number to be copied, but is typically a few milliseconds in length. The total program execution time (or cycle time) depends on the length of the control program. Each instruction takes 1–10 μs to execute depending on the particular programmable controller employed. So a 1K (1024) instruction program typically has a cycle time of 1–10 ms. However, programmable controller programs are often much shorter than 1000 instructions, namely 500 steps or less.

2.8 TYPES OF PLC SYSTEM

The increasing demand from industry for programmable controllers that can be applied to different forms and sizes of control tasks has resulted in most manufacturers producing a range of PLCs with various levels of performance and facilities.

Typical rough definitions of PLC size are given in terms of program memory size and the maximum number of input/output points the system can support. Table 2.6 gives an example of these categories.

Table 2.6 Categories of PLC.

PC size	Max I/O points	User memory size (no. of instructions)
Small	40/40	1K
Medium	128/128	4K
Large	>128/>128	>4K

However, to evaluate properly any programmable controller we must consider many additional features such as its processor, cycle time, language facilities, functions, expansion capability, etc.

A brief outline of the characteristics of small, medium and large programmable controllers is given below, together with typical applications.

Small PLCs

In general, small and 'mini' PLCs (Fig. 2.16) are designed as robust, compact units which can be mounted on or beside the equipment to be controlled. They are mainly

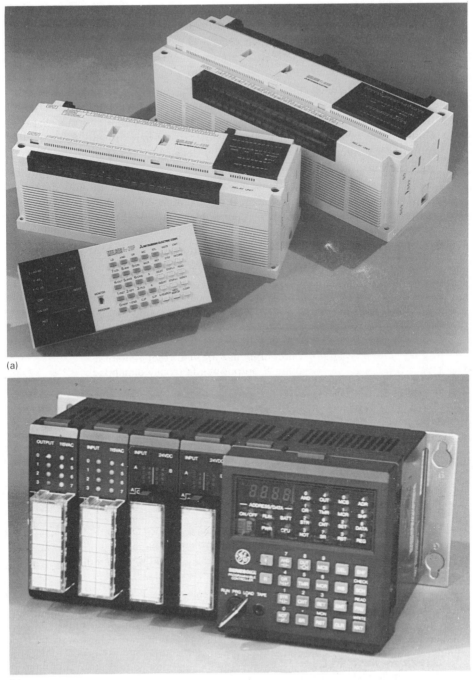

(a)

(b)

Fig. 2.16   Small PLCs: (a) Mitsubishi F40 (courtesy Mitsubishi Electric UK Ltd); (b) GE series 1 (courtesy General Electric).

used to replace hard-wired logic relays, timers, counters, etc. that control individual items of plant or machinery, but can also be used to coordinate several machines working in conjunction with each other.

Small programmable controllers can normally have their total I/O expanded by adding one or two I/O modules, but if any further developments are required this will often mean replacement of the complete unit.

This end of the market is very much concerned with non-specialist end-users, therefore ease of programming and a 'familiar' circuit format are desirable. Competition between manufacturers is extremely fierce in this field, as they vie to obtain a maximum share in this partially developed sector of the market.

A single processor is normally used, and programming facilities are kept at a fairly basic level, including conventional sequencing controls and simple standard functions: e.g. timers and counters. Programming of small PLCs is by way of logic instruction lists (mnemonics) or relay ladder diagrams. (Standard symbols to IEC 617-7 are being introduced.)

Program storage is by EPROM or battery-backed RAM. There is now a trend towards EEPROM memory with on-board programming facilities on several controllers.

Table 2.7   Features of a typical small PLC – Mitsubishi F20.

Electrical:	240 V a.c. supply;
	24 V d.c. on-board for input requirements;
	12 input, 8 output points (easily expanded);
	LED indicators on all I/O points;
	All I/O opto-isolated;
	Choice of output: Relay (240 V 2 A rated);
	                              Triac (240 V 1 A rated);
	                              Transistor (24 V d.c. 1 A);
	320-step memory (CMOS battery-backed RAM).

Programming:	Ladder logic or instruction set using hand-held or graphic LCD programmer, with editor, test and monitor facilities; (detachable for use on other PCs; provides interface to printer and tape store).
Facilities:	8 counters, range 1–99 (can be cascaded); 8 timers, range 0.1–99 s (can be cascaded); 64 markers/auxiliary relays; can be used individually or in blocks of 8, forming shift registers; Special function relays; Jump capability.

Medium-sized PLCs

In this range modular construction predominates with plug-in modules based around the Eurocard 19 inch rack format or another rack mounting system. This construction allows the simple upgrading or expansion of the system by fitting additional I/O cards into the rack, since most rack systems have space for several extra function cards. Boards are usually 'ruggedized' to allow reliable operation over a range of environments. (See Fig. 2.17.)

In general this type of PLC is applied to logic control tasks that cannot be met by small controllers due to insufficient I/O provision, or because the control task is likely to be extended in the future. This might require the replacement of a small PLC, whereas a modular system can be expanded to a much greater extent, allowing for growth. A medium-sized PLC may therefore be financially more attractive in the long term.

Communications facilities are likely to be provided, enabling the PLC to be included in a 'distributed control' system.

Combinations of single and multi-bit processors are likely within the CPU. For programming, standard instructions or ladder and logic diagrams are available. Programming is normally carried out via a small keypad or a VDU terminal. (If different sizes of PLC are purchased from a single manufacturer, it is likely that programs and programming panels will be compatible between the machines.)

Large PLCs

Where control of very large numbers of input and output points is necessary or complex control functions are required, a large programmable controller is the obvious choice. Large PLCs are designed for use in large plants or on large machines requiring continuous control. They are also employed as supervisory controllers to monitor and control several other PLCs or intelligent machines, e.g. CNC tools.

Modular construction in Eurocard format is standard, with a wide range of function cards available including analog input/output modules. There is a move towards 16-bit processors, and also multi-processor usage in order to efficiently handle a large range of differing control tasks (see Figs. 2.18 and 2.19), for example:

- 16-bit processor as main processor for digital arithmetic and text handling.
- Single-bit processors as co- or parallel processors for fast counting, storage, etc.
- Peripheral processors for handling additional tasks which are time-dependent or time-critical, such as:
  Closed-loop (PID) control
  Position controls
  Floating-point numerical calculations
  Diagnostics and monitoring
  Communications for decentralized (distributed) I/O (Fig. 2.20)
  Process mimics (screen graphics)
  Remote input/output racks.

This multi-processor solution optimizes the performance of the overall system as regards versatility and processing speed, allowing the PLC to handle very large

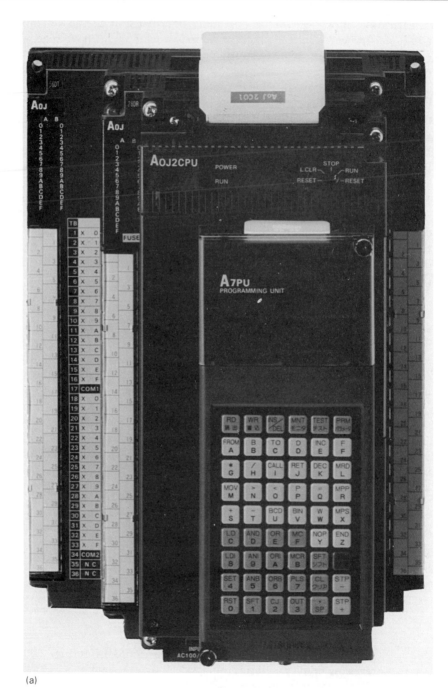

(a)

Fig. 2.17   Medium-sized PLCs: (a) Mitsubishi AO PLC; (b) SattControl
05–35 PC.

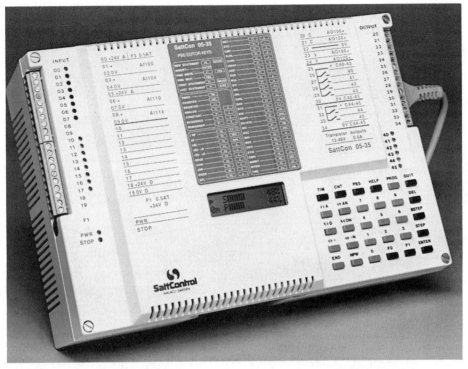

(b)

programs of 100K instructions or more. Memory cards can now provide several megabytes of CMOS RAM or EPROM storage.

Remote input/output

When large numbers of input/output points are located a considerable distance away from the programmable controller, it is uneconomic (and bulky) to run connecting cables to every point. A solution to this problem is to site a remote I/O unit near to the desired I/O points. This acts as a concentrator to monitor all inputs and transmit their status over a single serial communications link to the programmable controller. Once output signals have been produced by the PLC they are fed back along the communications cable to the remote I/O unit, which converts the serial data into the individual output signals to drive the process.

Programming large PLCs

Virtually any function can be programmed, using the familiar ladder symbols via a graphics terminal or personal computer. Parameters are passed to relevant modules

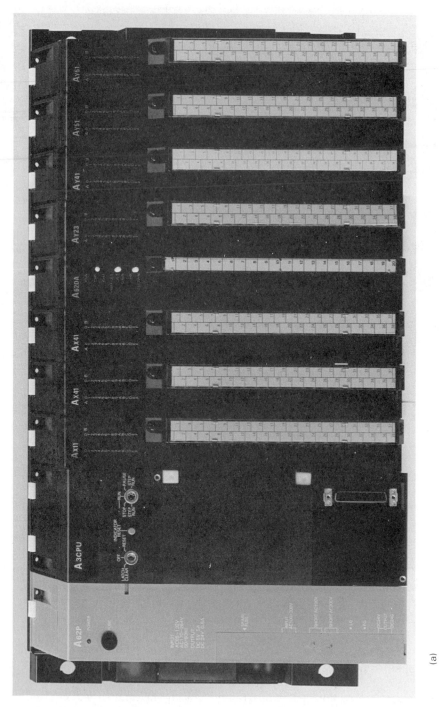

(a)

(b)

(c)

Fig. 2.18 Layout of medium and large PLCs: (a) Mitsubishi A range (courtesy Mitsubishi Electric UK Ltd); (b) GE series 6 (courtesy General Electric); (c) GE CIMSTAR I ('the truly industrial computer') (courtesy General Electric).

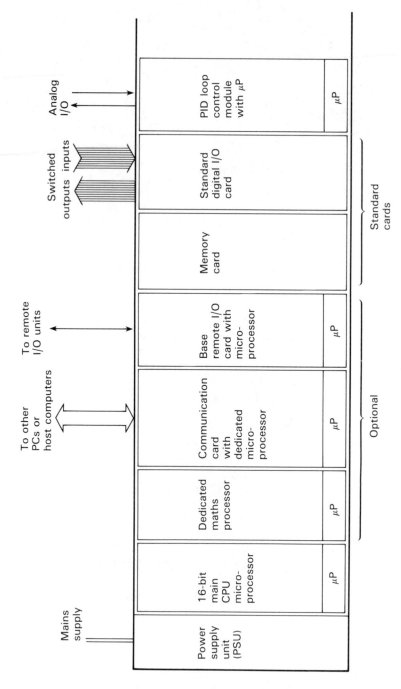

Fig. 2.19 *Multiprocessing structure in a large PLC.*

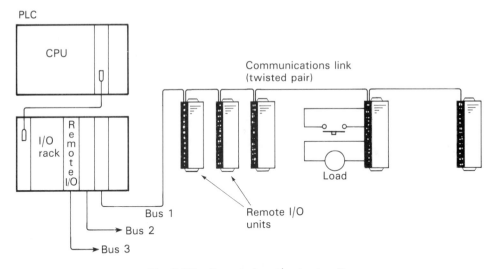

Fig. 2.20   Remote input/output units.

Prompt:	Response:
Controller no	1
Controller type (P, PD, PI, PID)	P
Auto selected	1 1400
Controller on	0 1401
Interlock and open	0 1402
Interlock and close	0 1403
Dev.alarm pos.	0 1404
Dev.alarm neg.	0 1405
Analog input	A 0100
Setpoint	A 0104
Deadzone: input-setpoint	0.0 DEG.C
Dev.limit pos.	5.0 DEG.C
Dev.limit neg.	5.0 DEG.C
Gain	2.5
Offset	50.0%
Hand-interlock speed	8M31S89
Max output signal	100.0 %
Min output signal	0.0 %
Direct control (Y/N)	N
Analog output (Y/N)	Y
Analog output	A 0300
*	

Fig. 2.21   Screen display of function module programming on a GE work-station.

(such as PID) either by incorporating constants into the ladder, or via on-screen menus for that module. (Fig. 2.21)

There may in addition be computer-oriented languages (such as dialects of BASIC, etc.) which allow programming of function modules and subroutines.

There is progress towards standardization of programming languages, with programs becoming easier to over-view through improvement of text (comment) handling, and improved documentation facilities. This is assisted by the application of personal computers as work stations.

Developments

Present trends include the integration of process data from a PLC into management databases, etc. This allows immediate presentation of information to those involved in scheduling, production and planning.

The need to pass process information between PCs, PLCs and other devices within an automated plant has resulted in the provision of a communications capability on all but the smallest controller. The development of local area networks (LAN) and in particular the recent MAP specification by General Motors (manufacturing automation protocol) provides the communication link to integrate all levels of control systems. (In-plant communications are investigated in Chapter 7.)

2.9 SUMMARY

The internal operation of any programmable controller is essentially similar to any other microprocessor-based system. Differences occur in the manner of input/output handling, and the interface hardware provided. PLCs are specifically designed to connect to most common industrial control systems, i.e. they are hardware-specific, but they offer great flexibility through programming.

Today virtually every manufacturer of electronics control equipment markets a range of programmable controllers, with facilities ranging from simple switched I/O through to sophisticated continuous control. Developments in this area are continuing at a rate almost equal to that in the field of personal computing. Because of this, the power and operating speed of all programmable controllers is constantly improving, whilst equipment prices at worst remain steady, and frequently fall!

2.10 EXERCISES

2.1 What is meant by the term 'programmable controller'?

2.2 Explain why programmable controllers were developed.

2.3 Describe the differences between a PLC and a conventional computer.

2.4 Draw a block diagram showing the main functional units of a programmable controller.

2.5 List the advantages and disadvantages of ladder programming.

2.6 What is the advantage of battery-backed RAM memory on a PLC system? When would EPROM or EEPROM be used as the program storage medium?

2.7 Describe the sequence of operations that occur during program scanning in a PLC.

2.8 Explain the relationship between memory size, program size and scan (cycle) time.

2.9 Explain the term 'associated contact', and outline, where such elements may be used in a ladder program.

2.10 Describe the different types of input and output unit that may be specified on a programmable controller. What is the purpose of providing opto-isolation on these units?

2.11 Outline the different facilities offered by small, medium and large PLCs, giving a typical area of application in each case.

## THREE

Programming of PLC Systems

In Chapter 2 we were introduced to logic instruction sets for programming PLC systems. The complete sets of basic logic instructions for two common programmable controllers are given below. Note the inclusion in these lists of additional instructions ORB and ANB to allow programming of more complex, multibranch circuits. The use of all these instructions and others is dealt with in this chapter. Some typical instruction sets for Texas Instruments and Mitsubishi PLCs are given in Table 3.1.

Table 3.1 Typical logic instruction sets.

| Texas Instruments | | Mitsubishi A series | |
Mnemonic	Action	Mnemonic	Action
STR	Store (start a new rung of a ladder diagram)	LD	Start rung with an open contact
OUT	Output	OUT	Output
AND	Series components	AND	Series elements
OR	Parallel components	OR	Parallel elements
NOT	Inverse action (used in conjunction with other instructions to invert their function, e.g. ANDNOT = NAND function.	..I	As for NOT, e.g; ORI meaning NOR function
		ORB	Or together parallel branches
		ANB	And together series circuit blocks

3.1 LOGIC INSTRUCTION SETS AND GRAPHIC PROGRAMMING

In the last chapter we introduced logic instructions as the basic programming language for programmable controllers. Although logic instructions are relatively easy to learn and use, it can be extremely time-consuming to check and relate a large coded program to the actual circuit function. In addition, logic instructions tend to vary between different types of PLC. If a factory or plant is equipped with a range of different controllers (a common situation), confusion can result over differences in the instruction sets.

RELAY LOGIC SYMBOLS: (MITSUBISHI PLC)

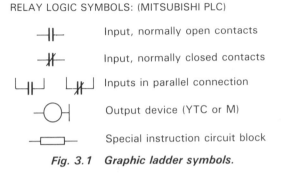

Input, normally open contacts

Input, normally closed contacts

Inputs in parallel connection

Output device (YTC or M)

Special instruction circuit block

Fig. 3.1   *Graphic ladder symbols.*

A preferable alternative is to use a *graphic programmer*, as available for several programmable controllers including the small Mitsubishi and Toshiba models from Japan. Graphic programming allows the user to enter his program as a symbolic ladder circuit layout, using standard logic symbols to represent input contacts, output coils, etc., as shown in Fig. 3.1. This approach is more user-friendly than programming with mnemonic logic instructions, and can be considered as a higher-level form of language.

The programming panel translates or *compiles* these graphic symbols into machine (logic) instructions that are stored in the PLC memory, relieving the user of this task.

Fig. 3.2   *Graphic programming panel.*

Different types of graphic programmer are normally used for each family of programmable controller, but they all support similar graphic circuit conventions. Smaller, hand-held panels are common for the small to medium-sized PLCs, although the same programming panel is often used as a 'field programmer' for these and larger PLCs in the same family. However, the majority of graphic programming for larger systems is carried out on terminal-sized units, as shown in Figs. 2.8 and 3.2. Some of these units are also semiportable, and may be operated alongside the PLC system under commissioning or test in-plant. In addition to screen displays, virtually all graphic programming stations can drive printers for hard copy of programs and/or status information, plus program storage via battery-backed RAM or tape/floppy disk. The facility to load (blow) resident programs into EPROM ICs may be available on more expensive units.

For the rest of this text, each example will be illustrated in symbolic ladder form, with the logic instructions included only in cases where the equivalence is not obvious.

Input/output numbering

It was previously stated that different PLC manufacturers use different numbering systems for input/output points and other functions within the controller. For continuity, in this chapter and Chapters 4 and 5 we will use the following range of

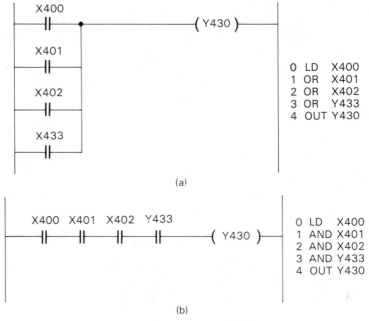

(a)

(b)

Fig. 3.3 (a) OR gate; (b) AND gate.

I/O assignments:

Inputs: X400–407; 410–413   (24 in total)
Outputs: Y430–437; 500–507, 510–513   (16 in total)

Elementary logic circuits

The basic logic gates that may be formed using ladder logic were introduced in Chapter 1 under relay systems. These gates and others are now constructed using ladder symbols and logic instructions to assist the reader in understanding the fundamentals of PC programming before moving on to more complex circuits.

Negation – NAND and NOR gates

These logic functions can be produced in ladder form simply by replacing all contacts with their inverse, i.e. AND becomes ANI; OR becomes ORI, etc. This changes the function of the circuit. For example, the AND circuit with normally closed contacts becomes a NOR circuit (see Fig. 3.4). Try checking this for yourself using truth tables.

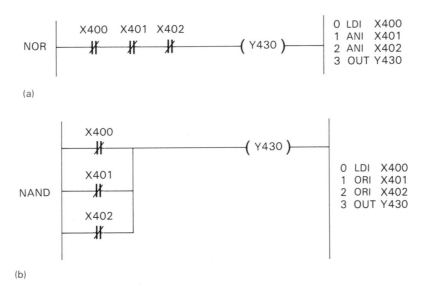

Fig. 3.4 (a) NOR gate – if any contact is opened, Y430 will de-energize;
(b) NAND gate – all contacts must be opened to de-activate Y430.

Exclusive-OR gate

This is different from the normal OR gate as it gives an output of 1 when either one input *or* the other is *on*, but not both. this is comparable to two parallel circuits, each with one *make* and one *break* contact in series, as shown in Fig. 3.5.

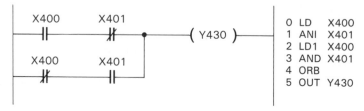

Fig. 3.5   EXCLUSIVE-OR gate.

Note the use of an ORB (OR branch/blocks together) instruction in this example. The programmable controller reads the first two instructions, then finds another rung start instruction (3 LDI X400) before an OUT instruction has been executed. The CPU therefore realizes that a parallel form of circuit exists and reads the subsequent instructions until an ORB instruction is found. This tells it to OR together the logical results of steps 0 and 1 with steps 2 and 3, then instruction 5 – OUT Y430 – specifies which output is to be driven with the result.

Programmed contacts

In the above example inputs X400 and X401 have *two contacts* in the circuit. This illustrates one of the main features of PLC programming:

> all physical and simulated functions may have as many associated contacts programmed as necessary. This includes different logical contacts (AND and ANI X400/401 above).

For example, a ladder circuit showing associated contacts for input X403 and Y433 is shown in Fig. 3.6.

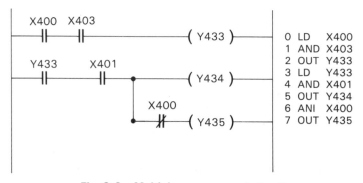

Fig. 3.6   Multiple programmed circuits.

Memory circuits (latches)

Memory elements are often required in logic control systems, being used to store brief command signals that must be transformed into continuous enabling signals. This type of operation is performed by a self-maintaining circuit or latch.

A latch circuit is shown in Fig. 3.7. It involves the output (Y430) having a bridge contact in parallel with (ORing) the initial operating contact(s). Thus, if and when the initiating contacts (X400 and X401) close, output Y430 operates and so do its associated contacts. Thus, the latch is now across X400 and X401. If either or both X contacts open, the latch path maintains power flow to output Y430, holding it on. In fact, the only way to release Y430 is by operating the normally closed contact X403.

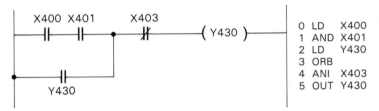

Fig. 3.7 Memory latch circuits.

3.2 FACILITIES

Standard PLC functions

In addition to the series and parallel connection of input and output (associated) contacts, the majority of control tasks involve the use of time delays, event counting, storage of process status data, etc. All of these requirements can be met using standard features found on most programmable controllers. These include *timers, counters, markers* and *shift registers*, easily controlled using ladder diagrams or logic instructions.

These internal functions are not *physical* input or output. They are *simulated* within the controller. Each function can be programmed with related contacts (again simulated) which may be used to control different elements in the program (see Fig. 3.8). As with physical inputs and outputs, certain number ranges are allocated to each block of functions. The number range will depend both on the size of a PLC, and the manufacturer. For example, for the Mitsubishi F 40-series, the details are as follows (octal numbering has been used):

Timers T	450–457⎫ 550–557⎬	16 points (elements)
Counters C	460–467⎫ 560–567⎬	16 points

The information in Fig. 3.9 illustrates the use of different number ranges assigned to each supported function. For example, the timer circuits for this

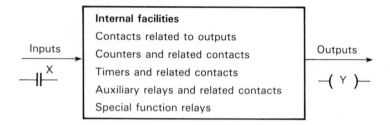

Fig. 3.8   Standard PLC functions.

programmable controller are addressed from 450 to 457 and 550 to 557, a total of 16 timers. It is the specified number that identifies a function and its point to the PLC, not the prefix letter (T in this case). These prefixes are included only to aid the operator.

The functions listed in Fig. 3.8 are provided on most programmable controllers, although the exact format will vary between manufacturers. Other functions may also exist, either as standard or by the selection and fitting of function modules to the PLC rack. Examples of a selection of these add-on functions are given in a later chapter.

PC (F40) I/O ASSIGNMENTS (BASE UNITS)		
		Key designation
Inputs: 24 points	400–407	
	410–413	X
	500–507	
	510–513	
Outputs: 16 points	430–437	Y
	530–537	
Timers: 16 points	450–457	T
	550–557	
Counters: 16 points	460–467	C
	560–567	
Auxiliary (control)	100–107	
Relays: 128 points	170–177	M
(internally	200–207	
simulated)	270–277	
Battery-backed: 64 points ⎱	300–307	
Special function ⎰	370–377	
Auxiliary relays; 5 points: 70, 71, 72, 75, 77		

Fig. 3.9   Typical number assignments to internal functions, using octal (base 8).

The operation and use of the listed standard functions is covered in the following sections.

Markers/auxiliary relays

Often termed *control relays* or *flags*, these provide general memory for the programmer, plus associated contacts. They also form the basis for shift-register construction. Normally a group of markers with battery back-up is provided, allowing process status information to be retained in the event of a power failure. These markers can be used to ensure safe start-up/shut down of process plant, by including them as necessary in the logic sequence.

Referring back to Fig. 3.9, the Mitsubishi F40 has:

128 auxiliary (marker) relays
64 battery-backed markers

Ghost contacts

In certain cases it will be necessary to derive an output from the combined logic of several ladder rungs, due to the number of contacts involved. The straightforward way of providing this is to common-up the respective circuit rungs and drive an internal relay or marker (M). This acts in the same manner as a 'physical' relay, in that it can have associated contacts – except for the fact that it is simulated by software within the programmable controller, and has no external appearance whatsoever! In common with other internal functions, auxiliary relays/markers can be programmed with as many associated contacts as desired. These contacts may be used anywhere in a ladder program as elements in a logic circuit or as control contacts driving output relays or other functions.

Example In Fig. 3.10 the two input contacts X401 and 402 are controlling output marker M100. There is an associated contact M100 across the X401 contact forming

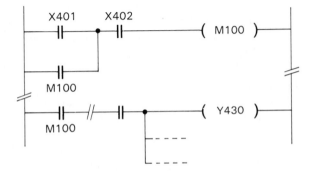

Fig. 3.10 Use of marker relay M100 and associated contacts.

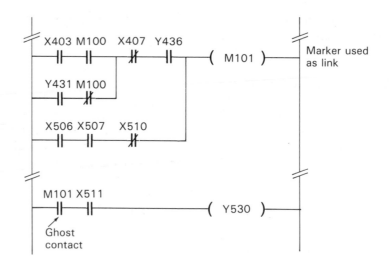

Fig. 3.11 Use of marker to link several program rungs.

a latch, whilst X402 remains operated. Elsewhere in the program, another M100 contact is used as part of a logic rung to control output Y430.

In Fig. 3.10 the use of markers and contacts has provided the means to link parts of a program together in a simple and readable manner, without the need for unnecessary use of (limited) input and output contacts or relays. Where several rungs are to be linked, the use of markers again provides the solution, as in Fig. 3.11. Here there are three rungs of logic that are to be combined to form an input to a later output circuit. Marker M101 provides the linking element, and we then place an associated contact in the appropriate circuit. In this case the M101 'ghost' contact forms an AND with X511 to control output Y530.

Typically, this may be necessary when the result of a logic 'network' is to be incorporated in a following rung or rungs. This simplifies the program and improves its readability.

Retentive battery-backed relays

If power is cut off or interrupted whilst the programmable controller is operating, the output relays and all standard marker relays will be turned off. Thus when power is restored, all contacts associated with output relays and markers will be off – possibly resulting in incorrect sequencing. When control tasks have to restart automatically after a power failure, the use of battery-backed markers is required. In the above PLC, there are 64 retentive marker points, which can be programmed as for ordinary markers, only storing pre-power failure information that is available once the system is restarted.

In Fig. 3.12 retentive marker M300 is used to retain data in the event of a power failure. Once input X400 is closed to operate the M300 marker, M300 latches

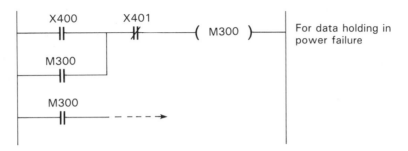

Fig. 3.12   Retentive marker used in a latch circuit.

via its associated contact. So, even if X400 is opened due to a power failure, the circuit holds on restart due to M300 retaining the 'operated' status and placing its associated contacts in the operated positions. Obviously X401 still controls the circuit, and if this input is likely to be energized (opened) by a power-failure situation, then a further stage of protection may be used.

There are many other applications of marker facilities, some of which will be shown in later sections as working elements in combination with other PLC functions.

Optional functions on auxiliary relays

From the above text it is apparent that auxiliary relays constitute an important facility in any programmable controller. This is basically due to their ability to control large numbers of associated contacts and perform as intermediate switching elements in many different types of control circuit.

In addition, many PLC manufacturers have provided additional, programmable functions associated with these auxiliary relays, to further extend their usefulness. A very common example is a 'pulse' function that allows any designated marker to produce a fixed-duration pulse at its contacts when operated, rather than the normal d.c. level change. This pulse output is irrespective of the duration of relay operation, thus providing a very useful tool for applications such as program triggering, setting/resetting of timers and counters, etc.

Pulse operation

The programming of this feature (and others like it) varies between controllers, but the general procedure is the same, and very straightforward.

A pulse-PLS instruction is programmed onto an auxiliary relay number. (Fig. 3.13). This configures the designated relay to output a fixed-duration pulse (equal to one cycle time of the program) when operated. The examples show how the relay may be used to output a pulse for either a positive or negative-going input.

The circuit in Fig. 3.14 uses a PLS instruction on auxiliary relay 101 to provide

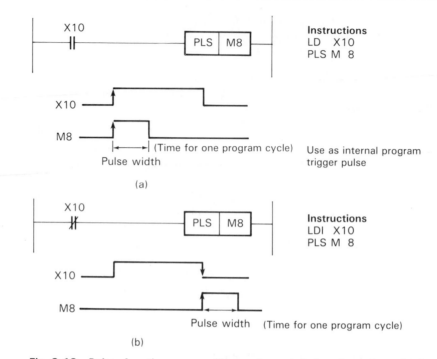

Fig. 3.13 *Pulse function on auxiliary relays: (a) rise detection circuit;*
(b) drop detection circuit.

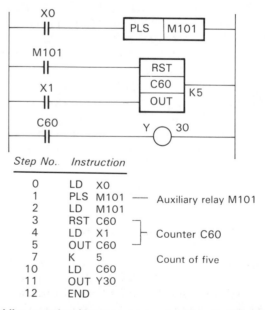

Fig. 3.14 *Providing a pulse input to a counter circuit. (See later in this*
chapter for details of counter circuits.)

a reset signal for a counter circuit C60. When input 0 is operated, a pulse is sent to relay 101, causing its contacts to pulse and reset counter C60. This is used here because counters and timers often require short duration resetting to allow the restart of the counting or timing process.

Set and reset

As with pulse-PLS, the ability to SET and RESET an auxiliary relay can often be produced by using appropriate instructions, as in Fig. 3.15. These instructions are used to hold (latch) and reset the operation of the relay coils.

The S-set instruction causes the coil M202 to self-hold. This remains until a reset (R) instruction is activated.

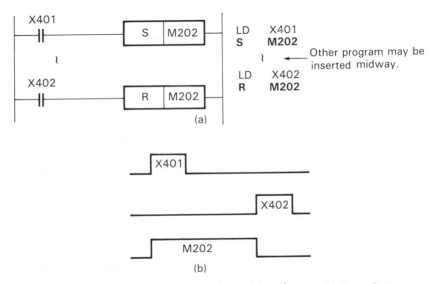

Fig. 3.15   Set and reset operations: (a) set/reset; (b) time chart.

Timers

In a large proportion of control applications, there is a requirement for some aspect of timing control. PCs have software timer facilities that are very simple to program and use in a variety of situations.

The common method of programming a timer circuit is to specify the interval to be timed, and the conditions or events that are to start and/or stop the timer function. As can be seen from Fig. 3.16, the initiating event may be produced by other internal or external signals to the controller. In this example the timer T450 is totally controlled by a contact related to output Y430. Thus, T450 begins timing only when Y430 is operated. This is caused by input X400 *and not* X401. Once

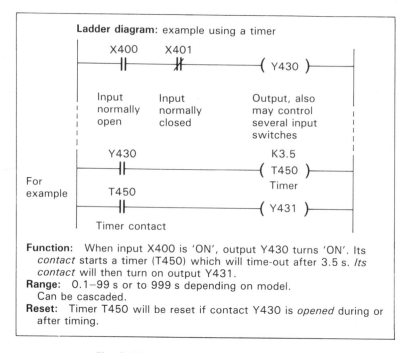

Ladder diagram: example using a timer

Function: When input X400 is 'ON', output Y430 turns 'ON'. Its *contact* starts a timer (T450) which will time-out after 3.5 s. *Its contact* will then turn on output Y431.
Range: 0.1–99 s or to 999 s depending on model. Can be cascaded.
Reset: Timer T450 will be reset if contact Y430 is *opened* during or after timing.

Fig. 3.16   Timer circuit (Mitsubishi PLC).

activated, the timer will 'time-down' from its preset value – in this case 3.5 seconds – to zero, and then its associated contacts will operate.

As with any other PLC contact, the timer contacts may be used to drive succeeding stages of ladder circuitry. Here the T450 contact is controlling output Y431. The enabling path to a timer may also form the 'reset' path, causing the timer to reset to the preset value whenever the path is opened. This is the case with most small PCs. The enabling path may contain very involved logic, or only a single contact.

Techniques for programming the preset time value vary little between different programmable controllers, usually requiring the entry of a constant (K) command followed by the time interval in seconds and tenths (or hundredths) of a second. The timers on this Mitsubishi controller can time from 0.1–999.9 s, and can be cascaded to provide longer intervals if required.

Example *Off-delay timer.* In the example of Fig. 3.17, timer T450 has been loaded with a preset of 19 s. When input X400 is operated, output Y430 is operated, placing a latch across the X400 contact. Timer T450 would also start, except for the normally closed X400 contact in its path. This contact opens whenever the initial start switch X400 is operated, not allowing timing to commence until X400 is released. This gives a delay from the time at which the switch is released rather than pressed. T450 will time down from 19 s to zero. On time-out, the T450 contact in the

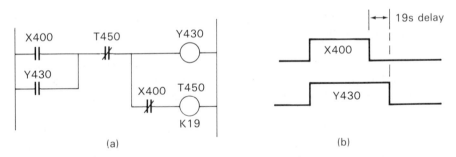

Fig. 3.17 Off-delay timer circuit: (a) ladder sequence; (b) time chart of operation.

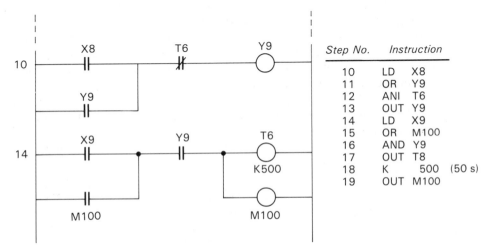

Step No.	Instruction		
10	LD	X8	
11	OR	Y9	
12	ANI	T6	
13	OUT	Y9	
14	LD	X9	
15	OR	M100	
16	AND	Y9	
17	OUT	T8	
18	K	500	(50 s)
19	OUT	M100	

This circuit is controlled by inputs X8 and X9, which both must be operated to start the release time (T6) on a preset 50 s period. This period ends with the release of Y9 by the opening of a T6 contact.

Y9 is operated as X8 is made, and holds until 50 s after X9 is operated.

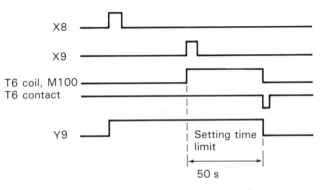

Fig. 3.18 Off-delay timer circuit.

first rung opens, breaking the operate path to Y430 and to T450, releasing the output and resetting the timer.

The examples in Figs. 3.18–22 show other applications of simple timer circuits. The numbering used for I/O and function contacts is for a Mitsubishi K-series PC – a different range from the F40 model. The reader should be able to follow the operation of these circuits by comparing them with the listings provided.

Note the frequent use of marker relays to provide latching and other functions.

In Fig. 3.20 the circuit is enabled when the X400 contact is closed, and will continue to 'flicker' until it is released. When X400 closes, this starts timer T450 which is loaded with a value of 1 s (K10). On time-out of T450, its associated contact in the next rung closes and starts timer T451. This timer is loaded with a K (constant) value of 15 to give a time period of 1.5 s. The associated T451 contact in the first rung is normally closed, and therefore opens on time-out. This action resets timer T450, causing its associated contact to open and hence reset timer T451. The timer 2 contact T451 then closes, re-activating timer T450. The above process then loops continuously until the X400 contact is released.

The effect of these operations can be seen from the timing diagrams below the figure. This shows the 1.5 s on/1.0 second off pulse routine that occurs at timer T451 and at the parallel output Y430.

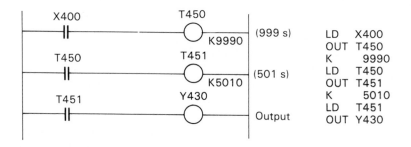

Here the two timers are used in series (cascade) to produce a time period of >999 s.

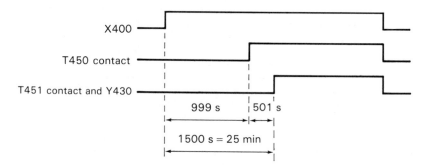

Fig. 3.19   Long-time timer by series use of two timers.

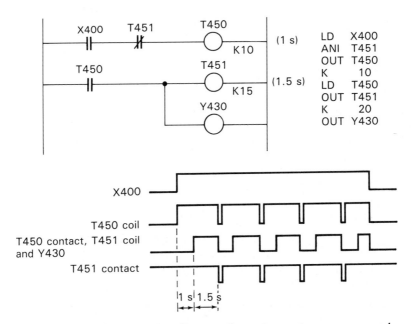

Fig. 3.20   Flicker circuit using two timers to create a programmed mark/space duration pulse train.

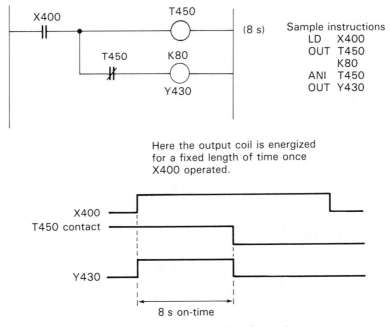

Here the output coil is energized for a fixed length of time once X400 operated.

Fig. 3.21   One-shot circuit using a timer.

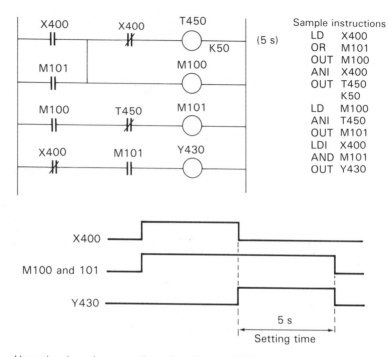

Here the timer is not activated until input X400 has been operated and released. This is via M101 contact. On time-out T450 breaks the hold path for relay M101, causing Y430 to release and breaking the operate path to T450.

Fig. 3.22   One-shot off-timer circuit.

One-shot circuits are used to obtain a fixed length of pulse or level from an initiating switch. Two examples are reproduced here: one for an ON time (Fig. 3.21) and the other for an OFF time period (Fig. 3.22).

Counters

Whenever the number of process actions or events are of significance, they must be detected and stored in some manner by the controller. Single or small numbers of events may be remembered by using latched relay circuits, but this is not suitable for larger event counts. Here programmable counter circuits are desirable, and are available on all PLCs.

Provided as an internal function, counter circuits are programmed in a similar manner to the timer circuits covered above, but with the addition of a control path to signal event counts to the counter block. Most PLC counters work as subtraction or 'down' counters, as the current value is decremented from the programmed set value (see Fig. 3.23). Here counter C460 has been programmed to reset when contact

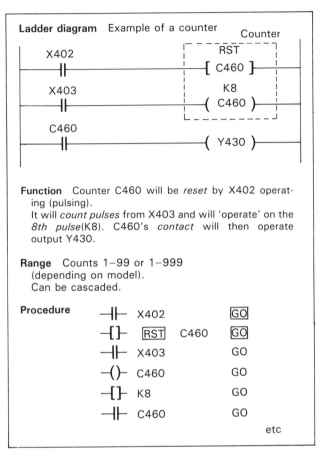

Ladder diagram Example of a counter

Fig. 3.23 Basic counter program.

X402 operates, and will count pulses from the X403 contact that is connected to its input line. The counter will operate after 8 pulses are received from X403, whereupon its associated contact C460 will operate output Y430. If the reset line is activated at any time during the count, the counter will reset to 8.

Further operation of the counter will be suspended whilst the reset line is activated.

The counters used here can be configured to count up to 999 events, and may be cascaded to provide larger values as necessary. In case of power failure, there will normally be specific counter circuits that are battery-backed to prevent loss of critical data.

Further examples of counter circuits and their applications are given in Figs. 3.24 and 3.25.

Fig. 3.24 uses two marker relays M100 and M101 driven by contacts associated with the same input, X405. The effect of the normally closed M101 contact in the operating path of M100 is to produce a short-duration pulse when X405 is closed.

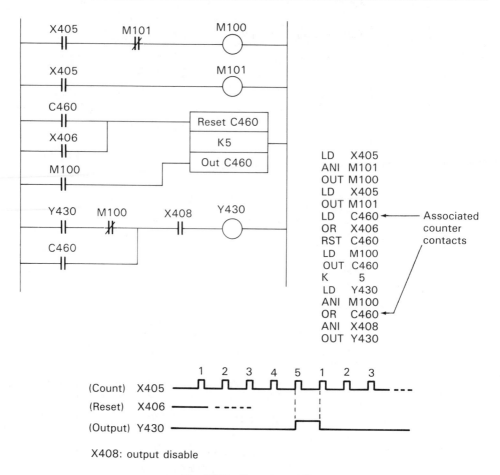

Fig. 3.24   Preset counter.

This pulse is offered to the counter input line via an M100 contact as shown. Thus, each closure of X405 passes a precise pulse to the counter. Counter C460 is loaded with the value of four, and will operate any associated contacts once this number of pulses has been received. In this case there are two such contacts: one controlling output Y430 as a parallel (OR) element across a Y430/M100 latch pair (with M100 normally closed), and a second C460 contact in parallel with the X406 contact on the reset line to the counter itself. The former C460 contact acts as the initial enabling path for output Y430: the other branch performs as an output latch (Y430 *and not* M100) once the counter is reset by its own contact. This latching holds only until a further X405/M100 pulse occurs, then releasing as a new count sequence begins.

X406 is a separate reset switch, and X408 acts as an output enable/disable switch.

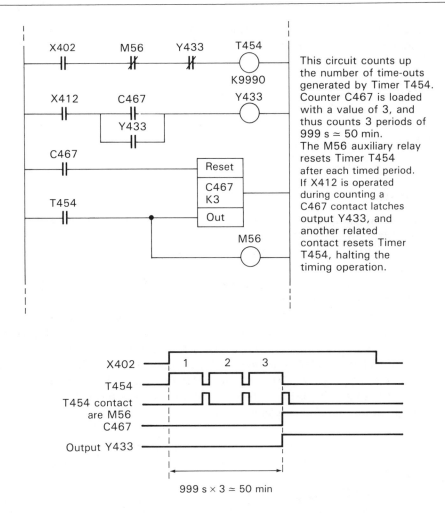

This circuit counts up the number of time-outs generated by Timer T454. Counter C467 is loaded with a value of 3, and thus counts 3 periods of 999 s ≃ 50 min. The M56 auxiliary relay resets Timer T454 after each timed period. If X412 is operated during counting a C467 contact latches output Y433, and another related contact resets Timer T454, halting the timing operation.

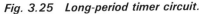

999 s × 3 ≃ 50 min

Fig. 3.25 Long-period timer circuit.

Counter circuits to operate after power failure

Where a controller is used in an application that involves event counting, it may be desirable to ensure the retention of counted information through a power failure, in order that the operation can continue correctly once power is restored. The solution is to use battery-backed counter and timer elements (if available). In order to ensure that a retentive counter/timer circuit is not *reset* by a power failure, it is wise to use normally open contacts in the reset line. This practice ensures that the return of power will not cause a false reset path purely by the inputs becoming energized. Were a normally closed contact to be used here, normal counter operation would require the reset contact(s) to be operated (open); reset would occur when contacts were off (closed) and the path was made. A power failure could result in the release of inputs and other internal contacts, causing the counter to be reset on repower-up.

Other applications

It is obviously possible to use many different combinations of counter/timer circuits, involving any other programmable controller functions and contacts in the control lines to these elements.

Registers

From using a single internal or external relay as a memory device to store a single bit of information, other PLC facilities allow the storage of several bits of data at one time. The device used to store the data is termed a register, and commonly holds 8 or 16 bits of information. Registers can be thought of as arrays of individual bit-stores – in fact many programmable controllers form the data registers out of groups of auxiliary marker relays (see Fig. 3.26).

Registers are very important for handling data that originates from sources other than simple, single switches. Instead of binary data in one-bit-wide form, information in a parallel data form may be read into and out of appropriately sized registers. Thus, data from devices such as thumbwheel switches, analog-to-digital converters, etc. can be fed into appropriate PLC registers and used in later operations that will generate other bit- or byte-wide (8-bit) data to drive switched outputs or digital-to-analog conversion units.

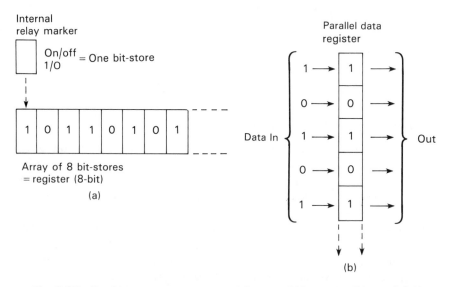

Fig. 3.26 *Register storage concepts: (a) array of bit-stores; (b) parallel data register.*

Shift registers

Before looking at parallel data registers (section 3.4), we will examine another form of storage unit provided on PLCs – the shift register.

A shift register provides a storage area for a sequence of individual data bits that are offered in series (one after the other) to its input line. The data are moved through the register under control of a 'shift' or clock line (see Fig. 3.27). The effect of a valid shift pulse is to move all stored digits one bit further into the register, entering any new data into the 'freed' initial bit position. Since a shift register will only be a certain size, for example 8 or 16 bits, then any data in the last bit of the register will be shifted out and lost.

The usefulness of a shift register (SR) lies in the ability to control other circuits or devices via associated SR contacts that are affected by the shifting data stream through the register. That is, as with marker relays, when a marker is ON any associated contacts are operated.

In programmable controllers, shift registers are commonly formed from groups of the auxiliary relays. This allocation is done automatically by the user programming a 'shift-register function', which then reserves the chosen block of relays for that register and prohibits their use for any other function (including use as individual relays).

The example in Fig. 3.28 shows a typical circuit for shift register operation on a Mitsubishi PLC. Here the register is selected by programming in the shift (SFT) instruction against the auxiliary relay number to be first in the register array – M160. This instruction causes a block of relays – M160–167 – to be reserved for that shift register. Note that only the first relay had to be specified, the remainder being implied by the instructions.

This shows the controlling contacts on the input lines to the register – RESET,

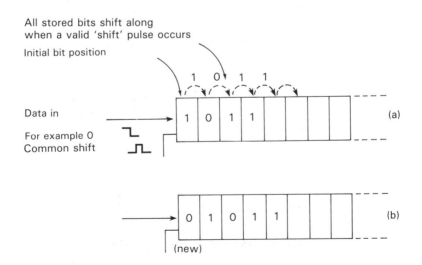

Fig. 3.27   Shift register operation: (a) before shift; (b) after 1 shift pulse.

The auxiliary relays can be grouped in blocks of 8 to form 8- or 16-bit shift registers. This feature is programmed as shown below using M160–177 internal relays (only M160 is keyed in, the other bits being transparent).

Shift register program

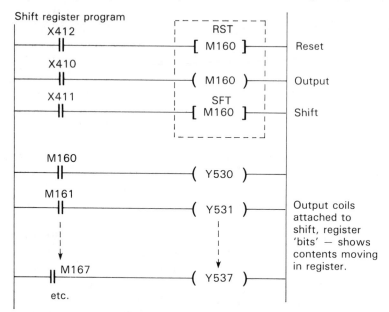

The shift register contacts perform as follows:

RST – a pulse or closure resets SR contents to 0
OUT – logic level (0 or 1) offered to register on this rung.
SFT – pulse moves contents along one bit at a time (eventually contents are lost off the final bit memory).

Fig. 3.28   Basic shift register circuit.

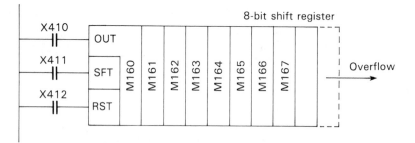

Fig. 3.29   Equivalent circuit of a shift register.

OUTPUT and SHIFT. Note the M-contacts below the SR circuit that are used to drive output coils (M160–167 driving Y530–537).

It is easier to understand the function of the register if we look at an equivalent circuit (Fig. 3.29). Here we can see the layout of other marker relays following M160. This helps us to visualize the shifting of data from bit to bit, affecting other parts of the circuitry as the data (1 or 0) in each bit changes.

Shift registers are commonly found as 8- or 16-bit, and can usually be cascaded to create larger shift arrays. This allows data to be shifted out of one register and into a second (cascaded) register, instead of being lost. Battery-backed markers can be selected as the register elements if it is necessary to retain register data through a power failure.

Applications of shift registers

There are many situations where the use of a shift register can save considerable amounts of conventional logic programming, both for interlock and sequential circuit requirements. (See later chapters for further details.)

A simple and common application is to use a shift register to keep track of particular items in a production system, as in Fig. 3.30. This shows the situation where finished goods are moving along a conveyor belt, with a photocell PH1 sensing for rejected items. This enters a 1 bit into the shift register for a faulty unit. There is a limit switch LS1 mounted on the conveyor mechanism to send a pulse to the SHIFT line of the shift register each time an item (good or bad) is moved on the conveyor. The requirement is for the reject items (found by PH1) to be dropped through a trap door into a reject bin at a certain point in the conveyor. Thus the shift register must track a faulty item as it moves along the conveyor and then cause the reject trapdoor to open at the correct time – a 1 in M144 – via a solenoid Y30. The trapdoor mechanism then opens provided the M101 auxiliary relay contact is not operated. A further photocell PH2 will detect reject units falling into the bin and will release the trapdoor solenoid mechanism via marker M101 to ensure that following 'good' units will not fall. M101 has a latch to ensure the trapdoor is closed, since X3 (PH2) provides only a fleeting pulse when an item falls past.

In addition to a unit pulse, the limit switch LS1 controls two auxiliary relays – M100 and M102 with a short pulse operation (PLS). The effect of the normally closed M100 contact in the data OUT line is to disable data input from the PH1 photosensor when a shift is in progress. This is to prevent possible sequence errors. The M102 pulse contact in the operate path to M101 is to ensure resetting of the trapdoor between each item.

There is obviously a requirement that the control system be able to hold data for up to 100% of item failures on the conveyor – a feature inherent to shift registers.

A switch X2 is provided to allow resetting of the register, switching all relays to 'off' and ignoring any shift or data inputs whilst the switch is on.

A further development of shift registers is the drum timer, which effectively combines an array of registers with counter/timer facilities. This function is examined later in Chapter 5.

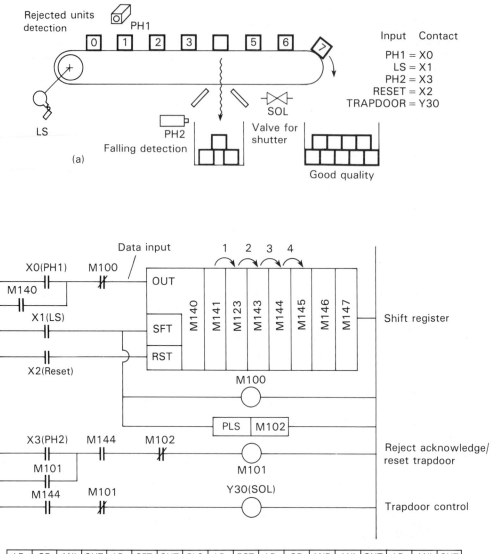

(a)

(b)

Fig. 3.30 Use of a shift register to track faulty items: (a) keeping track of
faulty items among good ones, e.g. a 1 entered and shifted each time a
faulty item is encountered; (b) trapdoor (Y30) to open when a reject (1) is
shifted into position M144 on the shift register (courtesy Mitsubishi Electric
UK Ltd).

Jumps

In the area of conventional computers and programming facilities, one of the most powerful facilities is the ability to move or JUMP to different parts of a program following a 'test' of some condition by a previous instruction or routine. This allows different actions to be taken depending on the condition or result of the test.

The instructions CJP and EJP (conditional jump and end jump) are provided in the Mitsubishi F40. These allow the skipping of sections of a program *if a condition* exists to activate the jump. The elements contained between JUMP and END are then held in the previous state. The instructions must be followed by a 3-digit octal number between 700 and 777. Up to 64 jumps can be employed.

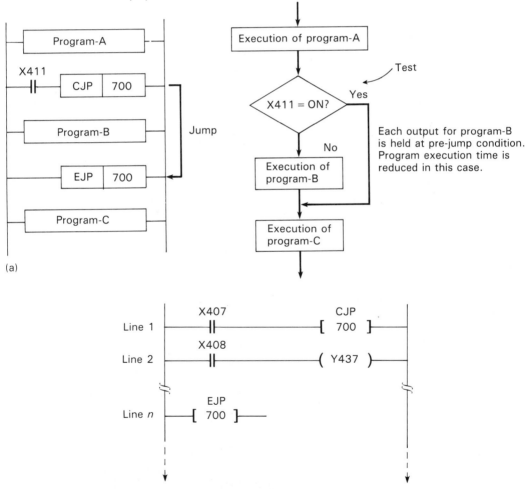

(a)

(b)

Fig. 3.31   *Conditional jump instructions: (a) concepts; (b) ladder example.*

A similar facility is often provided in programmable controllers, and may be used to great effect in control applications where there are several options of actions to be selected following a particular test or condition in the process. These operations are termed *conditional jumps*.

As with most PLC functions, the conditional jump can be programmed to activate from a logic circuit containing very simple or very involved logic or sequencing instructions. Fig. 3.31 gives an example of this operation for a Mitsubishi PLC, where a jump is started from a conditional jump instruction (CJP) and moves program execution to a corresponding end jump instruction (EJP).

These instructions and the associated ladder logic are simply placed in a user program to transfer execution to the position of the EJP code. In Fig. 3.31(b), if input X407 is operated (closed), the jump will occur to line *n* and line 2 will *not* be executed. If X407 is open, however, then line 2 is active, and if X408 closes, then output coil Y437 will operate. Here the 'test' for a jump is performed by the X407 contact, but a far more complex circuit could be used here.

As can be seen from the figures, a pair of CJP and EJP instructions are required to identify each jump operation, identified by the three-digit octal number that must follow each code. Thus, when a jump is selected during program execution, the PLC CPU looks for an end-of-jump instruction with the same octal code in order to recommence execution at that point.

Nested and multiple jumps

The use of jumps within other jumps is possible, and an example is given in Fig. 3.32.

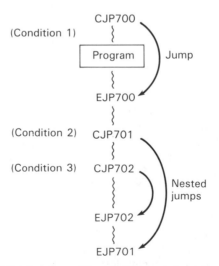

So if condition 2 is not met, condition 3 will be tested and, if met, CJP702 will occur.

Fig. 3.32   Nested jumps.

Multiple jump instructions with the same destination can usually be
programmed using a common identifier

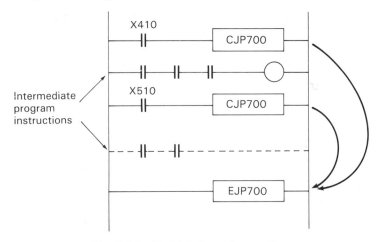

Fig. 3.33   Multiple jump instructions.

Another useful feature of jumps is the ability to program multiple 'start' or
'enter jump' instructions. This allows a jump to be entered from several different
locations in a section of program, ending at the same EJP instruction position
(Fig. 3.33).

As with all other instructions, care must be taken when laying out a program
that uses multiple or nested jumps, to ensure there is a logical pathway through *all*
jump operations both when a jump does occur and when it does not. Failure to
correctly structure this, and other aspects of program design can result in error-
prone, incomprehensible programs – all of which is unnecessary!

Applications of jump instructions

Fig. 3.34 illustrates a few useful applications of jumps, both as single control
elements and in combination with other functions.

3.3   SPECIAL FUNCTION RELAYS

Most programmable controllers are equipped with certain special function relays
that provide useful or essential facilities for the development of safe and optimal
control programs.

A common practice is to provide a reserved group of auxiliary relays for this
task, allowing the use of their associated contacts throughout any program to utilize
these 'special functions'. For example, Fig. 3.35 describes those of a Mitsubishi F40

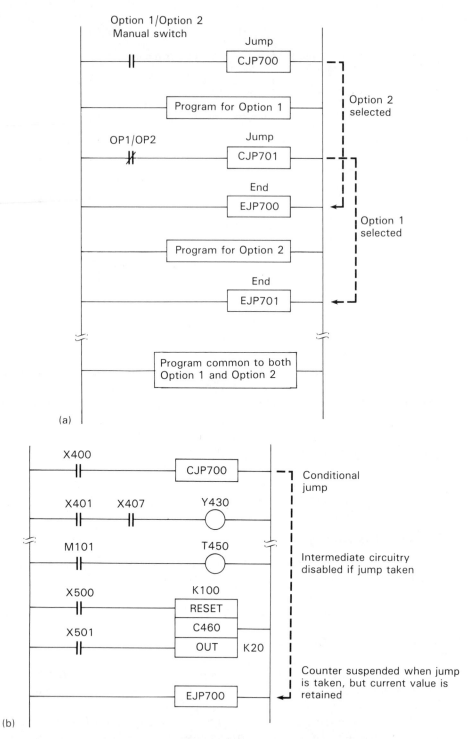

Fig. 3.34 *Applications of jump instructions: (a) switching between control options; (b) use with timer and counter circuits.*

PLC. Other functions are now being provided for more sophisticated applications, including facilities of high-speed counting, reversible registers and counters, etc.

These facilities can all be included in user programs either by operating a relay coil directly, or by using associated contacts. In this manner we can create such actions as auto-start, timing, fail-safe operation, emergency stops, etc.

- **Inner connection and application**

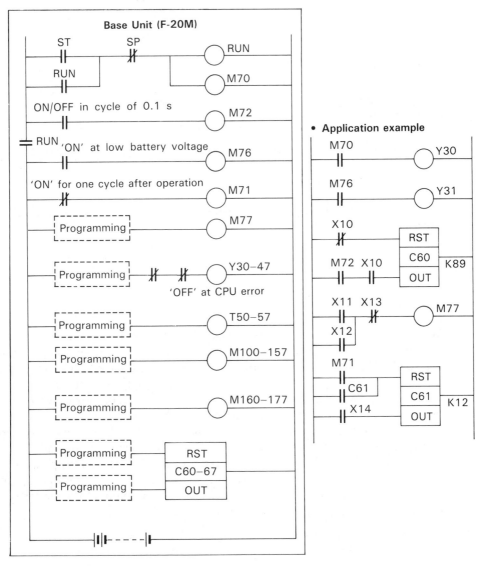

Fig. 3.35   (Continued overleaf)

In this example the relays are used as follows:

70 is used to display the RUN mode (in operation controlling a process or machine) with an external lamp, etc. through OUT 30.

71 introduces a single pulse at the commencement of a 'RUN' sequence, which can be used to reset a counter or shift register initially.

72 can be used to adapt battery-supported counters to timers, useful because the actual times do not have battery support and revert to the beginning of their time sequence when mains power is interrupted.

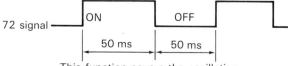

This function covers the oscillation circuit function.

The life of output relays will be shortened if the contacts of auxiliary relay 72 are used for their direct activation.

76 can be used to light external warning lamps, etc. in the case of low battery voltage through output 31 on the above example. A warning indicator is also provided through an LED on the base unit.

77 can be used to provide an emergency stop switch, etc. As shown in the inner connection above, all output terminal relays are switched off when relay 77 is activated. However, all other contacts such as input relays, auxiliary relays, timers and counters continue in operation according to the program.

The above special function relays (except relay 77) cannot be used for (out) instructions in a program.

As shown in the above inner connection diagram the base unit has another special internal relay, which is activated in the event of a CPU error, and switches output terminal relays 'OFF'.

Fig. 3.35 Special function relays.

Master control relays

When it is desirable to control large quantities of outputs or functions, a master control relay function is often used. This function allows the enabling of sections of logic and related outputs when certain major criteria are satisfied. Correspondingly, the same sections of program will be turned off (disabled) when the 'master' criteria are *not* met. This could be solved with conventional ladder logic by including a permissive contact in each rung controlling one of the output coils, but would require the use of many contacts and associated use of PLC memory. A preferable alternative is to divide the program into sections where this overall control is required, and enclose the sections with Master Control 'start' and 'stop' instructions as necessary. Fig. 3.36 illustrates this concept for a Mitsubishi F series PLC. In the 'F' series machine, the auxiliary relays M100–177 can be used for master control operations, by programming the following instructions:

MC (master control)   :   common series connection
MCR (master control reset) :   reset the MC instruction

These instructions are used with a marker relay from the range listed above.

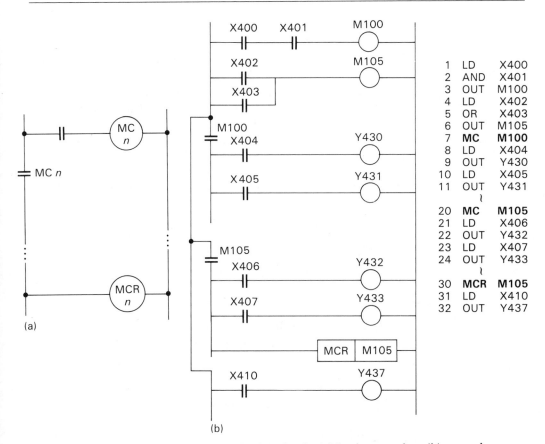

```
 1   LD    X400
 2   AND   X401
 3   OUT   M100
 4   LD    X402
 5   OR    X403
 6   OUT   M105
 7   MC    M100
 8   LD    X404
 9   OUT   Y430
10   LD    X405
11   OUT   Y431
        ⟨
20   MC    M105
21   LD    X406
22   OUT   Y432
23   LD    X407
24   OUT   Y433
        ⟨
30   MCR   M105
31   LD    X410
32   OUT   Y437
```

Fig. 3.36 Master control relay circuit; (a) basic operation; (b) example.

The program in Fig. 3.36 shows marker M100 with contacts X400 and 401 as the activating conditions, with M101 programmed in the next rung. A master control contact is entered, MC M100. This has the effect of creating a 'break' between the left-hand bus bar and any circuitry following the MC contact. Thus when the common series contact M100 is turned off, the succeeding rungs for Y430 and Y431 are also turned off. Further down the ladder, an M105 master control contact determines whether Y432, Y433, etc. are enabled. In all cases, the instruction after the master control must start a new rung, e.g. LD, LDI.

The end of a master control section is signified either by the occurrence of another MC series contact (lines 7 and 20 in Fig. 3.36), or by programming an MCR reset instruction after the section of logic that is to be controlled. The MC and MCR are like brackets or quotations around the logic to be controlled. At line 30 of the program an MCR instruction is used to reset the M105 master control, since it is not immediately followed by another MC section.

In this example, unless X400 *and* X401 are closed, Y430 and Y431 will be off. Likewise, X402 *or* X403 must be operated to activate master control contact M101, otherwise Y432 and Y433 will be off. The remaining rung with an X410 contact and Y437 coil is unaffected by any master control instructions.

Fig. 3.37 shows a master control program for a General Electric Series 1 PLC. Note the similarity of the instructions for this application compared to the Mitsubishi code, except that in the GE program a master control start or reset is not associated with a marker relay, but is a separate relay provided for this purpose.

The use of multiple master control functions are possible in any logic program, each controlling as many logic elements as are required. The MC functions may also be interleaved if necessary, as shown in Fig. 3.38 for a GE controller. In this example the first group of outputs under master control of contacts 010 OR 011 is 304–313 inclusive. This includes two subgroups of master control: outputs 307–310 are

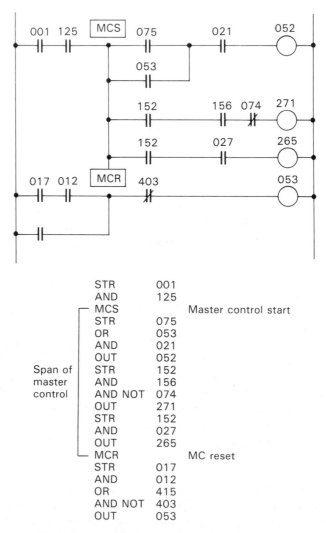

Fig. 3.37   *GE Series 1 master control example.*

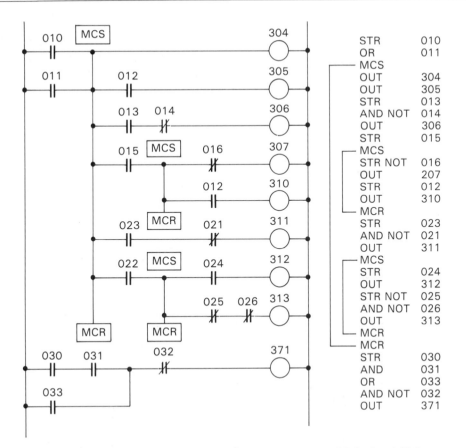

Fig. 3.38 Sample multiple master control logic on a GE Series 1 PLC.

master controlled by contacts 010 or 011 AND 015, whilst outputs 312 and 313 are dependent on contacts 010 or 011 AND 022. Notice the use of two successive MCR reset instructions to complete both the subgroup and the major group of master control.

NOTE: care must be taken when using conditional jump instructions in conjunction with master control relays. Any programmed jumps completely outside MCR blocks are unaffected, and jumps confined within any one MCR block will be enabled/disabled with the MCR, but jumps that cross MCR boundaries are affected in a variety of ways depending on the direction of the jump. The latter case must be checked using manufacturer's data or programming guides.

3.4 DATA-HANDLING INSTRUCTIONS

With the exception of the shift register, the previous instructions and elements in this chapter have all dealt with the handling of individual bits of information, rather than complete bytes or words of data. For a large proportion of logic control applications, controller operation on bit information is perfectly adequate. However, when a control task involves operations on related groups of bits – for example a block of 8 input points – then it is convenient to be able to reference the bits as a data word or byte (depending on the number of bits involved).

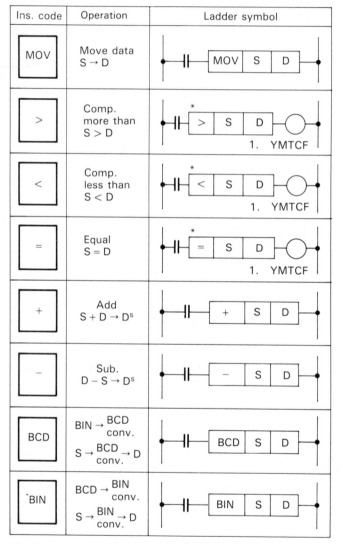

Ins. code	Operation	Ladder symbol
MOV	Move data $S \rightarrow D$	⊣⊢ MOV S D
>	Comp. more than $S > D$	* ⊣⊢ > S D —○ 1. YMTCF
<	Comp. less than $S < D$	* ⊣⊢ < S D —○ 1. YMTCF
=	Equal $S = D$	* ⊣⊢ = S D —○ 1. YMTCF
+	Add $S + D \rightarrow D^s$	⊣⊢ + S D
–	Sub. $D - S \rightarrow D^s$	⊣⊢ – S D
BCD	BIN → BCD conv. $S \rightarrow$ BCD conv. $\rightarrow D$	⊣⊢ BCD S D
BIN	BCD → BIN conv. $S \rightarrow$ BIN conv. $\rightarrow D$	⊣⊢ BIN S D

S means source; D, destination   * Cannot handle negative value.
1. Instructions may be applied to: outputs, markers, timers and counters.

Fig. 3.39   *Typical data-handling instructions (Mitsubishi K Series PC).*

The operations that may be performed on data words will normally include:

- data move
- magnitude comparison $(>,<,=)$
- addition and subtraction
- conversion between number systems (BCD, binary, octal).

These operations allow a programmable controller to handle signals other than discrete switched digital inputs and outputs. For example, analog signals may be offered to the PLC once they have been converted into a digital representation of 8, 10 or 12 bits. (See Chapter 6 for further details of analog I/O.) An example list of the instructions used to program these operations is given in Fig. 3.39.

The data move operations are also used to enhance the existing facilities, for example reading in data from manual thumbwheel switches to alter set values for timers and counters. This type of application is very common, allowing operators to alter various control parameters before or during a program cycle. Other devices interfaced to the PLC such as high-speed counters, can be read using MOVE instructions. Outputs include LED numerical displays, analog signals and counter presets.

Most of the data operations listed above are supported in a similar form by the majority of small programmable controllers, with larger machines (and an increasing number of smaller machines) providing more sophisticated mathematical functions. These include multiplication and division, square roots, etc., through to more advanced scientific operations in certain cases. Programmable controllers with comprehensive instruction sets such as these are used in applications that previously required a much more expensive 'control computer' to handle the complete range of process signals and operations.

Operation of data-handling instructions

As with previous instructions and operations, the various makes of PLC deal with data operations in different ways. The following sections look at general methods and procedures, with specific examples included to illustrate particular examples or points.

Memory

In common with other microprocessor-based systems, programmable controllers store data in memory locations that are each identified by a unique address. This allows the CPU to read or write data at a specific location, using the data for operations within the control program. As we discussed in previous sections, PLCs have areas of memory designated to specific functions. These include:

- Input memory   X ⎱
- Output memory   Y ⎰   or I/O RAM areas

- Timer and counter
  memory   T, C
- Auxiliary relays   M
- Constant value   K   used to hold preset values for T, C, etc.

We do not tend to think about memory usage when creating programs with the above elements, but they occupy a significant portion of PLC memory space. This may be automatically reserved for each function, always leaving a certain area of RAM for user programs. However, some modular systems require RAM to be designated for each function (including I/O points), with the result that memory available for user programs is reduced as more functional RAM area is taken up.

Previous examples have used number codes to specify the various elements, such as X400, M100, T450, etc. These refer to memory addresses for the chosen element. Data instructions also require memory addresses to be specified so that data words may be moved from one location to another. Locations in PLC memory are allocated for the storage of data words, commonly termed 'data registers'.

- data registers D
  e.g. D 0–99.

As for any other function, there is a numbered array of these registers, providing say, 100 locations in total. Each data register can store a binary word of data, usually 8 or 16 bits in size. The length of data register determines the maximum value of any stored number. For a 16-bit register, each word is regarded as a binary number between 0 and 65 535 ($2^{16} - 1$). In comparison, an 8-bit word can represent up to only $2^8 - 256$ in decimal. Therefore 16-bit operation provides greater precision. However, two 8-bit words may be used to form an effective 16-bit word for many applications, by performing operations on each pair of 8-bit words in turn.

The example in Fig. 3.40 shows the layout of a typical 16-bit register, together with details of the bit-positions. The most significant bit (MSB) in a word is bit 15, which represents 2^{15} (32 768) when set to a 1. Bit 0 is the least significant bit, representing 2^0 (1) when set. To obtain the total value of the word we add the values of all the bits that are set to a 1.

The CPU of the programmable controller operates with either 8-bit or 16-bit data words, depending on the size and age of the system. Smaller, older PLCs tend to have 8-bit microprocessors, with larger controllers and most modern systems using 16-bit operation or multiprocessor configurations to produce high-capacity/high-speed operation. Processors that can handle data in 16-bit words have

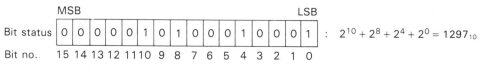

Fig. 3.40   16-bit register example.

a much greater potential throughput than their 8-bit counterparts, performing arithmetic operations directly on the 16-bit data words.

Programming data operations

To effect data handling in a PLC program we must give both the source and destination addresses (Fig. 3.41). Here we see a typical instruction containing three areas of information:

1. The data-handling instruction.
2. The source address S to obtain the data for the operation.
3. The destination address D for storage of the result – (not to be confused with the D for data register).

Depending on the data operation to be performed, source and destination addresses may be referenced to inputs, outputs, auxiliary relays or the other supported functions such as timer and counter: it is not always desirable or possible to transfer (or operate on) a data word in certain combinations of source and destination types, without using a data register as an intermediate storage area. For example the MOV (move data) instruction described in Fig. 3.42 allows data to be moved only from Dn to X, Y, M, T, C, K, D and from X, Y, M, T, C, K, D to Dn. The ladder circuit for a MOV instruction is also shown.

This type of instruction is provided to get information into and out of the PC, allowing the reading of input data and the output of resulting signals via physical and programmed devices (Y, T, C, etc.). Once data has been transferred into data registers it can be operated on by other arithmetic instructions. These normally use data registers for source and destination purposes.

Data-handling instructions are composed of three steps, and their notation method is shown in the following data.

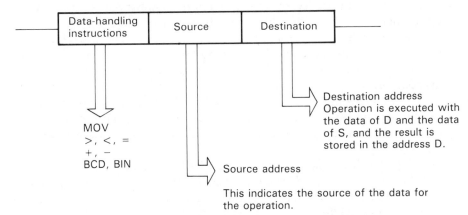

Fig. 3.41   Data-handling operations.

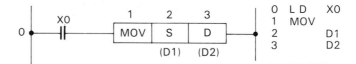

When X0 becomes ON, the data of S (source register) D1 are
transmitted to D (D2) (detination). The possible operation S/D
combinations include the following:

Constant to register	MOV	K	D
Register → register	MOV	Di	Dj
Timer/counter to data register	MOV	T or C	D
Input to data register	MOV	X	D
Data register → timer or counter	MOV	D	T or C
Data register → output	MOV	D	Y
Data register → marker	MOV	D	M
Marker → data register	MOV	M	D

Fig. 3.42   Data move instruction (MOV).

3.5   ARITHMETIC INSTRUCTIONS

BCD numbering

All internal CPU operations are performed in binary numbers. Since it may be
necessary to deal with decimal inputs and outputs in the outside world, conversion
using binary-coded-decimal (BCD) numbering is provided on most PLCs. BCD
numbering is briefly described in Fig. 3.43. Readers wanting further information are
referred to the many texts dealing with number systems. When data is already in
binary format, such as analog values, it is placed directly in registers for use by other
instructions.

Decimal input/output is often required for operator input via thumbwheel
switches or similar devices, with a decimal display of certain information back to the
operator. For example, the count-down status of a counter or timer (see the unit in
Fig. 3.34(b)).

Fig. 3.39 listed the common instructions for data handling in a programmable
controller. Since the actual layout of each instruction is not dissimilar to the above
MOV example, full descriptions are not provided here. The remaining part of this
chapter gives a brief discussion the uses of these functions.

(1) BIN (pure binary)

	X7	X6	X5	X4	X3	X2	X1	X0	
Upper digit	1	1	0	1	0	0	1	1	Lower digit
	128	64	32	16	8	4	2	1	

$$128 + 64 + 16 + 2 + 1 = 211$$

In the data made up of 8 bits from X0 to X7, the number 211 is expressed when X0, X1, X4, X6 and X7 are turned on (= 1 in the above figure) and the others are turned off (= 0 in the above figure).

(2) BCD (binary-coded decimal)

M113	M112	M111	M110	M107	M106	M105	M104	M103	M102	M101	M100
1	0	0	1	0	1	1	0	0	1	1	1
800	400	200	100	80	40	20	10	8	4	2	1

100s digit	10s digit	1s digit

$$(800 + 100) + (40 + 20) + (4 + 2 + 1) = 967$$

BCD data is such that each digit of the decimal number is expressed in 4-bit binary. No digit will exceed 9.

E.g.: If both M103 and M102 are turned on (= 1) in the BCD data shown in the above figure it will result in an error.

The values of timers or counters may be treated as BCD

(a)

(b)

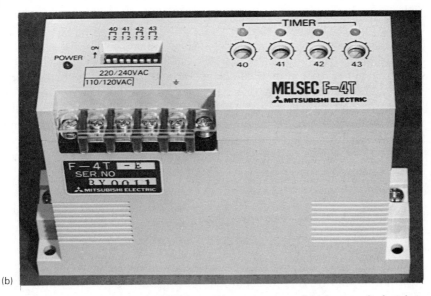

Fig. 3.43 *(a) Binary and BCD number systems; (b) timer unit for data operations (courtesy Mitsubishi Electric UK Ltd)*

Magnitude comparison

Magnitude comparison instructions are used to compare a digital value read from some input device or timer, etc., with a second value contained in a destination data register. Depending on the instruction – more than, less than, or equal – this will result in a further operation when the condition is met. For example, a temperature probe in a furnace returns an analog voltage representing the current internal temperature. This is converted into a digital value by an analog-to-digital converter module on the PC, where it is read from input points by a data-transfer instruction and stored in data register D10. The process requires that if the temperature is less than $200°C$, then the process must halt due to insufficient temperature. If the temperature is greater than $200°C$ and less than $250°C$, then the process operates at normal rate (e.g. items are baked for 5 minutes each). If the temperature is between 250 and $280°C$, then baking time is to be reduced to 3 minutes 25 seconds, and once temperature exceeds $280°C$ the process is to be suspended.

This is the type of area where magnitude comparison can provide the necessary control, in conjunction with other circuitry to drive the plant equipment.

Other common applications include the checking of counter and timer values for action part-way through a counting sequence.

Addition and subtraction instructions

These instructions are used to alter the value of data held in data registers by a certain amount. This may be used simply to add/subtract an offset to an input value (in order to place it within range) before it is processed by other instructions. For example, when two different sensors are passing values to the controller and one sensor signal has to be compared against the other, but is a fundamentally smaller signal with a narrower output swing. It may be possible to add an offset to the smaller signal to bring it up near to the level of the larger one, thus allowing comparison to take place. The alternative would be to use signal conditioning units to raise the sensor output before the PLC – an expensive option.

Other uses of + and − include the alteration of counter and timer presets by programmed increments when certain conditions occur.

Extended arithmetic functions

All programmable controllers are being given more advanced features as the demand dictates, including more advanced arithmetic and data-handling facilities. Most larger PLC systems possess these facilities, along with a growing number of smaller controllers.

Extended arithmetic functions may include:

- Double-precision add and subtract
- Multiplication and division
- Scientific functions.

In many cases these arithmetic functions are carried out on signed values, where any value can be a positive or negative number. This can be very useful when dealing with analog values of between, say, -10 and $+10$ V, thermocouple input signals of between $-120°$F and $+60°$F, or any other parameter that can produce an output which swings between positive and negative values.

The double-precision functions normally use a pair of consecutive 16-bit registers to represent each value, allowing integer values of up to plus or minus 2 147 483 647 2^{31} (-1) to be handled. Apart from the size of the data words, double-precision arithmetic operates in a similar fashion to conventional signed arithmetic.

Multiply operations normally use two single-precision (single register) signed values and multiply the contents. This results in a signed double-precision product that occupies two registers. Division operates on a double-precision signed number (pair of registers) and divides it by a single-precision signed value, resulting in a signed single-precision quotient together with a remainder.

Multiplication and division facilities allow the construction of relatively sophisticated mathematical algorithms that can be used to (amongst other things) process signals and values to perform direct closed-loop control. This topic is considered again in Chapter 6 under advanced features.

3.6 DATA MANIPULATION

Tables

Further to the word-move instructions introduced above, the development of data tables now allows the manipulation of whole groups of consecutive storage locations – data registers. As shown in Fig. 3.44 a data table is made up from a number of consecutive registers (from 1 to 256) selected by the user program. There is a pointer register at the start of the table that points to the next register to be accessed.

Tables are constructed by transferring in register contents that contain data required to build up indexes or recipes on, say, production data or machine diagnostics, etc. This data can then be accessed either as a total table or as individual locations via table-to-table or register-to-table instructions.

Table-to-table moves transfer the entire contents of the specified table into a second table. This type of operation is used for transferring data between PLCs via communications modules, and for passing data between function cards on a single system.

Other table functions exist that allow the addition, removal and sorting of the table contents, making this a very powerful tool for data manipulation.

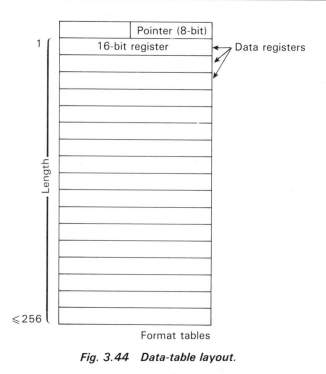

Fig. 3.44 Data-table layout.

Matrix bit operations

A further variation on the idea of data tables – that of logic matrices – is provided on many larger controllers, including the General Electric Series 6 PC – a large modular system. The Series 6 supports logical operations on a matrix basis, where arrays of bit-storage locations are arranged and accessed in a similar manner to data tables. Instead of 16-bit registers containing numerical values to be processed, the matrices are concerned with the logical status of the discrete bits that make up the 16 bits. Each bit in the matrix has a single identifying number or name made up of the register name and the bit location within that register. With a possible size of 16×255 bits, a matrix can be up to 4080 bits in total. The bits of a matrix can be compared against those of a second matrix using normal logic instructions – AND, OR, EOR, etc., with the results being placed in a third matrix for examination. A compare function is a very powerful facility when using logic matrices, replacing hundreds or thousands of relay logic instructions. The compare operation forms the basis for several maintenance diagnostic functions. If any miscomparisons are detected, the bit-addresses of the error source are output by the matrix logic, activating other logic elements that may drive alarm lamps, message annunciation or any other operation.

All bit-storage locations in a matrix can be operated on at very high speed during a single program scan, giving fast PC operation even with a large number of logic operations. One matrix can represent inputs from a variety of field inputs such as switches, auxiliary contacts, etc., with another matrix used to drive outputs to the plant equipment.

3.7 PROGRAM SUBROUTINES

Where a section or routine of program performs a function that is required several times in the course of a main program execution, it should ideally be written only once in the program and be 'called' from each part of the program that wants to use it. This is termed a *subroutine*, and is normally stored after the main program. Use of subroutines avoids having to program the same group of instructions each time that function is required, with resultant savings in programming time and memory usage.

Subroutines have been used in computer systems and some larger controllers since their introduction, but have only recently been provided on small PLCs. A subroutine is normally a user-defined program that performs a certain function not provided by the standard PLC instruction set. For example, the arithmetic functions described above can be termed *resident subroutines*, because they are provided as standard for use in user programs. Here the user program will pass values or parameters to the function, and this same technique is adopted for user-defined subroutines. Thus, when a subroutine is required at any point in a program, it will normally be called and have the current associated parameters input to its data registers, in a similar way to activating say, a timer circuit with a desired time preset.

Once a subroutine is called, program execution moves from the main program to the subroutine, which is then carried out. On completion, control is passed back to the next instruction in the main program by a RETURN from subroutine instruction at the end of the subroutine. Any resulting data values will be stored in data registers, and can then be used in the main program. A subroutine can be passed different parameters each time it is called, often from a different place in the main program.

Fig. 3.45 below shows the main program occupying memory area 0000–1000, with subroutine #3 from 1100 to 1150. When initiating conditions for the subroutine do not exist, the PC only scans the main program area – the subroutine is *not* scanned unless it is enabled by one of the programmed contacts. This gives a useful speed advantage in program execution.

Fig. 3.46(a) shows the general format of subroutine instructions, where different conditions can be programmed to initiate the routine. For example, external inputs may be used to start a subroutine where a fast controller response to the input is required. Applications of this nature include: counting pulses from tachometers or flowmeters; reacting to halt commands; reading data values within a short time-window.

In Fig. 3.46(b), an example from a GE series 6 programmable controller handbook shows the ladder program for a logic subroutine, possibly for the control

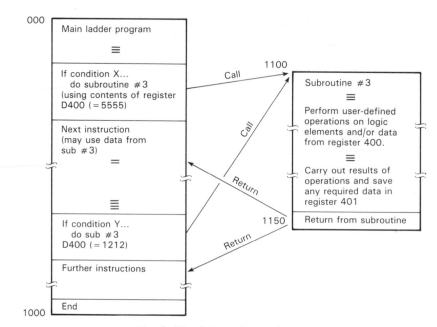

000

| Main ladder program |
| ≡ |
| If condition X...
do subroutine #3
(using contents of register
D400 (=5555) |
| Next instruction
(may use data from
sub #3)
= |
| ≡ |
| If condition Y...
do sub #3
D400 (=1212) |
| Further instructions |
| End |

1000

Call → 1100

Subroutine #3

≡

Perform user-defined operations on logic elements and/or data from register 400.

≡

Carry out results of operations and save any required data in register 401

Call

Return

Return → 1150 Return from subroutine

Fig. 3.45 Subroutine operations.

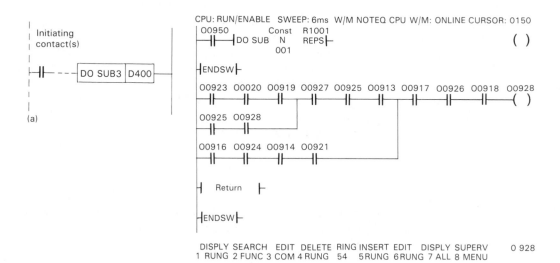

(a)

Initiating contact(s)

---||---- [DO SUB3 | D400]----

CPU: RUN/ENABLE SWEEP: 6ms W/M NOTEQ CPU W/M: ONLINE CURSOR: 0150

```
00950          Const   R1001
--||--DO SUB  N    REPS|-                        ( )
              001
-|ENDSW|-

00923 00020 00919 00927 00925 00913 00917 00926 00918 00928
--||---||---||----||---||---||----||---||---||----( )
00925 00928
--||---||--
00916 00924 00914 00921
--||---||---||----||--

-|    Return   |-

-|ENDSW|-
```

DISPLY SEARCH EDIT DELETE RING INSERT EDIT DISPLY SUPERV   0 928
1 RUNG 2 FUNC 3 COM 4 RUNG 54   5 RUNG 6 RUNG 7 ALL 8 MENU

Outputs 901–915 are loaded with the individual head's input status (possible with a simple data-move function) each time the subroutine is activated. The output (00916) is then copied to the head's output after the subroutine. For a 20-head transfer line, this technique saved 180 words of memory with this simple logic. In addition, if changes to the head-servicing logic are required, only one area need be changed to affect all heads. Subroutines are also useful where different logic is required by the control system to build different parts, for example. The logic is stored for each type of part (up to 16) and activated by a small main program that determines which one of sixteen parts is to be built.

(b)

Fig. 3.46 Typical ladder program for subroutine operation: (a) format for subroutine operation; (b) subroutine for GE series 8 PLC.

of a single workhead in an automated component transfer line containing 10–20 workheads. The subroutine is activated with different values as each head in turn is to be operated.

3.8 SUMMARY

In this chapter we have looked at a fairly broad range of PLC instructions, functions and circuit applications. These are the fundamental building blocks for all logic and interlock control programs, and in some cases can constitute almost the complete solution. Even so, not all of the possible facilities that can be provided by programmable controllers have been introduced – several others exist for use mainly in sequential control and other specialist applications. The most important of these will be covered in later chapters. Chapter 4 deals with the problems and solutions of ladder-circuit design.

3.9 EXERCISES

3.1 Explain the difference between simulated and physical contacts and coils. Give examples to illustrate your answer.

3.2 Describe the relationship between ladder symbols and the logical instructions that can be used to program a given programmable controller. What format of instruction does the PLC microprocessor execute at the machine level?

3.3 What is a battery-backed relay or function used for?

3.4 Describe the operation and use of the following:
  (a) set and reset on an auxiliary relay;
  (b) a timer circuit;
  (c) a cascaded counter circuit.

3.5 Describe how a 'one-shot' timer circuit may be implemented using ladder symbols.

3.6 With the aid of a diagram, explain how a basic shift register is formed within a programmable controller. Outline two common applications of this function.

3.7 State three common facilities provided by special function relays in a PLC.

3.8 Describe the use and effect of master control relays in ladder program operation.

3.9 In a programmable controller, what is the advantage of being able to handle data in *words* as well as bits?

3.10 Give four examples of data-handling instructions and explain their operation.

3.11 What facilities are offered by matrix bit-operations on a PLC?

3.12 Explain the difference between jump and subroutine functions in ladder programming.

## FOUR

Ladder Program Development

In Chapter 3 we examined the facilities and internal functions supported by most common programmable controllers. These are the component parts that allow the construction of tailor-made control programs for virtually any application. In many cases certain facilities may be mostly or completely unused.

Once a PLC with the necessary facilities has been selected for a particular application, the tasks of software design and development have to be considered. This task is only part of the total design procedure, but is of vital importance as it constitutes the means for controlling the target process. Fig. 4.1(a) outlines the overall design strategy for programmable control systems. This allows the development and installation of both hardware and software by separate but parallel routes, providing the program designer with the maximum possible time to consider the control requirements and generate the software solution. The other aspects of overall system design are examined in Chapter 9.

4.1 SOFTWARE DESIGN

When ladder programs are being developed to control simple actions or equipment, the amount of planning and actual design work for these short programs is minimal, mainly because there is no requirement to link with other actions or sections within the program. (Most of the early examples in Chapter 3 fall into this category.) The ladder networks involved are small enough to be easily understood in terms of circuit representation and operation. In practice, of course, circuits are not limited to AND or OR gates, etc., often involving mixed logic functions together with the many other programmable functions provided by modern programmable controllers.

When larger and more complex control operations have to be performed, it quickly becomes apparent that an informal and unstructured approach to software design will only result in programs that are difficult (if not impossible) to understand, modify, troubleshoot and document. The originator of such software may possess an understanding of its operation, but this knowledge is unlikely to remain after even a short period of time away from that system.

In terms of design methodology, then, ladder programming is no different from conventional computer programming. Thus, considerable attention must be

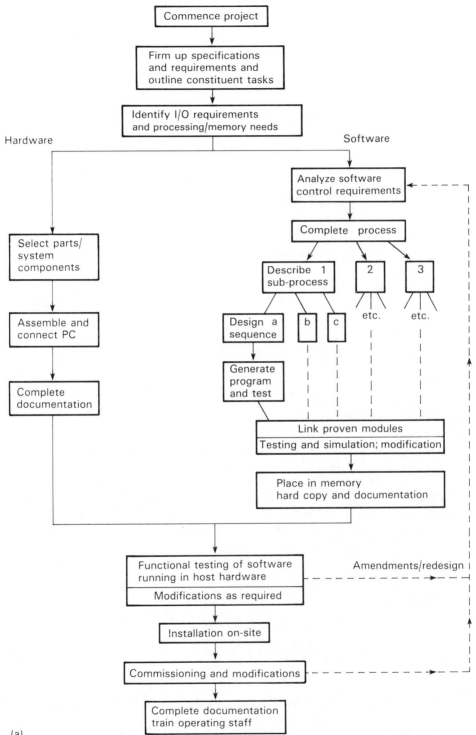

(a)

Fig. 4.1   (Continued overleaf)

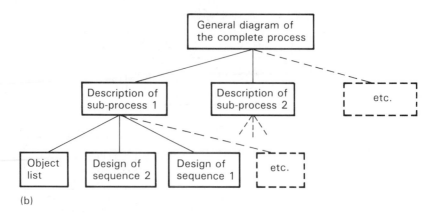

(b)

Fig. 4.1 (a) PC system design procedure; (b) describing the functional structure of a process.

given to:

- Task definition/specification
- Software design techniques
- Documentation
- Program testing

System functions

Most industrial control systems may be considered as a set of functional areas or blocks, in order to aid the understanding of how the total system operates. For example, each machine in a plant unit can be treated as a separate sub-process. Each machine process is then broken down into blocks that may be described in terms of basic sequences and operations. Fig. 4.1(b) illustrates this approach.

A functional block could, for example, consist of all actions required to control a certain machine in the process.

Block 1

Block 2

Block 3

etc.

The division of programming tasks into functional blocks is an important part of software design.

In logic programming, there are two different types of network that may be used to implement the function of a given block:

- Interlocks or combinational logic, where the output is purely dependent on the *combination* of the inputs at any instant in time.
- Sequential networks, where the output is dependent not only on the actual inputs but on the *sequence* of the previous inputs and outputs (so this involves memory of events).

Before examining the design techniques used in each case, let us look at examples of interlock and sequential systems.

Interlock example

A conveyor belt is to be used for transporting loaded pallets from a loading machine to a pick-up area (see Fig. 4.2). When a pallet is loaded onto the start of the belt, a manual switch X1 signals to start the conveyor motor. The belt now runs until the pallet trips a limit switch X2 at the far end of the conveyor. Once the pallet is removed, the belt starts running again, transporting the next pallet along to the trip-switch X2, and so on. Switch X1 must be able to stop the conveyor at any time to allow the loading of subsequent pallets as necessary.

For the motor drive to operate, X1 must be switched ON, and the limit switch X2 must *not* be operated. This system can be controlled by an *interlock network* (Fig. 4.3) with X1 and X2 inputs and an output which activates the motor drive, since the on/off status of the output is determined purely by the combination of the input signals at any instant in time, rather than their order or sequence.

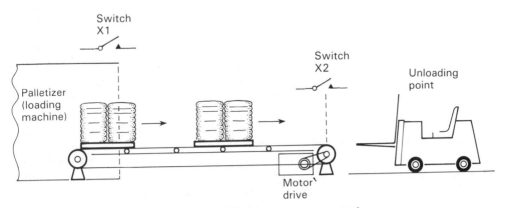

Fig. 4.2 Simple conveyor control.

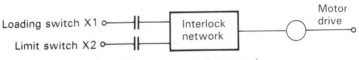

Fig. 4.3 An interlock network.

Sequence control example

The conveyor system given in the previous example can be extended to provide two separate belts, each with an independent motor drive. A common requirement in this type of system is for a final weighing of a loaded pallet, so an automatic weigher is installed beneath the second belt. An additional limit switch X2 is provided to indicate when a pallet is positioned directly above the weigher.

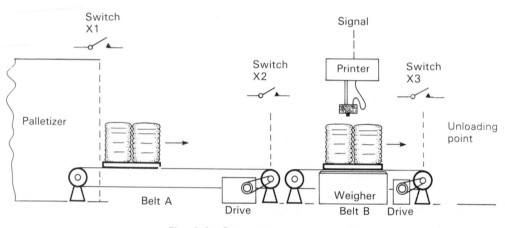

Fig. 4.4   Conveyor sequence control.

System operation

Both conveyor belts A and B are to be started by switch X1. When a pallet has traveled from belt A onto B and is positioned above the weigher, both belts stop by X2 operating. The weighing operation takes 4 s, then the measured weight and other relevant data is printed onto the package by an ink-jet printer. Belt B now starts up and moves the pallet to the end of the conveyor line, where it trips switch X3 and stops the belt. Once the pallet is removed, X3 opens and both belts start running to move other pallets up to the weigher. The above sequence is then repeated.

This description involves a sequence of states and events that must always occur in the same order. Also, the system must always start up (initialize) at the same point in the sequence if reliable operation is to be achieved:

State 1:   both belts stopped, switch X1 off
State 2:   belts A and B running, pallet moving towards limit switch X2
State 3:   both belts stopped; pallet weighed (timer running)
State 4:   both belts stopped; data printed onto package
State 5   belt B on, belt A stopped, pallet moving towards X3
State 6:   both belts stopped; pallet to be removed by fork-lift truck.

As described in Yohansson (1985), we can more easily understand how the

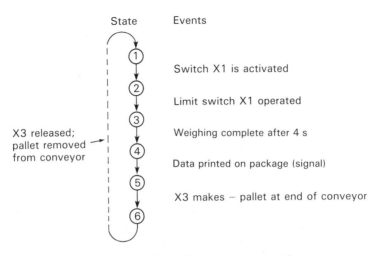

State   Events

Switch X1 is activated

Limit switch X1 operated

X3 released;   Weighing complete after 4 s
pallet removed
from conveyor   Data printed on package (signal)

X3 makes – pallet at end of conveyor

Fig. 4.5 A flowchart of a sequence operation.

system is to function by using a *flowchart* or diagram that illustrates all states the system may be in, and the events that are to cause a transition between these states (Fig. 4.5). We can also represent the system as a sequential network having four inputs and four outputs (Fig. 4.6)

Note that the timers involved are internal functions, and do not appear on the external diagram. However, this diagram does not supply enough information for us to calculate the combination of input signals necessary at any instant to produce a required output. In fact to obtain a certain output, a complete sequence of input signals is necessary. For example state 5, where belt B alone is running, can be active only if states 1–4 have occurred *and* inputs X1, X2 and the printer signal have also taken place.

In order to hold the status of previous events, the controller logic system must contain memory. This allows the recording, for example, of the operation and release of the limit switch X3 when a pallet moves to the end of the conveyor and is then removed. The system must remember this even though the pallet that operated the switch has passed out of the system, to ensure it takes the correct route through the operating sequence. In a programmable controller this memory is usually provided using marker relays or registers.

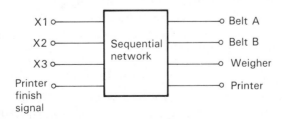

Fig. 4.6 A sequential network.

In a programmable controller, the logical elements deal with interlock/ combinational requirements, in conjunction with counters, timers and shift registers, etc. for sequential functions. Examples of interlock and sequential systems follow in the next sections.

Many control functions can be implemented using either combinational *or* sequential networks, or a combination of both. A solution based on sequential design tends to be a much smaller and simpler program than an interlock equivalent, plus it has a greater resistance to noise or erroneous signals. This is due to the sequential system being made up of a number of stable states. Once a certain state is entered, the system will only leave that state *if* the exit conditions are met. It will not respond to any changes in entry conditions to its present state.

This aspect of stable states is also of assistance when troubleshooting a sequential system. When a false or erroneous signal occurs, the system is likely to remain in the current state, which allows the fault to be localized to the input signals and devices that caused entry to that state or would initiate the next.

4.2 COMBINATIONAL LOGIC DESIGN

We have already been introduced to the standard logic instructions as provided on all programmable controllers – AND, OR, NOT, etc., and also some simple examples of how these functions may be combined to create logic networks. This section builds on this and considers the use of Boolean algebra as a tool to assist in the design of logic networks.

Consider the following problem:

Open valve Y1 if limit switches A and B are operated (closed), and level switch C is not operated (open).

This is an elementary example which illustrates the concept of an output determined purely by a combination of inputs. Logic circuits can be usefully and easily described using Boolean algebra, which provides a framework of standard rules and relationships concerning binary logic. Any logic circuit or program may be converted into a Boolean equation, which shows how the inputs are related to derive the output for that circuit.

This conversion is carried out by starting at the top of the ladder diagram and moving through each branch in turn, contact by contact by contact. In the above example the output Y1 is energized when all contacts are closed (see Fig. 4.7), i.e. when A AND B AND NOT C are activated. (C is a normally-closed contact.) In Boolean terms this is stated as:

$$Y1 = A.B.\bar{C}$$

This example can be expanded to illustrate circuits with additional inputs (Fig. 4.8). Here valve Y1 will operate if switch C is open AND valve X is operated AND valve Z is NOT operated, together with switches A AND B closed OR switch D

Fig. 4.7   Example of a logic switching circuit.

closed. In Boolean:

$$Y1 = A.B.\bar{C}.X.\bar{Z} + D.\bar{C}.X.\bar{Z} \text{ which can be reduced to:}$$
$$Y1 = (A.B + D).\bar{C}.X.\bar{Z}$$

This expression is produced by identifying common terms in the equation $(\bar{C}.X.\bar{Z})$ and then using brackets to simplify the components that act on the common terms.

Similar expressions would be produced for every output device in the system, based on the specification. At this level of ladder design you can virtually sketch out the circuit operations and then produce the ladder logic version. This would then be encoded and programmed into the target PLC. When dealing with fairly complex combinational logic tasks many program designers would express the requirements in terms of Boolean equations before translating this into ladder logic. As can be seen from Fig. 4.8, we can easily express a logic network using equivalent logic functions or symbols. Thus the above circuit can be described with:

- 1 two-input AND gate
- 1 two-input OR gate
- 1 four-input AND gate
- inverters (NOTS) as necessary.

Again, this form of expressing logic requirements promotes ease of understanding and implementing a given circuit. Logic symbols are often used in conjunction with Boolean algebra, as they group inputs into logical functions that are then easily transferred into Boolean terms. The techniques of Boolean algebra

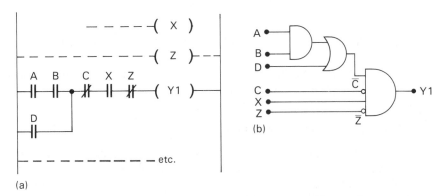

Fig. 4.8   (a) Ladder diagram; (b) logic function. X and Z are other actuators in the same system; D is an input from a level switch.

can often result in a solution that is more economical or elegant in terms of logic functions used, when compared to the original 'rough' program.

Before getting involved in the overall design of system software, we will look at other examples of combinational logic.

Example of multiple branch networks

This circuit comprises six inputs and one output. Output Y2 is activated when:

Input X0 OR X1 OR X2 AND X3 AND X4 OR X5 on its own,
is/are made

Notice the use of brackets, in the caption to Fig. 4.9 to show the relationship between all the inputs. We can in fact consider the circuit as a gate system, with two OR-gates plus a single AND gate, as shown in Fig. 4.9.

If you are unsure of the manner in which the Boolean equation was found, investigate all the individual ways that output Y2 may be energized in the above circuit.

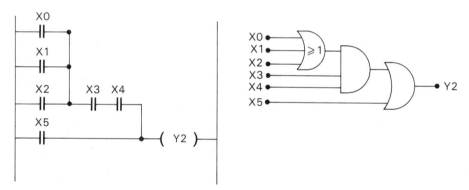

Fig. 4.9 *(a) Ladder diagram; (b) logic function. The Boolean equation is:*
$$Y2 = ((X0 + X1 + X2).X3.X4) + X5.$$

Converting designs to ladder logic and logic instructions

Once we have obtained the expressions for each output, they have to be translated into a form that can be entered into a PLC.

Method 1: Using logic instructions

Taking the above equation for Y2, we can convert this to logic instructions as shown in Table 4.1. The relationship between the Boolean terms and the PLC logic instructions can clearly be seen in this example. Coding of this type of circuit is straightforward providing you remember the need for, and the effect of, the ORBRANCH and ANDBRANCH type of instructions when building up larger

Table 4.1 **Coding Boolean expressions.**

	Instructions	
Boolean	(Mitsubishi)	(GE)
((X0	LD X0	STR X0
+ X1	OR X1	STR X0
+ X2)	OR X2	
.X3	AND X3	STR X0
.X4)	AND X4	
+ X5	OR X5	
= Y1	OUT Y1	output

networks (see Chapter 3). For the type of PLC that does not support these branch instructions, the normal practice is to use marker relays to hold the result of each logic path, then to pass on this information to succeeding rungs via related contacts.

Method 2: Ladder logic programming

When the PLC system in use provides a graphic programming panel or terminal, the logic network appears on the screen as each symbol key is pressed.

The details in Fig. 4.10 show the sequence of key entry for the previous program. Most graphic programmers have a form of automatic circuit checking built-in, which can detect and/or correct errors in ladder layout and structure. In this example, the logic circuit must be rearranged as shown before it can be programmed.

Exercise 4.1 Transfer the ladder logic shown in Fig. 4.11 to a Boolean equation.

Exercise 4.2 For Exercise 4.1, write a logic instruction program for the programmable controller used in this text (or one of your own choice).

Interlock systems – second example

A machine work cell (see Fig. 4.12) is bounded by an array of four sensors to detect the presence of people or objects entering the work area. When any sensor is tripped, this must stop the motion of all machines in the cell to ensure that no injury or damage to the 'intruder' occurs. To allow the process to continue, all sensors must signal 'all clear'.

The type of sensor used may detect proximity, the presence of a light beam or ultrasonics, etc., but whatever system is employed, it should be fail safe in nature: i.e. if a sensor or related component fails, this event should result in the process being halted. This can be achieved by designing the PC control program and connections so that the machines will be halted if any sensor malfunctions *or* signals an 'intrusion'. Thus we require a control system that incorporates these two features in the design.

Revised ladder

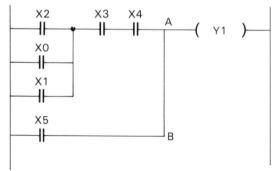

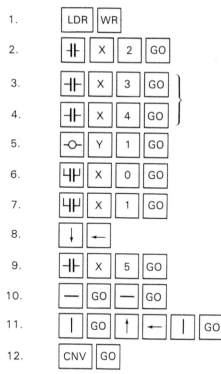

Entry procedure	*Key operation*	*Comment*
1.	LDR WR	Mode and function key
2.	⊣⊢ X 2 GO	Symbol element execution (cursor moves to next position)
3.	⊣⊢ X 3 GO	Where the ladder symbol is the same as the one used previously, it need not be rekeyed
4.	⊣⊢ X 4 GO	
5.	─○─ Y 1 GO	Output coil, cursor moves to next line
6.	ЧⱵ X 0 GO	Parallel contact across X2
7.	ЧⱵ X 1 GO	Parallel contact across X2
8.	↓ ←	Move cursor to next line
9.	⊣⊢ X 5 GO	Parallel contact across previous contacts
10.	─ GO ─ GO	Horizontal circuit links to point B
11.	│ GO ↑ ← │ GO	Vertical links up to point A, using cursor keys
12.	CNV GO	Editing and writing code to RAM (compiled into logic instructions)

Fig. 4.10 Graphic programming (Mitsubishi HGP).

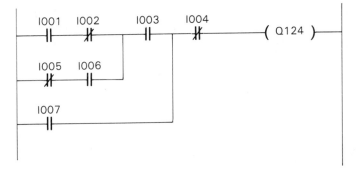

Fig. 4.11 Ladder logic for Exercise 4.1.

The system requires 11 inputs and 4 outputs:

4 sensors – active and functioning signals (X0–X3)
4 sensors – intrusion signals (X4–X7)
3 manual inputs – start/stop, reset alarm (X8–X10)
3 output paths to the drives of robot, CNC drill and conveyor (Y0–Y2), plus an alarm (Y3).

To understand the requirements, we verbally stated the specific conditions that allowed the plant to function. Now we shall express this in a logical form. The initial equation describes the conditions necessary for the drives to operate:

$$Y0\text{–}Y2 = X0.X1.X2.X3.\overline{X4}.\overline{X5}.\overline{X6}.\overline{X7}$$

outputs   functions   intrusion

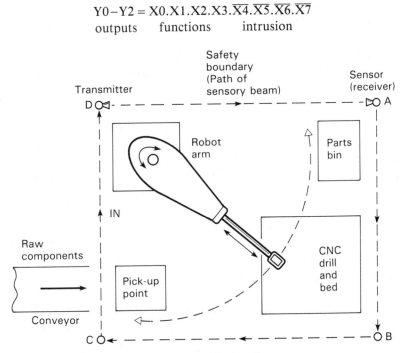

Fig. 4.12 Work cell.

This states that the drives are enabled when (X0–X3) sensors return an active functioning signal, whilst other sensors return a negative intrusion signal (X4–X7). To create the expression for halted machines, we simply invert this equation:

$$\overline{Y0-Y2} = \overline{X0.X1.X2.X3.\overline{X4}.\overline{X5}.\overline{X6}.\overline{X7}}$$

The alarm output Y3 is activated if the machines are halted because of a sensor failure or intrusion, so we can say that:

$$\overline{Y0-Y2} = Y3$$

We shall label this condition as M1; thus

$$M1 = \overline{X0.X1.X2.X3.\overline{X4}.\overline{X5}.\overline{X6}.\overline{X7}}$$

This expression can be simplified using De Morgan's law for removing inverting bars in logic equations: 'break the bar and change the logical sign'. In this case the signs change from AND to OR:

$$M1 = \overline{X0} + \overline{X1} + \overline{X2} + \overline{X3} + \overline{\overline{X4}} + \overline{\overline{X5}} + \overline{\overline{X6}} + \overline{\overline{X7}}$$

The double bars above X4 – X7 cancel out, giving:

$$M1 = \overline{X0} + \overline{X1} + \overline{X2} + \overline{X3} + X4 + X5 + X6 + X7$$

Consideration of this arrangement will reveal a problem in the event that a transition through the sensor field occurs, rather than an object remaining in the path of a sensor. This will cause the work cell to halt and then to restart once the intruder leaves or moves inside the field. To deal with this we need to insert a latch or memory circuit that will retain the alarm until reset manually by an operator (contact X10). A related contact from M1 is used for the latch in this case, giving:

$$M1 = \overline{R}.(M1 + \overline{X0} + \overline{X1} + \overline{X2} + \overline{X3} + X4 + X5 + X6 + X7)$$

Elsewhere in the program there are also manual start/stop switches (X8 and X9). These latch in the start position once the start push button is depressed, and enable all outputs via associated normally open contacts. The stop button breaks this latch as shown in Fig. 4.13. The logic for this complete start/stop circuit is thus:

$$S = (X8 + S).\overline{X9}.\overline{M1}$$

In practice, the control logic will not act directly on the output coils, but via an intermediate marker or auxiliary relay. This approach simplifies the ladder layout and aids its readability. Here we have used a marker M1 as the intermediate. (Remember that internal relays and functions can have as many related contacts as necessary, unlike the hardwired equivalent.)

It should be realized that the complete control program for an application like this would have a greater amount of intermediate logic to deal with other tasks and conditions that will occur in the work cell: for example, machine faults that require the cell to halt. Also, the sequencing of the elements of the work cell (robot, CNC

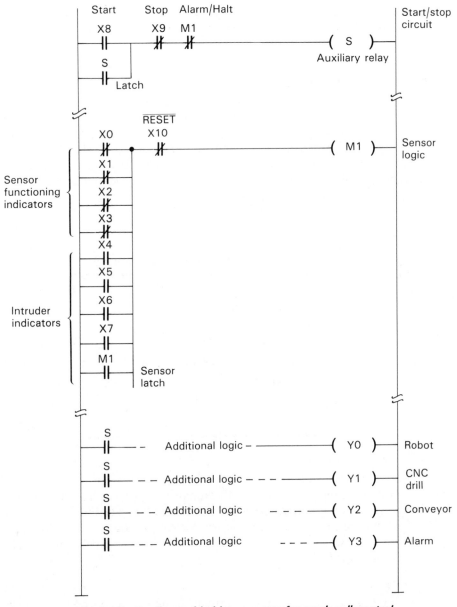

Fig. 4.13 *Sections of ladder program for work-cell control.*

drill and conveyor) are controlled from this same program, resulting in more involved ladder logic for the output drives than is shown in Fig. 4.13.

Exercise 4.3   Explain the need for a comprehensive and careful treatment of automatic stop conditions in any plant- or process-control system.

4.3   PROGRAM STRUCTURE

At this stage in the investigation of design techniques, it is appropriate to discuss the layout and structure of PLC programs. It is sound practice to base any program layout on the general operating structure of all process-control systems. This means having definite sections dealing with operating modes, basic functions, process chain or sequence, signal outputs and status display, as indicated in Table 4.2.

Table 4.2   Sections of a PLC program.

Start
Operating modes and basic functions starting (basic) position
Enabling/reset conditions
Process operation/sequence logic
Signal outputs
Status/indicator output
Finish

(a) Operating modes
Basic position.   The controlled equipment is likely to have a basic or normal position, for example when all actuators are off and all limit switches are open. All these elements can be combined logically to signify and initialize a 'basic position', which may be programmed as a step in a sequential process (see Section 4.2).

Enabling/reset conditions.   Most industrial processes have manual start and stop controls that may be incorporated into the PLC program structure at this point. These would be included as enabling and reset contacts, having overall control of the PLC in terms of 'run or stop', etc. There may also be a manual switch to enable the system outputs, which would allow the program to run without driving the physical outputs connected to the PLC, i.e. a test function.

(b) Process operation/sequence

This is the main topic of this chapter, involving the design and programming of combinational and sequential networks as necessary. The resultant outputs do not normally drive actuators directly, but instead are used to operate intermediate marker relays.

(c) Signal output

Output signals to process actuators are formed by interlocking the resulting operation sequence outputs (markers) with any enabling conditions that exist in (a) above.

(d) Status/indicator outputs

Process status is often displayed using indicator lamps or alarms, etc. Such elements are programmed in this section of the software.

By adopting this systematic approach to program structure, we can create reliable, easily understood software, which will allow rapid fault location and result in short process down times. The programs that are developed in this chapter deal mainly with the topic of process operation, but will be structured in this manner where possible. Examples given in later chapters will be laid out as described above.

4.4 SEQUENCE CONTROL

At the beginning of this chapter the need for sequential control systems was introduced, where the output is dependent not only on the actual inputs but on the sequence of the previous inputs and outputs, i.e. this involves memory of events that have occurred.

Sequential problems have long been solved using conventional logic gates as 'building blocks', but using certain techniques to express and identify the sequence logic equations that control the system outputs. This approach is still often used today and is covered in the next section. However, there are several other programmable controller functions that are provided specifically to simplify the design and implementation of sequential systems, and several of these were introduced in Chapter 3, e.g. shift registers, master control relays, drum timers. Example applications of these functions are to be found later in this chapter. To fully appreciate the operation and problems of sequential control programming, it is desirable to have some knowledge of sequential logic program design.

Programming sequences

As with digital electronics, sequential logic program design requires a slightly different approach to pure combinational design. For a programmable controller, the software design procedure is as follows:

1. The process is verbally described.
2. This description is translated into a flowchart or function diagram.
3. From this stage the conditions are easily identified, then converted into Boolean equations representing each state of the sequential process.
4. Finally the Boolean equations are simply converted into ladder logic format, then programmed into the PLC.

The verbal or written description of an automatic process is usually long, difficult to follow and imprecise. As was stated earlier, the complete process is more easily understood when divided up into a number of self-contained sub-units or sub-processes. Each sub-unit can then be constructed from sequences and interlocks to create the required function. Other methods of describing such systems in clear, easily followed layout are necessary.

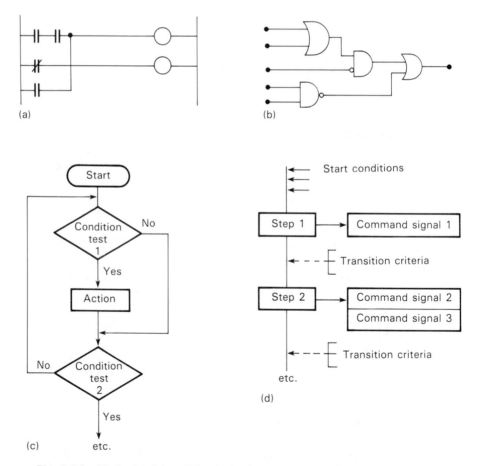

Fig. 4.14 *Methods of describing logic control requirements: (a) relay ladder logic; (b) logic schematic; (c) flow chart; (d) function chart.*

Several options exist: relay logic diagrams, logic schematics, flowcharts and function charts. These methods do not replace, but greatly supplement the verbal description. Choice of a particular method depends mainly on previous experience and the nature of the application. Staff who have a background in digital electronics or computing may tend to favor the latter three methods, leaving the relay ladder logic approach to those with considerable current or relay experience. However, the engineers working closely with a process always have the best knowledge of its operation, and the method chosen to describe the control requirements should always reflect their opinion.

Relay ladders and schematic logic

Both these methods relate directly to physical circuit layouts, and are therefore ideal for applications in which a programmable controller is replacing a conventional relay or logic system, since the original system drawings may be used as the basis for programming the PLC. However, these methods are normally used only for combinational control systems or small sequential systems, since the diagrammatic layout tends to become complex and very difficult to follow when used in larger sequential applications.

Flowcharts

These are more often associated with computer programming, but are a common method of displaying the sequential operation of a control system. Flowcharts have a direct relationship with the verbal description of a control sequence, showing each test and resultant action within a series of interlinked symbols (Fig. 4.14(c)). Actions within the flowchart are contained in rectangular symbols, whilst conditions or tests are entered in diamond-shaped symbols. Unfortunately, when used to describe larger control systems, the flowchart can become difficult to lay out, often resulting in large, cumbersome diagrams.

Function charts

Over the last few years function charts have become a popular way of representing sequential control tasks, being suitable for portraying the fine detail of the system as well as the outline process operation (Fig. 4.14(d)). Using a condensed form of symbolic description, the function chart combines most of the advantages of the other methods, showing the control sequence coherently and clearly. For all separate stages, it shows the setting and resetting conditions, transition criteria and all necessary control signals. Function charts are also valuable aids for testing and commissioning the control system, and for automatic fault diagnosis. The symbols and labels used in function charts are standardized under DIN and IEC specifications.

Boolean algebra

Whichever method of system description is to be used, once the functions have been described they have to be converted into a form that can be programmed into a PLC. This is normally carried out by converting the functions into a series of Boolean equations, and hence into the programming language used by the target PLC. Once familiar with the techniques, one can easily convert a functional description into Boolean equations, irrespective of the original format.

It is also possible to describe a logic control system completely using Boolean expressions, although this is often less effective in terms of design time and degree of friendliness to the inexperienced user. The Boolean solution, however, takes up much less space on paper during the design process.

These topics are best explained by considering suitable examples. This will also highlight various techniques and procedures used in the implementation of PLC control programs.

Simple sequence control of an assembly machine

Verbal description

A feed belt moves components in succession towards the the assembly machine. When a component is in position in front of the assembly jig, limit switch LS1 operates to stop the feed belt. A pneumatic ram is then operated to push the component firmly into the assembly jig, when LS2 is operated. This halts the ram

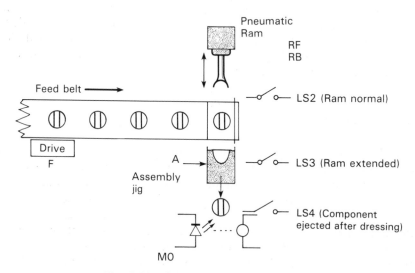

Fig. 4.15 Component assembly machine.

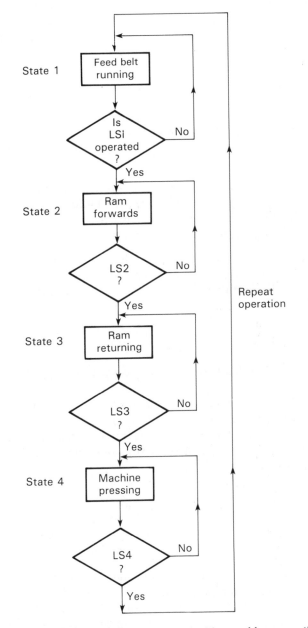

Fig. 4.16 Flowchart of component assembly machine operation.

and moves it back into the normal position, operating LS3. This limit switch also signals that the ram is clear of the assembly jig, and the machine now presses the raw component into the desired shape. The pressing operation is completed and the resulting artifact is ejected from the press, which activates a light sensor LS4. This indicates that the next component can be moved up to and then into the jig. The process then continues.

Flowchart

We can draw up a flowchart representation of this sequence quite easily from the verbal description. From Fig. 4.16 you can clearly see the four states of the sequence:

1. Feed belt on, pneumatic ram normal, pressing machine inactive.
2. Ram extending, feed belt off, press inactive.
3. Ram returning, belt off, press inactive.
4. Press operating, belt off, ram normal.

We can also see the relationship of the input signals to these states. What is occurring is that changes of state (from 1 to 2, 2 to 3, etc.) are initiated by the input signals from the four limit switches. Also, the four different states always occur in the same sequence, i.e.

$$1 \xrightarrow{\quad LS1 \quad} 2 \xrightarrow{\quad LS2 \quad} 3 \xrightarrow{\quad LS3 \quad} 4 \xrightarrow{\quad LS4 \quad} 1 \quad \text{etc.}$$

It must be realized that there can only be one active (current) state in the system at any instant.

From this we can say that (a) You will *enter* state 1 if in state 4 when LS4 is operated, *or* (b) if in state 1, you will remain in 1 while LS1 does *not* operate. Inspection of these statements will reveal that they (a) form *entry* criteria (or *initialization*) to a state, and (b) *self-maintaining* (*latching*) and *reset* (*exiting*) criteria of that state.

This can be stated in Boolean terms as:

$$1 = 4 \cdot LS4 + 1 \cdot \overline{LS1}$$

entry   self-   reset (exit)
maintaining

The Boolean expression for a state must include all permitted combinations of entry, holding and exiting that state, if it is to be reliable as a control program. In this case, the input LS4 causes the transition from state 4 into state 1. Since we are now in state 1, the 'entry' section of the equation (4 . LS4) is no longer effective, so there must be another part of the equation to do this. The $1 \cdot \overline{LS1}$ provides this self-maintaining aspect, providing LS1 does not operate. When LS1 operates, it causes the AND function of $1 \cdot \overline{LS1}$ to release, since $\overline{LS1}$ now equals 0 (as $1.0 = 0$).

Entry, hold and reset

An understanding of these three elements of state definition is essential for the designer of sequential control systems.

The other states can be similarly expressed:

$$2 = 1.LS1 + 2.\overline{LS2}$$
$$3 = 2.LS2 + 3.\overline{LS3}$$
$$4 = 3.LS3 + 4.\overline{LS4}$$

Before considering the translation of these state equations into PLC programs, another aspect of their operation must be examined – when the system changes from one state to the next, and the effect on the logic involved.

In the above logic equations, the same input signal is used to *reset* the current state as is used to *enter* the subsequent state. For example, LS2 is used to exit state 2 and to enter state 3. What has to be considered is that each of these equations will be implemented in one or more lines of a ladder program, and the host programmable controller will execute each line or rung sequentially one after the other. Thus one line of program causes state 2 to reset before the line to enter state 3 is executed. The outcome of this is that LS2 correctly causes exiting from state 2, but since this means state 2 is no longer active, it cannot be used to gain entry to the next state (3) via $3 = 2 . LS2$, the entry section of this equation.

This problem is overcome by resetting the previous state with information from the state that has just been initialized, rather than by the input signal which caused this transition. That is, the system *enters* the next state before *exiting* from the previous one – there is a slight overlap! In practice, this is usually negligible, as the overlap exists for only one PLC memory cycle of a few milliseconds.

The equations for the above application become:

Feed belt: $\quad\quad\quad 1 = 4 . LS4 + 1 . \overline{2}$
Ram extending: $\quad 2 = 1 . LS1 + 2 . \overline{3}$
Ram returning: $\quad\; 3 = 2 . LS2 + 3 . \overline{4}$
Machine pressing: $4 = 3 . LS3 + 4 . \overline{1}$

Initial system start-up

Consideration has to be given to this aspect of system operation. Since start-up will require the system to enter state 1 (the first state) in this case, an additional entry condition has to be added to the equation for state 1. Also, this entry path should only be valid on the first pass through the program, otherwise state 1 would be constantly maintained. If initiated from a manual start button this would cause state 1 to be entered irrespective of what other state was active at that time. An automatic 'boot' into state 1 can be achieved by ORing a normally closed marker relay contact (M1) onto the state 1 equation, and programming the unconditional activation of this marker relay further down the listing.

$$\text{State } 1 = 4.\,LS4 + 1.\,\overline{2} + \overline{M1}$$

plus

$$M1 = 3 + M1$$

This provides a latching path for M1 which is permanently activated once state 3 is entered.

Input/output and internal assignments

Before the Boolean equations are converted into a control program, all physical I/O lines have to be related to I/O points on the controlling PLC. Also, it is standard practice to assign internal memory elements or auxiliary relays to each state of the process, in order to simplify the program layout. A table or tables would be drawn up to show all assignments. This forms a vital part of the system documentation. (See Section 4.5) The assignments for this example application are given in Table 4.3.

When the previous equations are rewritten with the appropriate input/output code, the following expressions are formed:

State 1:	M101	$= M104.X4 + M101.\overline{M102} + \overline{M1}$
State 2:	M102	$= M101.X1 + M102.\overline{M103}$
State 3:	M103	$= M102.X2 + M103.\overline{M104}$
State 4:	M104	$= M103.X3 + M104.\overline{M101}$
Start:	M1	$= M103 + M1$
Belt:	Y10	$= M101$
Ram out:	Y11	$= M102$
Ram in:	Y12	$= M103$
Press:	Y13	$= M104$

Table 4.3   Assignments for the example application

Signal	PC address	State number
LS1	X1	I/P
LS2	X2	I/P
LS3	X3	I/P
LS4	X4	I/P
State 1	M101	1
State 2	M102	2
State 3	M103	3
State 4	M104	4
Belt	Y10	O/P
Ram out	Y11	O/P
Ram in	Y12	O/P
Press	Y13	O/P
Start	M1	Marker

The resultant ladder diagram is given in Fig. 4.17. Note the use of the intermediate auxiliary relays (markers) to distance the control logic from the physical outputs for clarity, etc. It must be appreciated that this program only covers the aspects of sequence design, and in practice far more attention would be given to the areas of safety circuits, start/stop, etc.

The last example was represented by the flowchart given in Fig. 4.16. Alternatively, the operation sequence can be described by constructing a function

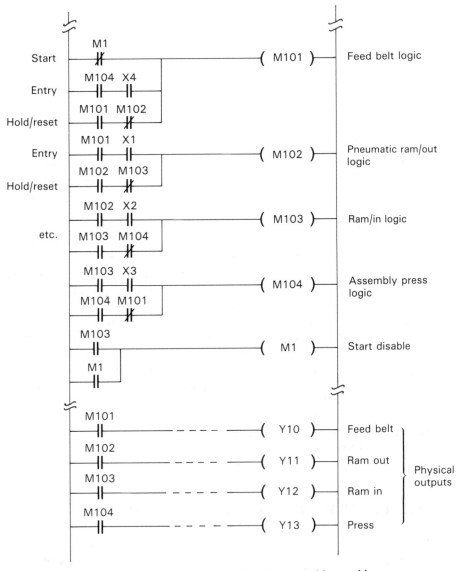

Fig. 4.17   Ladder diagram for the assembly machine.

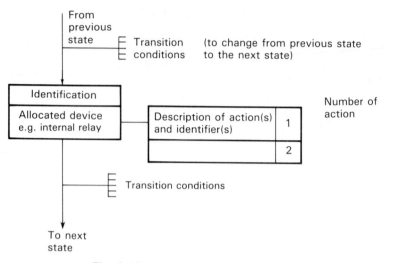

Fig. 4.18   Function chart description.

chart. This offers advantages in terms of the amount of operational information that may be displayed. The usual format of each block in the function chart is shown in Fig. 4.18.

The sequence operation of the assembly machine can then be described in a function chart as illustrated in Fig. 4.19(a). Again, like the flowchart, the function chart involves a series of steps or states represented by consecutive symbols. The steps relate to memory stages, and in this example are programmed using the self-maintaining logic circuitry in conjunction with marker relays. Only one step at a time is active, and the sequence is 'pulsed' forwards from one step to the next in the program once the necessary transition (entry) conditions are met. Note that the function chart does not show the self-hold details of a circuit, or those necessary to exit a state. These are assumed to be included by the designer, since they depend on the method used to implement the sequence (e.g. logic, shift register, drum sequencer, etc.) Some designers use a modified function symbol that incorporates the self-hold or reset conditions. Reset may also show details of a total system reset if this is provided (see Fig. 4.19 (b)).

Exercise 4.4   For the example given in Fig. 4.4, draw up a flow chart or function chart to describe the process sequence.

4.5   FURTHER SEQUENTIAL CONTROL TECHNIQUES

In many practical applications, a control system has to deal with a process sequence that requires the concurrent operation and control of more than one step. Also, steps in a sequence may require a time delay or event count as entry criteria for a

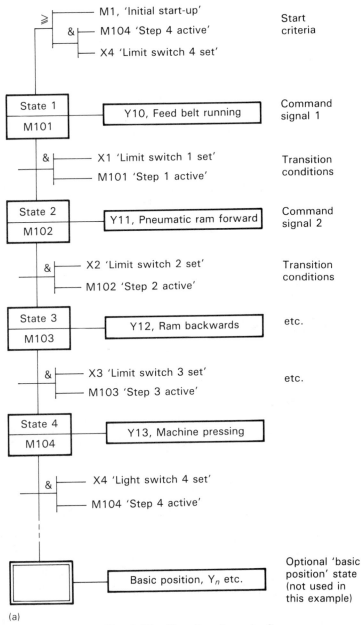

(a)

Fig. 4.19 (Continued overleaf)

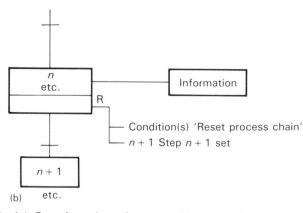

(b)   etc.

Fig. 4.19   (a) Function chart for assembly task; (b) modified function symbol.

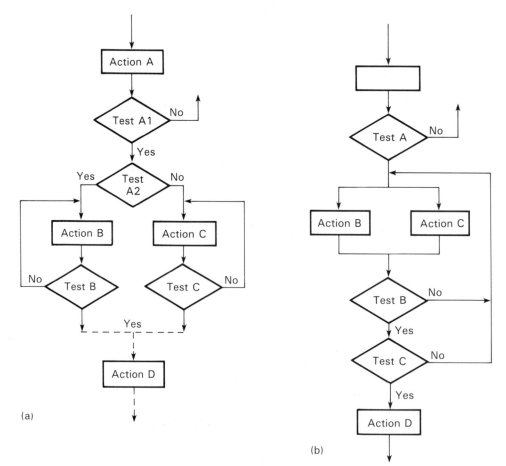

Fig. 4.20   (a) OR branching in a flowchart; (b) AND branching.

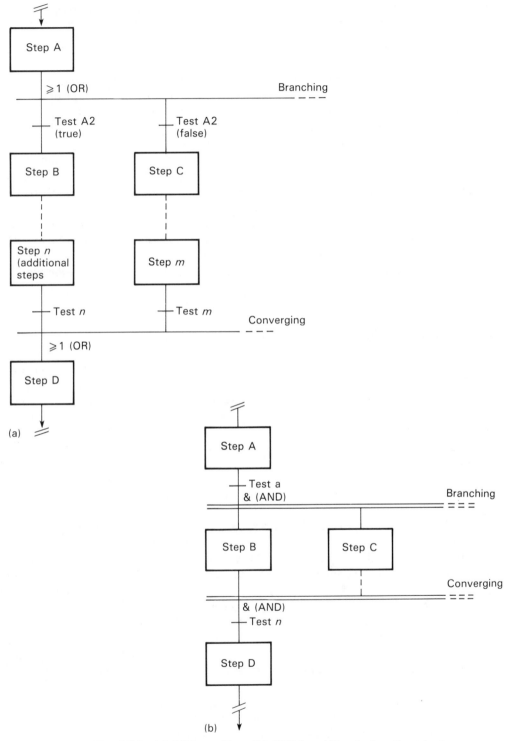

Fig. 4.21 *(a) OR branching; (b) AND branching in function charts.*

succeeding step. To describe the different types of parallel operation, we use the conventions in Fig. 4.20.

In Fig. 4.20(a), actions B OR C are taken, depending on the result of test A2. Either action will (on completion) allow entry to action D. Fig. 4.20(b) shows the format for a process where two actions A AND B are initialized once test A is true; also both tests B AND C must be true before progression to action D.

The equivalent function chart descriptions are illustrated in Fig. 4.21. The number of parallel activities may be extended via the branching and converging rails. The chart in Fig. 4.21(a) shows the tests that allow entry to steps B OR C, and also the individual tests or conditions that will allow resetting of the chosen step (tests *n* and *m*). Notice the OR signs at each branch rail.

In Fig. 4.21(b) the ANDing of steps is signified by the double connecting rails after test A and before test *n*. This means that all parallel steps (in this case B and C) are set once state A is active *and* test A is fulfilled.

These descriptive conventions are demonstrated in the following example.

4.6 CHEMICAL PROCESS PLANT CONTROL

The plant consists of a four-tank system with pumps to transfer the liquid contents through the system. Each tank is fitted with sensors to detect 'empty' and 'full' states, and tank 2 has an integral heating element with associated temperature sensor. Tank 3 is equipped with a stirring arm to mix the two constituent liquids when they are pumped in tanks 1 and 2. The lower tanks, 3 and 4, have twice the capacity of tanks 1 and 2, and will therefore be filled by the contents of tanks 1 and 2 (alkali plus polymer).

Operation

Tanks 1 and 2 are to be filled from supply reservoirs of alkali and polymer respectively, via pumps 1 and 2. Pumps 1 and 2 turn off as each tank-full sensor operates. The heating element in tank 2 is activated, raising the polymer temperature up to $60°C$, when the temperature sensor closes. This should turn off the heater and turn on pumps 3 and 4 to transfer the liquids into the reaction vessel, tank 3. The stirring arm must also run when this tank is being used, and for a minimum of 60 s. Once tank 3 is full, pumps 3 and 4 are stopped. If the stirring time is greater than 60 s, pump 5 is to transfer the mixture to tank 4 (the product silo) via a filter unit. Pump 5 is stopped once tank 4 is full or tank 3 is empty. Finally the product is run off into storage using pump 6. This marks the completion of one cycle, and the process can begin again.

This sequence may be represented using a function chart or flow chart. In this case a function chart has been constructed, and appears in Fig. 4.23.

The sequence function chart follows the symbolic conventions described earlier, illustrating the use of AND branching rails to link steps 4–6. These steps

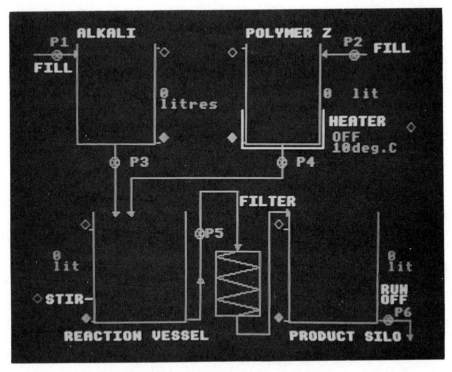

Fig. 4.22   Chemical tank schematic.

must *all* be completed before step 7 (pump 5 on) is initiated. Notice also the two actions that stem from step 6: 1, the stirring arm is activated and 2, the associated timer T is started. This step will be reset on time-out of timer T. An additional step 9 has been introduced at the bottom of the chart. This *basic position step* is often used to form an overall check on the system by testing several conditions before allowing entry to the next process loop. For example, all tanks must be empty and adequate raw materials must exist before continuing. It is not used in the following design exercise, being included here for information only.

Input/output assignments

Table 4.4 provides a typical set of I/O data for a small programmable controller to be used as the process controller.

We can draw up the boolean equations to represent the process steps as follows. Firstly each set of conditions is stated in the form that is most readily understood from the verbal description. Once this is done, and all elements of the sequence are understood, this set of equations is amended to eliminate any state-transition problems as described in the previous examples.

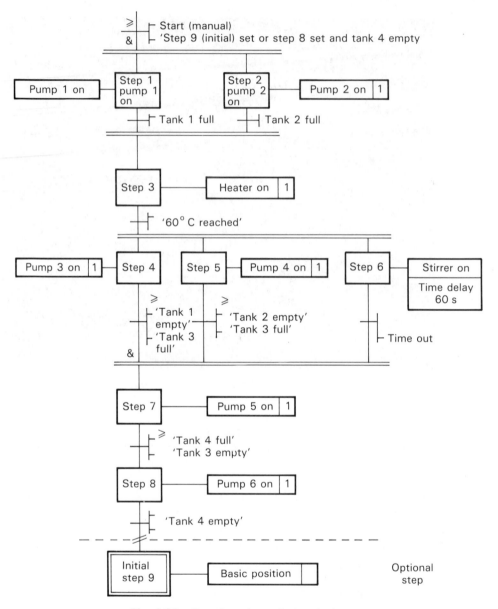

Fig. 4.23   Function chart of chemical process.

(a) Initial equations

Step   Equation
$1 = 8 \cdot T4E + \overline{INIT} + 1 \cdot \overline{T1F}$

Here step 1 is entered either when step 8 is active *and* tank 4 is empty, *or*

Table 4.4   Typical I/O data.

Signal and description		PC address	Status
tank 1 empty,	T1E	X401	input
tank 1 full,	T1F	X402	input
tank 2 empty,	T2E	X403	input
tank 2 full,	T2F	X404	input
tank 3 empty,	T3E	X405	input
tank 3 full,	T3F	X406	input
tank 4 empty,	T4E	X407	input
tank 4 full,	T4F	X410	input
temp sensor,	TEMP	X411	input
initial start relay		M70	special marker operates when PC in run mode
	STATE 1	M101	marker
	STATE 2	M102	marker
	STATE 3	M103	marker
	STATE 4	M104	marker
	STATE 5	M105	marker
	STATE 6	M106	marker
	STATE 7	M107	marker
	STATE 8	M110	marker
pump 1,	P1	Y430	output
pump 2,	P2	Y431	output
pump 3,	P3	Y432	output
pump 4,	P4	Y433	output
pump 5,	P5	Y434	output
pump 6,	P6	Y435	output
heating element,	HEAT	Y436	output
stirring arm,	STIR	Y437	output
timer time-out	TIME	T450	internal timer

because the initial cycle is in progress and the INIT signal has occurred. Step 1 is maintained until the T1F sensor is on (true). The other equations are formed in a similar manner.

$$2 = 8 \cdot T4E + \overline{INIT} + 1 \cdot \overline{T2F}$$
$$3 = 2 \cdot T2F + HEAT \cdot \overline{TEMP}$$
$$4 = (HEAT \cdot TEMP + 4 \cdot \overline{T3F}) \cdot \overline{T1E}$$
$$5 = (HEAT \cdot TEMP + 5 \cdot \overline{T3F}) \cdot \overline{T2E} \quad \rbrace \quad \text{branches}$$
$$6 = (HEAT \cdot TEMP + 6 \cdot \overline{TIME}) \cdot \overline{T3E}$$
$$7 = (6 \cdot TIME + 7 \cdot \overline{T4F}) \cdot \overline{T3E}$$
$$8 = 7 \cdot T4F + 8 \cdot \overline{T4E}$$

The next step is to modify these equations to eliminate any transition problems, if they exist. Here this need only be done where the steps follow each other directly. If there is a delay as in step 6, where the timer is controlling the stirring arm and hence the initializing of pump 5, it is imperative that step 7 is *not* used to reset steps 4 and 5. Otherwise pumps 3 and 4 would continue to run after tank 3 was filled, until the 60 s timer had timed out, creating overflow problems! Therefore several equations are unaltered.

(b) Final equations

$1 = 8 . \text{T4E} + \overline{\text{INIT}} + 1 . \overline{\text{T1F}}$ Step 3 is not used to reset step 1 as not entered until step 2 is completed.

$2 = 8 . \text{T4E} + \overline{\text{INIT}} + 2.\overline{3}$ Step 3 used for reset as it follows 2.

$3 = 2 . \text{T2F} + 3.\overline{5}$ Step 5 for reset as it signals TEMP reached.

$4 = (3 . \text{TEMP} + 4 . \overline{\text{T3F}}).\overline{\text{T1E}}$ Unchanged

$5 = (3 . \text{TEMP} + 5 . \overline{\text{T3F}}).\overline{\text{T2E}}$ Unchanged

$6 = (3 . \text{TEMP} + 6.\overline{7}).\overline{\text{T3E}}$ Step 7 used to reset 6 as it follows on.

$7 = (6 . \text{TIME} + 7.\overline{8}).\overline{\text{T3E}}$

$8 = 7 . \text{T4F} + 8.\overline{1}$

$\text{INIT} = 1$ For automatic start in cycle 1.

The physical input/output assignments are now inserted into the equations, and then the PC program or ladder diagram prepared. (Fig. 4.24.)

Exercise 4.5 Reconstruct the above equations using the listed I/O and internal functions. Once complete, relate your equations to the ladder diagram given in Fig. 4.24.

The operating sequence described for the chemical process could of course be changed in various ways. One alteration that would result in a greatly increased throughput is to refill tanks 1 and 2 after they have been emptied by pumps 3 and 4. This allows tank 2 to be heated towards the target temperature of 60°C whilst tank 3's contents are being stirred under timer control. Likewise, once the contents of tank 3 have been transferred to tank 4, tank 3 may be refilled from tanks 1 and 2 (once the contents of 2 have reached 60°C).

Exercise 4.6 One of the great advantages of any programmable controller is having the ability to easily amend existing control programs. For the above chemical process changes, prepare a revised function chart, logic equations and the resulting ladder diagram.

4.7 LIMITATIONS OF LADDER PROGRAMMING

1. Ladder programs are ideal for combinational/interlock tasks and simple sequential tasks. However, the lack of 'comment' facilities on most small

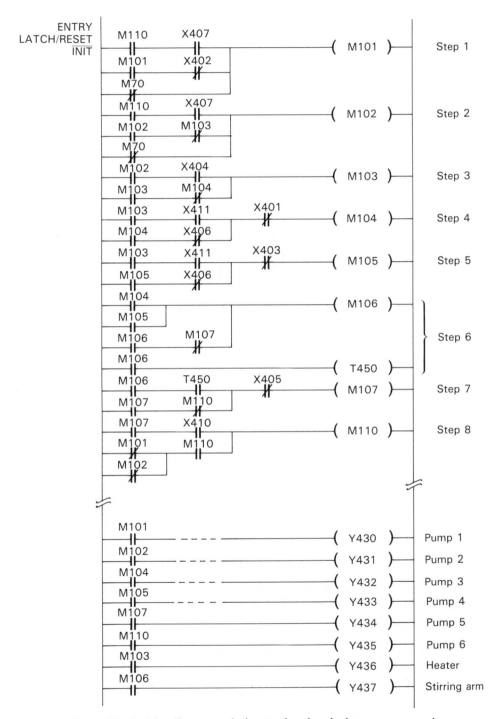

Fig. 4.24   Ladder diagram solution to the chemical process example.

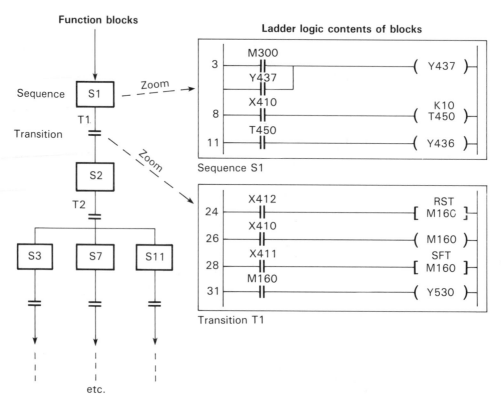

Fig. 4.25   Function block programming and zoom facility used to show contents of a block.

programmers makes interpretation of any program extremely difficult (thus creation of full documentation is essential).

2. When applied to complex sequential tasks, ladder programs become cumbersome, difficult both to design and debug. This is mainly due to having to provide entry, hold and reset elements in every stage to ensure no sequence errors occur (see section on sequence program design).

Several manufacturers are adopting a *function block* style of programming that removes most of this complexity. This employs basic programming symbols that are closely related to the function-chart symbols used for program-design purposes, as used earlier in this chapter.

Advanced graphic programming languages

The facility for programming using *functional blocks* is currently available on a few larger programmable controllers, such as those from Siemens and Telemechanique.

This approach uses *graphic blocks* to represent sections of circuitry related to a particular task or part of a process. Each function block is user-programmed to contain a section of ladder circuitry required to carry out that function. The sequential operation of the control system is obtained by progression from one block to the next, where a step is entered only if its entry conditions are fulfilled, in which case it becomes active and the previous step becomes inactive.

Thus, there is no need to reset the previous step – an important advantage over conventional ladder programming.

To examine or program the contents of each block the user would zoom in on the block in question. This 'windows in' on the contents as shown, which are displayed on the programming panel. The necessary details are then entered in normal ladder format.

Provision for displaying simultaneous sequences is a further important feature of these programming methods, displaying the multi-tasking ability of PLCs in an easy-to-follow manner.

Workstations

The traditional tool for programming PLCs is the small, hand-held panel which can provide only limited monitoring and editing facilities. Most manufacturers are now using personal computers as workstations on larger programmable controllers, in order to fully exploit these features, and those of graphic function blocks.

4.8 SUMMARY

This chapter has examined the basic design techniques used in solving combinational and sequential control tasks. Boolean algebra formed the basis of constructing the

Table 4.5   Good programming practice – guidelines.

1. Plan your program on paper first! Don't just power up your
   PLC programmer and start keying in elements.
   90% of your time should be spent working out the program,
   and only the last 10% keying it in.
   (When familiarizing yourself with a PLC, however, this is only
   partially valid.)
2. Keep documentation of all elements used in the program –
   add comments as necessary.
3. Assume the program will find every error sequence possible –
   design safety into it!
4. Avoid 'smart' programs: keep them 'readable'. Documentation helps here.
5. Use of printer and program storage (tape and disk) can aid
   development. Try sectional development and testing if
   possible.
6. Use 'forcing' and 'monitoring' facilities (if provided) to
   observe program operation *in situ*.

logic networks required to implement each step in the design solution. System operation has been described using flowchart and function-chart techniques.

Besides standard logic functions, all but the smallest PLCs are equipped with other facilities that allow the straightforward implementation of programs to control sequential systems. These are examined in the following chapter. Whatever application control sequence is to be developed, it is worth following a few simple, common-sense guidelines during the software development process. Table 4.5 gives examples of these.

4.9 FURTHER EXERCISES

(Exercises 4.1–4.6 are included within the body of the chapter.)

4.7 Explain why a structured approach to software design is often necessary when dealing with ladder programs.

4.8 What are the main differences between combinational and sequential control tasks? Why are function charts and flowcharts used in the design sequence?

4.9 With the aid of a diagram, describe the functional structure of a typical, well-designed program.

4.10 State the advantages and disadvantages of using Boolean terms to describe control requirements.

4.11 What is meant by 'entry, hold and reset' elements in a ladder program? Give an example to illustrate your answer.

4.12 Compare function-block programming techniques with conventional ladder programming.

## FIVE

Sequential Control Facilities

In Chapter 4 the nature and requirements of PLC programming for sequential control systems were investigated. This revealed certain key points:

1. The division of each control system into several steps or states.
2. A logic network was required to implement each of these steps.
3. Each step required entry, self-hold and reset elements to allow progression through a sequence of steps, as certain conditions were fulfilled.
4. The need for memory elements to hold information on sequence position.

It was shown that these requirements can be met using conventional logic networks. However, programmable controllers provide several other facilities that may be used to simplify the design and construction of all types of sequential control problem. These include shift registers, master control relays, stepladder functions, drum timers and sequencers.

5.1 SHIFT REGISTERS

The operation and programming of a shift register in a typical PLC was examined in Chapter 3, where an example used the register to hold the status of items on a conveyor belt. The shift register is equally capable of holding the status of a sequential system, and of controlling that system via associated marker contacts. Also, the shift register solution to a given control problem will be in simpler format than the equivalent logic network.

The basic principles of applying shift registers to multistep sequential problems are as follows:

1. The shift register must possess at least as many 'bit stores' as there are steps in the control task (8- or 16-bit registers are the most common). Thus each bit represents a process step.
2. A single logic 1 is entered in the first bit of the shift register on start-up, using a logic network on the DATA (output) line (see Fig. 5.1).
3. This logic 1 is moved from bit to bit through the shift register, enabling the step associated with the bit containing the 1, and disabling the bit-step it has just left.
4. This bit-shifting is accomplished by programming the necessary step transitions onto the shift (SFT) line of the register.

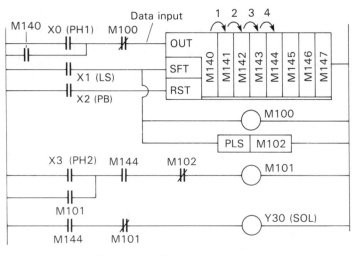

Fig. 5.1   Shift register contents.

5.   When the process cycle is complete, this can RESET the entire register contents to 0 by programming the necessary logic conditions onto the reset (RST) line.

Example of shift-register control – a small robot arm

The following example illustrates a typical shift register solution for controlling a limited sequence robot arm, using a small PLC.

Task   Move the actuator using the solenoid valves; UP, DOWN, RIGHT, LEFT (Fig. 5.2(a)), to follow a sequence, viz:

When in the normal position, move the arm DOWN when the START button is

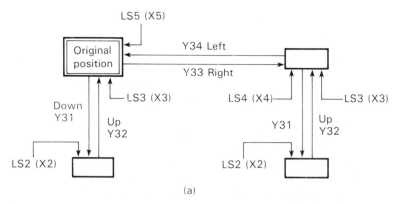

(a)

Fig. 5.2   (a) Schematic of robot path; (b) function chart of robot sequence.

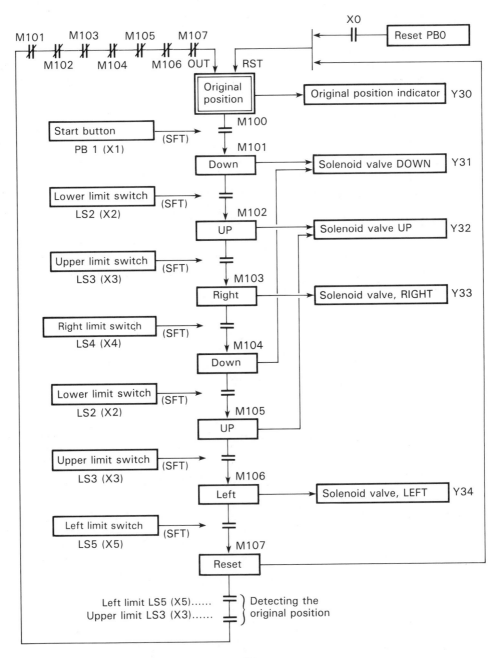

Start button
PB 1 (X1)

Lower limit switch
LS2 (X2)

Upper limit switch
LS3 (X3)

Right limit switch
LS4 (X4)

Lower limit switch
LS2 (X2)

Upper limit switch
LS3 (X3)

Left limit switch
LS5 (X5)

M101 M103 M105 M107
M102 M104 M106 OUT RST

X0
Reset PB0

Original position
Original position indicator Y30

M100

M101
Down
Solenoid valve DOWN Y31

M102
UP
Solenoid valve UP Y32

M103
Right
Solenoid valve, RIGHT Y33

M104
Down

M105
UP

M106
Left
Solenoid valve, LEFT Y34

M107
Reset

Left limit LS5 (X5)......
Upper limit LS3 (X3)......
} Detecting the original position

(b)

pressed, until it operates LS2. Then move the arm UP until LS3 is operated, which moves the arm RIGHT until it reaches LS4. Now swing the arm DOWN until LS2 is reached, then raise the arm UP to LS3. Move it LEFT to LS5, then reset. The arm should now be back in the normal position.

A flow chart or function diagram is constructed in Fig. 5.2(b) to help clarify the desired process, and to show all relevant input/output points that have been assigned on the PLC as in Fig. 5.3.

Examination of the above sequence description will reveal that several limit switches are used repeatedly during the process, but at different times in the cycle. If control of the arm were based purely on combinations of these inputs, the required sequence of movements would not occur. This process is sequential and needs a form of memory to determine which 'stage' in the process is currently active, and thus what the next stage will be when a particular input occurs. Obviously just one solenoid should be energized at any moment in time. Note the different exit paths from 'arm UP' – firstly to 'right', and later to 'left'.

This may be achieved by using a shift register. Its function is to shift in data bits presented on the 'out' line whenever the 'SFT' line is pulsed. Thus if a data pattern of 1001 is offered over successive shift pulses, the pattern will move through the register as shown in Table 5.1.

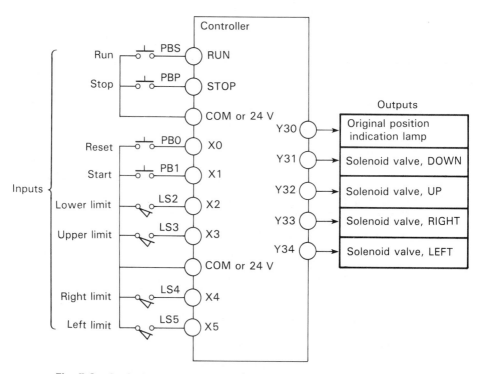

Fig. 5.3 Assignment of PLC input/output points (Mitsubishi F Series).

Table 5.1   Operation of a shift register.

Shift pulses	Data offered	Shift reg. contents
0	0	0000 0000
0	1	1000 0000
1	0	0100 0000
2	0	0010 0000
3	1	1001 0000
4	0	0100 1000
5	0	0010 0100
6	0	0001 0010
7	0	0000 1001

The RST line will reset the shift register contents to 0 if pulsed. In most PLCs, shift registers are made up of 8 or 16 individual bit-stores, or marker relays as they are often termed. Each marker (M) can have associated contacts which can be programmed to operate any other devices within the programmable controller. In this example markers M100–M107 form the shift register, and can be seen in Fig. 5.4 controlling output coils to solenoids (Y30–34), and also normally open and closed contacts to the shift-register control lines.

Operation

From Figs. 5.2 and 5.4 it can be seen that a single 1 bit will enter the register when the arm is in the normal position, caused by the limit switches LS3 and LS5 completing a path to the DATA line via the normally closed contacts of M101–107. This rung is designed to offer a logic 1 only when the robot is in the normal position. At every other step an M contact is open, offering a logic 0 to the line. M100 is ON in the basic position, energizing Y30. The START button X1 will activate the 'shift' line, causing the 1 to be shifted from M100 into M101, which turns on (M100 is turned off). This operates the DOWN solenoid via the related M101 contact to Y31. Note that the first parallel rung to the SFT line (M100.X1) is now disabled.

Step transition

The shifting of the controlling logic 1 from one marker bit to the next eliminates any possible timing problems, since the enabling 1 leaves one step bit then enters the next step bit. In this example the shift-register marker relays, M100–M106, provide interlocking contacts on the 'shift' logic rungs. Each condition for changing from one step to the next is programmed here, with the limit switches ANDed to enabling contacts from the M relays. These interlocks ensure that only one 'shift path' will be enabled at one particular stage in the process.

It is worth noting that the 'exit' conditions for a given step are also the 'entry' conditions for the next step. For example, the exit criterion for the basic position is M100 AND X1. This is also the entry criterion for step 1 (M101). With a shift register no holding or self-maintaining logic is necessary, since the data is automatically retained until shifted or reset. (Battery-backed shift registers are normally provided for use where system status has to be retained through power failure.)

It is left to the reader to follow through the remainder of the robot control program, as this will highlight the design details.

Once the control bit 1 is shifted into M107, this activates the reset step. M107 does not drive a physical output, but has contacts in the OUT and RST lines to the shift register. Thus M107 resets the register to 0 and holds the OUT line at zero. The logic 1 in M107 is also reset, and this re-establishes the path to the OUT line if the robot is back in the normal position. So M100 operates and the process recommences.

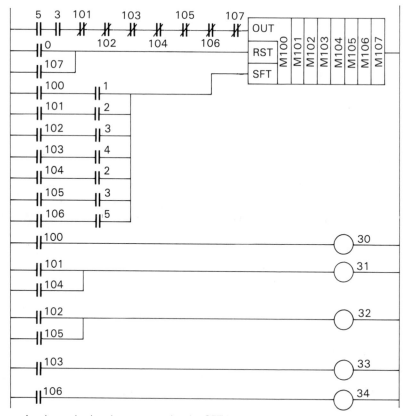

As shown in the above example, the SFT instruction circuit should have an interlock contact (M100 above) to prevent a sequence error in the process.

The output circuits should be programmed collectively at the end of a program to prevent any duplication of the same output relays.

(a)

Fig. 5.4 (a) Ladder diagram; (b) coding sheet of shift-register control for a point-to-point robot.

The other means of resetting the register is by using the X0 manual push-button. This is programmed as an OR contact with the M107 reset contact.

Program design

In the majority of cases the logic requirements of shift, reset and data entry lines in the shift register are very straightforward, reducing (if not removing) the need for formal design techniques. However, most designers use logic equations to define *all* elements of program layout, if only to provide complete documentation of the system. For the robot example the logic equations are as follows:

$$OUT = X5.X3.\overline{M101}.\overline{M102}.\overline{M103}.\overline{M104}.\overline{M105}.\overline{M106}.\overline{M107}$$
$$RST = X0 + M107$$
$$SFT = M100.X1 + M101.X2 + M102.X3 + M103.X4 + M104.X5 + M105.X6$$

Y30 = M100	Y31 = M101 + M104	Y32 = M102 + M105
Y33 = M103	Y34 = M106	

Mitsubishi F series program

A ladder diagram is not necessarily required to complete a coding sheet, which can be made directly from a flowchart of the process.

0	LD	0	Resetting circuit	31	LD	104	Shift at lower limit
1	OR	107		32	AND	2	
2	RST	100		33	SFT	100	
3	LD	5	Detection of original position	34	LD	105	Shift upper limit
4	AND	3		35	AND	3	
5	ANI	101		36	SFT	100	
6	ANI	102		37	LD	106	Shift at left limit
7	ANI	103		40	AND	5	
10	ANI	104	Starting condition of the shift register	41	SFT	100	
11	ANI	105		42	LD	100	OUT 30 by M100
12	ANI	106		43	OUT	30	
13	ANI	107		44	LD	101	OUT 31 by M101 and M104
14	OUT	100		45	OR	104	
15	LD	100	Interlock Shift by start button	46	OUT	31	
16	AND	1		47	LD	102	OUT 32 by M102 and M105
17	SFT	100		50	OR	105	
20	LD	101	Shift at lower limit	51	OUT	32	
21	AND	2		52	LD	103	OUT 33 by M103
22	SFT	100		53	OUT	33	
23	LD	102	Shift at upper limit	54	LD	106	OUT 34 by M106
24	AND	3		55	OUT	34	
25	SFT	100		56	END		
26	LD	103	Shift at right limit				
27	AND	4					
30	SFT	100					

As shown in the above example, the shift instruction can be programmed by dividing SFT instructions according to the conditions in parallel.

(b)

A ladder logic diagram is then drawn up using the above description and data (Fig. 5.4). The instruction list to implement this ladder is also given for a Mitsubishi F series programmable controller.

This example has introduced the principles of shift register sequence control. For comparison, the example from Chapter 4 on chemical-process control is now solved below, using shift-register techniques.

5.2   THE CHEMICAL PLANT REVISITED – SHIFT-REGISTER CONTROL

Referring to Figs. 4.22 and 4.23, the operation of the process is to be similar in terms of tank sequencing, but with the inclusion of a 'basic position' that marks the end of a cycle. Referring to Fig. 5.5, this step is entered when all tanks are empty *and* all other steps in the process are inactive (off). This forms the logic circuit for the data OUT line to the shift register.

Input/output assignments

The shift register is formed by marker relays M160 to 167, but M167 is unused since the process now consists of seven steps. Inputs and outputs are as before, with the inclusion of an automatic start contact, M70. This is related to a special function relay that operates on power-up of the programmable controller (Mitsubishi F model). For clarity, all I/O and internal functions used are listed in Table 5.2. The marks shown in Table 5.3 comprise the internal shift register.

Notice that the number of states or steps in the process has decreased to seven, although an extra state has been added. This is accomplished by combining the steps that were previously operated in parallel branches using logic control (see Figs. 4.23 and 5.5). For simple shift-register control, it is necessary to control all the elements

Table 5.2   Input/output and internal functions.

Signal and description		PC address	Status
Tank 1 empty,	T1E	X401	Input
Tank 1 full,	T1F	X402	Input
Tank 2 empty,	T2E	X403	Input
Tank 2 full,	T2F	X404	Input
Tank 3 empty,	T3E	X405	Input
Tank 3 full,	T3F	X406	Input
Tank 4 empty,	T4E	X407	Input
Tank 4 full,	T4F	X410	Input
Temp sensor,	TEMP	X411	Input
Initial start relay		M70	Special marker operates when PC in run mode

Table 5.3   Markers comprising the internal shift register.

BASIC POSITION STATE		M160	Initial marker
	STATE 1	M161	Marker
	STATE 2	M162	Marker
	STATE 3	M163	Marker
	STATE 4	M164	Marker
	STATE 5	M165	Marker
	STATE 6	M166	Marker
Pump 1,	P1	Y430	Output
Pump 2,	P2	Y431	Output
Pump 3,	P3	Y432	Output
Pump 4,	P4	Y433	Output
Pump 5,	P5	Y434	Output
Pump 6,	P6	Y435	Output
Heating element,	HEAT	Y436	Output
Stirring arm,	STIR	Y437	Output
Timer time-out	TIME	T450	Internal timer

at a particular stage in a sequence from the single 1 that is moved through the register.

(It might be assumed that a better technique would be to place a two- or three-bit stream of 1s in a larger shift register, and reconnect the output-enabling lines to ensure that only the required actuators are enabled as the 'word of 1s' moves along, i.e. several marker bits would have no connection where only a single device has to be activated. This approach is intrinsically problematical, however, and is inferior to other mechanisms provided for complex sequence control.)

The modified function chart in Fig. 5.5 identifies all the conditions that are to cause transition between states in the process. When fulfilled, each condition is to shift the controlling 1 along the register one bit, de-activating the last step and activating the next. These criteria are programmed onto the SHIFT line as parallel OR branches. Again, each branch consists of the shift condition ANDed with a contact related to the step that must be current for that shift condition to be valid – an interlock to prevent sequence errors.

The shift register RESET criteria are shown, in this case the activation of M166 – the reset step. The provision of a manual reset in this process might, if activated, result in a system 'crash'. Since the tanks would be part full at any stage except the first and last, a reset would not take account of any part-completed batch of product. This would result in a spoilt product, and must therefore be avoided. It is worth noting that an overflow will not occur, due to the design of the logic controlling the pump operations.

From the above descriptions of all the shift-register inputs, the ladder logic program can be constructed as in Fig. 5.6. The reader should compare this control program with the more complex pure logic equivalent in Fig. 4.24. When dealing with larger and more complex control sequences, a logic solution becomes even more difficult to design and interpret, whilst the shift register version requires only additional control logic on the relevant lines.

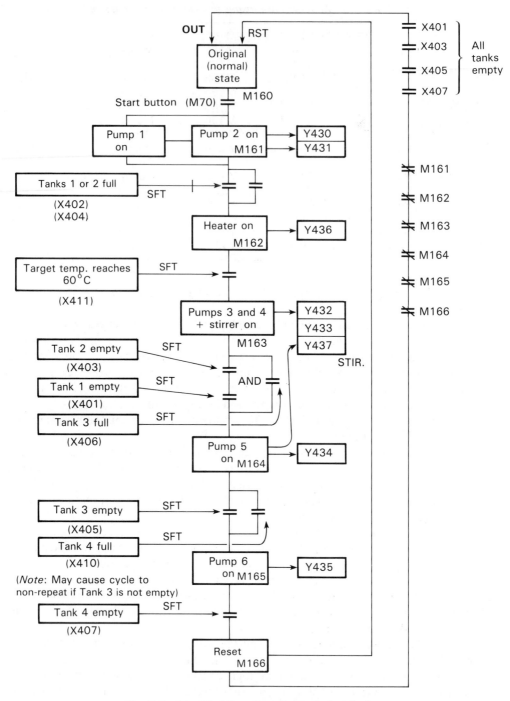

Fig. 5.5 Modified function chart of chemical process.

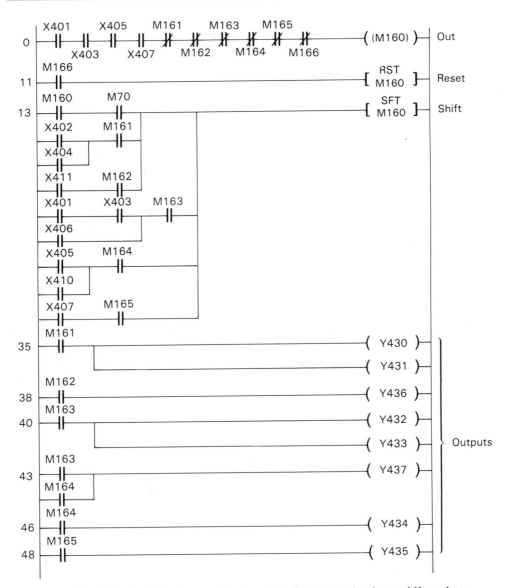

Fig. 5.6 Ladder diagram for chemical plant control using a shift register.

Shift-register facilities are provided on virtually all programmable controllers, including even the smallest models. The provision of alternative mechanisms for handling sequence control varies between manufacturers and their ranges of PLCs. Several of these facilities are examined in the following sections.

Exercise 5.1 For the above example of shift register control, describe the logic inputs to all register lines using Boolean equations.

Exercise 5.2   Describe the advantages of this form of sequence control over pure logic solutions.

5.3   STEPLADDER FUNCTIONS

The stepladder instruction is provided by certain manufacturers to facilitate the programming of process step control. The stepladder function closely resembles the function-chart description of a sequential stepping system, eliminating the need for most of the control logic necessary with the previous methods. It achieves this simplicity of control through certain key features:

1. Self-latching state memory elements that are normally battery-backed.
2. Automatic reset of the previous state – the 'transfer origin'.
3. Ease of creating parallel branch networks for both OR and AND cases.

In the case of Mitsubishi, these facilities are used in conjunction with the internal states S of simulated relays (from the block S600–647 in an F series PLC). Like ordinary auxiliary relays or markers, the state S may be used for output instructions such as OUT, Set, Reset, etc. with related contacts used in conventional logic networks. (When used in this way their function is identical to a marker relay, Fig. 5.7.) Before examining the nature and techniques of stepladder programming, a simple example is used to show the relationship to a function-chart description. See Fig. 5.8.

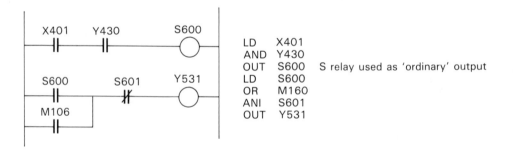

Fig. 5.7   *General format of S (state) relay program.*

Simple example of stepladder circuit

The function chart in Fig. 5.8(a) describes part of a sequence that is to be controlled, showing the process steps, actions and transition conditions as normal. This sequence may be implemented with stepladder functions as detailed in Fig. 5.8(b).

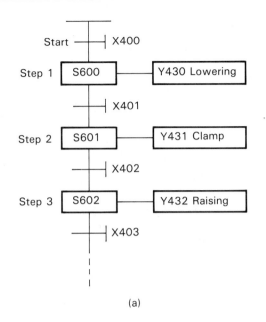

(a)

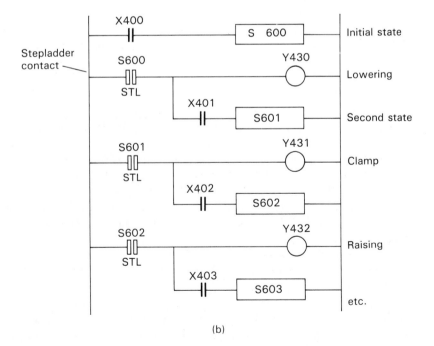

(b)

Fig. 5.8 (a) Function chart of simple process; (b) stepladder program.

Outline circuit operation

The state memory S600 will be set when contact X400 closes. This is an initial state, and there is no previous state to be reset. S600 has a stepladder contact programmed in the next rung, which acts as a master control over any following circuitry. As S600 is set as the active (current) state, the contact closes, operating output Y530. This is the action associated with state S600. When contact X401 closes due to this action, state S601 is set. This automatically *resets* state S600, with S601 now the active state. It also has a stepladder (STL) contact which controls the succeeding rung, etc. The reader may expect this action to result in state S601 being released, since the 'entry' contact of S600 is now opened. It must be appreciated that the state is latched in the 'on/set' position once initially activated. As stated above, the reset and self-latching facilities are handled automatically by the stepladder/state S instructions. The order of programming the state numbers is optional, so for example S610 may be substituted for S601 in the above example.

This brief example has introduced the use of stepladder control. The following sections consider the topic in greater detail.

Stepladder control function

The STL instruction is similar to a master control function in that it connects a section of logic to the bus bar, only allowing it to function when the stepladder contact is closed. The STL instruction is only applicable to the state S, transforming it into a store for holding the current operation stage of a controlled process. Fig. 5.9(a) shows a basic stepladder circuit, with a conventional logic equivalent in Fig. 5.9(b).

In Fig. 5.9 the output Y1 is operated (set) when contacts Sa *and* Xb close. If the Xc contact is then closed, the state Sd is set on and its related contact will reset (R) state Sa. This opens the original Sa contact, breaking (resetting) the original transfer state. However, the use of 'set' prefixes on the outputs Y1 and Sd (S Y1 and S Sd) makes them self-latching, and they will remain active unless reset elsewhere in the circuit. This does in fact happen to Y1, as the M105 relay at the foot of the circuit is pulsed by the Sd contact, which resets Y1 at the top of the circuit via contact M105 pulsing. Note the use of a master control relay, M106, in the program for Fig. 5.9(b). When MC M106 is operated, the logic enclosed by its contacts will operate normally.

In simpler terms, a stepladder circuit is a development of the master control relay: enabled via a contact associated with the previous state S, once any intermediate contacts have closed to give power flow to the outputs, the stepladder section activates. This in turn resets the previous stepladder section and possibly enables the next, i.e. automatic reset and set up of transfer states. It is these features that make the stepladder an ideal method for controlling sequential step systems.

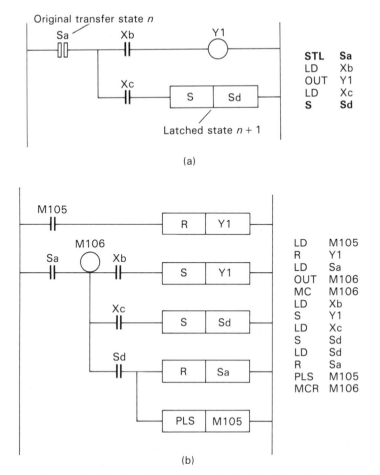

Fig. 5.9 Stepladder functions: (a) basic circuit; (b) equivalent circuit.

Programming stepladders

The circuit in Fig. 5.10 illustrates the basic principles for creating stepladder programs. When the STL instruction is used, the next commencing circuit element (LD or LDI) will effectively appear after the STL contact. This continues as other circuit rungs are entered, for example:

$$\ldots \text{LD} \quad \text{M102}$$
$$\text{S} \quad \text{S605}$$
$$\text{STL} \quad \text{S605}$$
$$\text{OUT} \quad \text{Y533} \ldots$$

This gives further parallel branches off the stepladder contact, similar to a master control relay function.

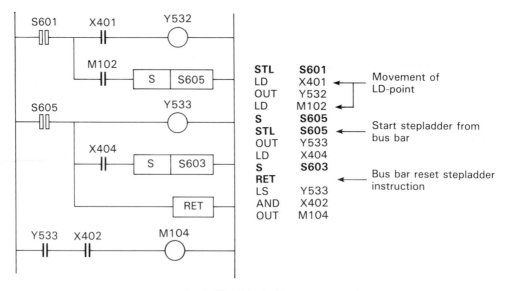

Fig. 5.10 Stepladder programming.

If the next circuit block starts with an STL instruction, the block is connected to the left-hand bus bar. However, when the next block is of conventional logic, the preceding stepladder must be terminated with a return (RET) instruction.

Restrictions

Each state using stepladder facilities may control only *one* associated stepladder contact. If this is not observed, unpredictable output switching may result. The exception to this rule is when parallel branches are being programmed (see later in this chapter). Care must also be exercised when combining stepladders with other PLC functions, such as JUMP and Master Control, to ensure correct program operation.

Selective (OR) branching

Figure 5.11 shows a function chart and ladder diagram for selecting between two paths A and B, depending upon the test conditions, X401 or X404. Both paths A and B culminate with entry to the common path via state S606.

(a) Branching Here the selective (OR) branching is achieved by programming the two ORed test conditions after the S601 stepladder contact, so if X401 is operated after stepladder S601 then S602 will be set, enabling path A. Alternatively, if X404 is operated, S604 will be set and enable path B in the process. The transfer origin, state S601, will be reset automatically when either S602 or S604 is set.

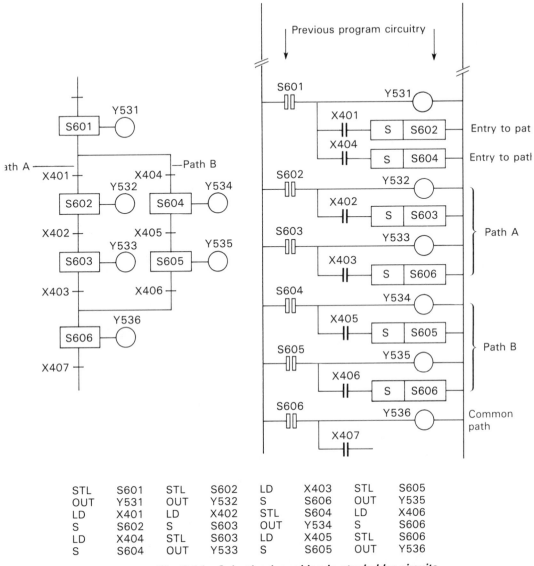

STL	S601	STL	S602	LD	X403	STL	S605
OUT	Y531	OUT	Y532	S	S606	OUT	Y535
LD	X401	LD	X402	STL	S604	LD	X406
S	S602	S	S603	OUT	Y534	S	S606
LD	X404	STL	S603	LD	X405	STL	S606
S	S604	OUT	Y533	S	S605	OUT	Y536

Fig. 5.11   *Selective branching in stepladder circuits.*

(b) Joining   The common state S606 is set from state S603 in path A, or
from state S605 in path B. The transfer origin state will be automatically reset
by S606 under stepladder operation. This joining operation is achieved by
programming an entry condition to S606 into each branch of the process,
ensuring state transfer to S606 from the end of either path A when X403
operates, or from path B when X406 operates. Additional selective paths may
be added in the same manner.

Parallel (AND) branching

The example in Fig. 5.12 illustrates the construction of multiple paths A and B, which operate simultaneously (in parallel), whilst starting and culminating in a single common path. When the transfer origin state S601, is active, the operation of X401 sets both S602 and S604, resetting state S601 and executing paths A *and* B concurrently.

Joining   The common state S606 can only be entered when states S603 (path A) *and* state S605 (path B) are active, *and* X404 is operated. That is,

$$S606 = S603 \cdot S605 \cdot X404$$

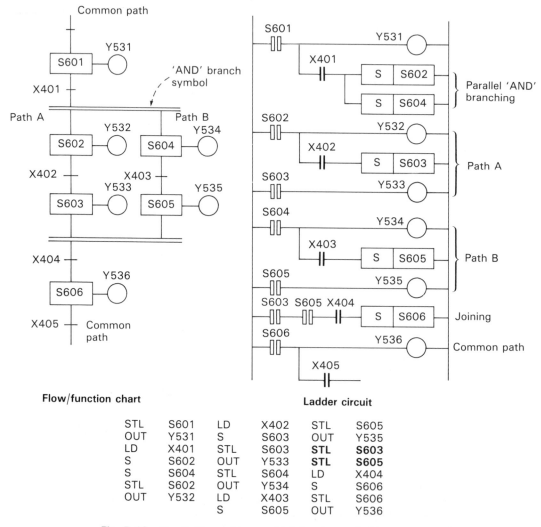

STL	S601	LD	X402	STL	S605
OUT	Y531	S	S603	OUT	Y535
LD	X401	STL	S603	**STL**	**S603**
S	S602	OUT	Y533	**STL**	**S605**
S	S604	STL	S604	LD	X404
STL	S602	OUT	Y534	S	S606
OUT	Y532	LD	X403	STL	S606
		S	S605	OUT	Y536

Fig. 5.12   Parallel branching and joining in stepladder programming.

Once this condition is met, S606 is set and both the transfer origins S603 and S605 are reset. Additional parallel paths may be programmed in a similar fashion.

Combinations of branch types

Many industrial processes involve situations where one operation is always to be performed, and, if certain conditions exist, a second parallel operation is to be carried out simultaneously. This type of operation may be programmed using stepladder functions as shown in Fig. 5.13.

(a) Operation   In this example, the main process path consists of states S600, S601 and S603. State S601 has a parallel branch containing S602. Under normal conditions when X401 operates, the state S601 is set and the branch to

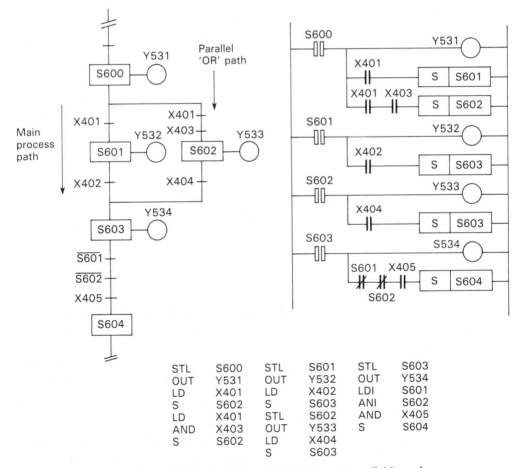

STL	S600	STL	S601	STL	S603
OUT	Y531	OUT	Y532	OUT	Y534
LD	X401	LD	X402	LDI	S601
S	S602	S	S603	ANI	S602
LD	X401	STL	S602	AND	X405
AND	X403	OUT	Y533	S	S604
S	S602	LD	X404		
		S	S603		

Fig. 5.13   Combination of selective and parallel branches.

S602 enabled. However, S602 becomes concurrently active only if X403 is operated, i.e. contact X403 determines whether the parallel path is executed.

(b)   Joining   The state S603 is set from states S601 or S602 when inputs X402 or X404 operate respectively. When only S601 has been active, i.e. the mandatory path, this criterion is adequate. However, if the parallel path has also been operating, then a route for entering state S603 exists from each path, regardless of whether the other path's operation is complete or not:

$$S601 . X402 + S602 . X404$$

This situation is undesirable, since the operations associated with states S601 and S602 are unlikely to be completed at the same instant in time, and further consideration of the problem is necessary to ensure the process cannot continue with an operation incomplete. State S603 resets the state that caused the transition, but it can reset the initial transition state, say S601, then once the second (S602) transition signal comes after a delay, this tries to 'set' S603 again, and so S602 can be reset accordingly. This is possible only if the process does not proceed beyond S603 before both previous paths are complete. By ensuring that transition from state S604 to S605 can only occur when both states S601 *and* S602 are turned off (completed), this problem is overcome. That is,

$$S604 = S603 . \overline{S601} . \overline{S602} . X405$$

Process repetition

It is often desirable to repeat a section of process operation a certain number of times, under manual or automatic setting. In the example shown in Fig. 5.14, a programmable counter is incorporated in the operation sequence to allow this facility.

Operation

The counter circuit is pulsed each time state S604 is activated. The value of the counter is preset either manually by thumbwheel switches connected to the programmable controller inputs, or by instructions earlier in the program. In this example it is assumed to be preset to a value of 8. Down to state S604 the process is in normal stepladder operation, then beyond this a selective branch is programmed, with a choice of S602 *or* S605 as the next state, depending on the transfer criteria:

$$S602 \text{ (loop)}   = S604 . C460 . X505 \qquad \text{(a)}$$
$$S605 \text{ (continue)} = S604 . C460 . X504 \qquad \text{(b)}$$

Therefore until the counter C460 reaches the preset value of 8, equation (b) to enter state S605 is not met, since the C460 contact will not be operated. Equation (a) however, is true if the counter has operated less than four times, *and* if X505 is

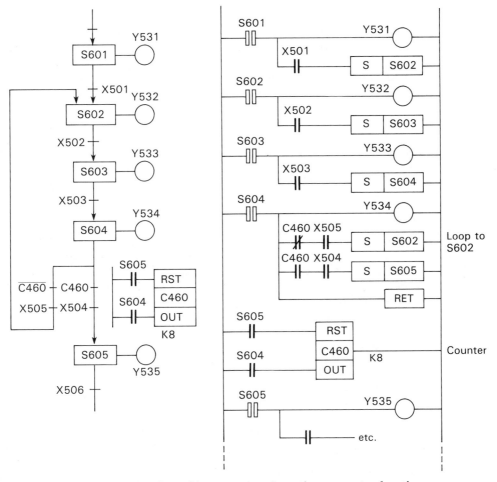

Fig. 5.14 Repetition programming using a counter function.

closed. This will cause the section of process S602 to S604 to repeat eight times, as the counter is incremented once each time S604 is set. On the eighth pass, S605 is enabled instead of S602, as the counter contacts operate. This also resets the counter via an S605 contact, preparing C460 for re-use when the whole process is repeated.

5.4 EXAMPLE OF STEPLADDER PROCESS CONTROL – A 'PICK-AND-PLACE' MACHINE

This example illustrates the use of stepladder techniques as part of a total control solution, in conjunction with the conventional logic and counter/timer facilities used to provide various control options (original source: Mitsubishi UK Ltd).

Machine function

Fig. 5.15 shows the layout of a 'pick-and-place' machine, consisting of a horizontal/vertical mechanism which is used to transfer workpieces from the left-hand table to the right-hand table. The horizontal and vertical movements are each driven by a double solenoid powered by air cylinders — a raising/lowering unit and a leftward/rightward unit. Once a solenoid is energized, the current position is maintained *if* the reverse operation solenoid is also energized. The clamp/unclamp action is driven by a single air solenoid, which clamps the workpiece onto the table when energized, and releases it when the power is removed. Limit switches detect the position of the mechanism, with a photoelectric switch which, for safety, checks for the presence of workpieces remaining on the right-hand table.

The basic sequence of the machine is as follows: from the home position (signified by the upper and left limit switches), the lowering solenoid Y430 is operated, moving the mechanism down until X401 operates. This stops the lowering motion and activates the clamp for a period of 1.7 s, to grip the workpiece. The mechanism now moves upwards to limit switch X402, when a horizontal movement starts, taking the mechanism over to X403 and initiating another downward path towards the right-hand table — providing no previous workpieces remain on the table (tested by the X405 photoelectric switch). The clamp then releases over 1.5 s, leaving the workpiece on the table as the mechanism returns to the home position.

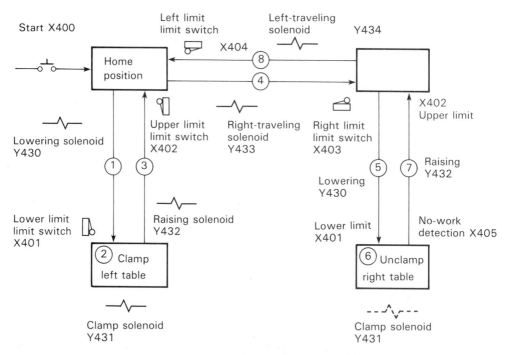

Fig. 5.15   Pick-and-place machine.

Operation modes

There is a requirement for the machine to be operated in four different modes, selected by a manual switch on the control panel shown in Fig. 5.16. The status of these manual switches is to be presented to programmable controller inputs as X406 and upwards.

The different modes are to function as follows:

(a) single operation mode (X407 on);
(b) step mode (X410 on);
(c) one-cycle mode (X411 on);
(d) continuous operation mode (X412 on).

(a) *Single operation mode* Here the action or load selected by the 'load selection switch' operates by depressing the start/stop button. For example, if the start button is pressed when the load selected is 'clamp', then the clamp will close, and if the stop button is pressed instead, then the clamp opens. This mode can be used to set the machine to the home position.

(b) *Step mode* The operation is advanced one step each time the start button is pressed, for example, the movement from the home position downwards to the left-hand table.

(c) *One-cycle operation mode* One complete cycle is to be executed automatically when the 'start button' is pressed with the mechanism in home position.

(d) *Continuous operation mode* When the start button is pressed at the home position, the operation cycle is performed continuously. If the stop button is pressed, the machine moves to the home position and is stopped.

All these requirements must be incorporated in the program design, together with the sequential control needs of the process.

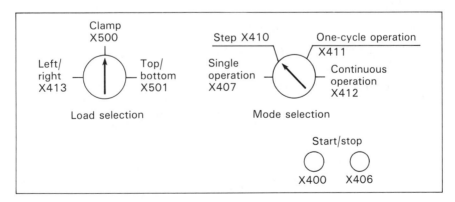

Fig. 5.16 Operating panel switches.

Input/output assignments

From the schematic diagram in Fig. 5.15 and the verbal description of the process, a set of I/O allocations can be made for the chosen programmable controller as shown in Fig. 5.17 (Mitsubishi F series PLC).

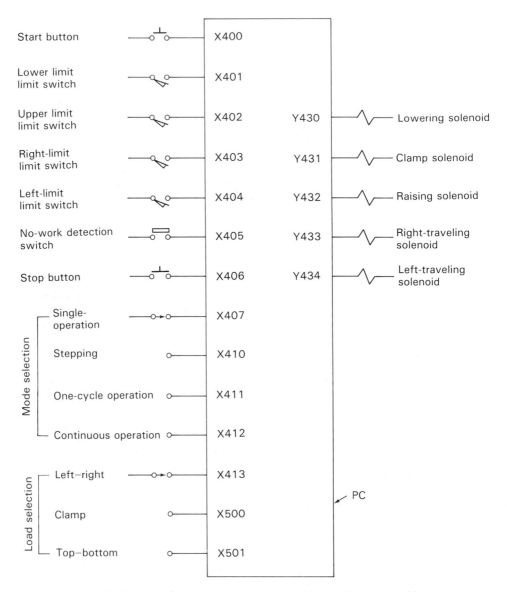

Fig. 5.17 Input/output assignments for pick-and-place machine.

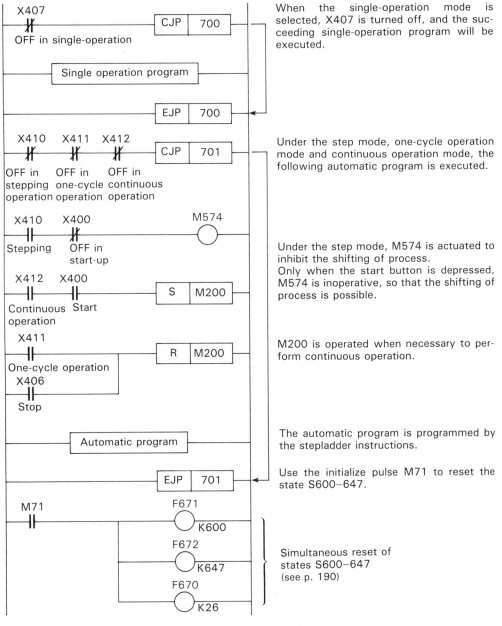

When the single-operation mode is selected, X407 is turned off, and the succeeding single-operation program will be executed.

Under the step mode, one-cycle operation mode and continuous operation mode, the following automatic program is executed.

Under the step mode, M574 is actuated to inhibit the shifting of process.
Only when the start button is depressed, M574 is inoperative, so that the shifting of process is possible.

M200 is operated when necessary to perform continuous operation.

The automatic program is programmed by the stepladder instructions.

Use the initialize pulse M71 to reset the state S600–647.

Simultaneous reset of states S600–647 (see p. 190)

Fig. 5.18   Overall program.

Program solution

The overall program layout is initially described, to show the relationship between the different modes that may be selected and their effects on the program operation. For clarity, the constituent programs for single operation and automatic operation are described by name only in the overall program layout of Fig. 5.18. Notice the use of conditional jumps to either select or bypass different sections of the program, depending on the selected mode of operation. The two separate subroutines for single operation and automatic operation can then be designed independently, in a modular fashion (Figs. 5.19–5.21). For details of the operation of each program, refer to the comments on the relevant figure.

The final rung in Fig. 5.18 uses facilities provided with most modern PLCs – an initial reset. Here a contact from a special function relay M71 in a Mitsubishi F40 PC provides an initial pulse when the controller is switched to run mode. This enables a function F670–672 that has been programmed to simultaneously reset all states S600–647 (note the K values).

In the single-operation mode, it is necessary to include various interlock circuits in order to prevent incorrect operation in certain positions. Since sequence control is not necessary in this part of the program, the interlocks can be designed using ordinary ladder logic. The X402 interlock contact in the first rung of Fig. 5.19

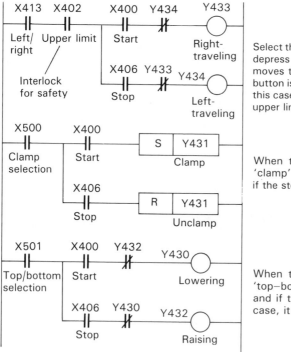

Select the load selector switch 'right–left' and depress the start button, so that the machine moves to the right; if, alternatively, the stop button is depressed, it will move to the left. In this case, this operation is allowed only at the upper limit position for safety.

When the start button is depressed with 'clamp' selected, the machine clamps, and if the stop button is depressed, it unclamps.

When the start button is depressed with 'top–bottom' selected, the machine lowers, and if the stop button is depressed in this case, it raises.

Fig. 5.19   Single-operation program.

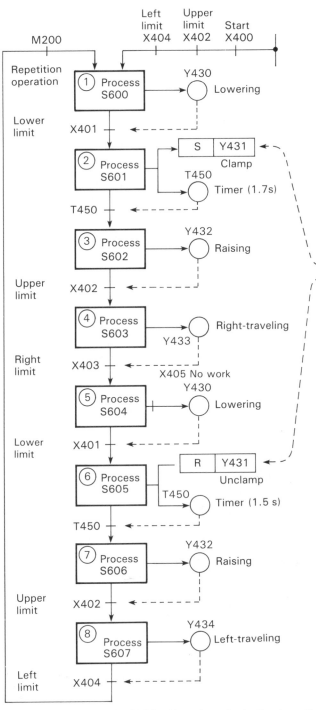

When the start button is depressed at the home position (upper limit/left limit) the state S600 is set, and the lowering solenoid Y430 is operated. Then, when the lower limit switch X401 is turned on, the machine moves to the state S601, setting the clamp solenoid Y431, and then moves on to the succeeding process 1.7 s after waiting for clamping.

The output of a set instruction such as Y431 is held until the reset instruction is activated.

Since the timer T450 is used in state S601 and state S605 but is not operated at the same time, the same timer can be programmed with different setting times. (The same timer cannot be used in consecutive processes.)

Fig. 5.20 Function chart of automatic operation.

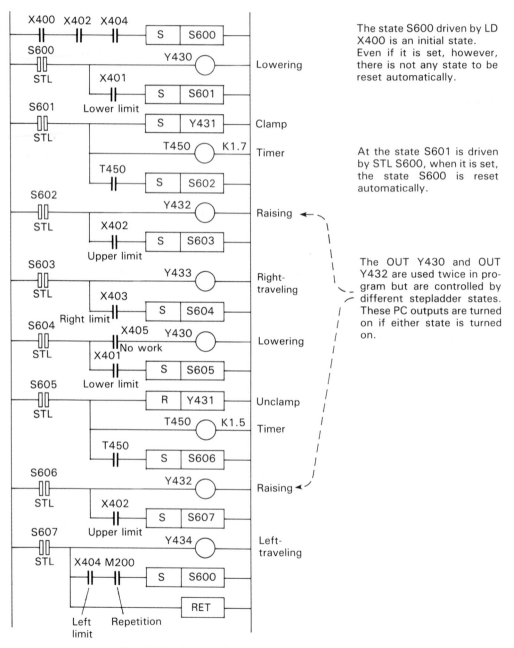

The state S600 driven by LD X400 is an initial state. Even if it is set, however, there is not any state to be reset automatically.

At the state S601 is driven by STL S600, when it is set, the state S600 is reset automatically.

The OUT Y430 and OUT Y432 are used twice in program but are controlled by different stepladder states. These PC outputs are turned on if either state is turned on.

Fig. 5.21 Automatic program using stepladder functions.

is to ensure that a 'right–left' movement can occur only when the mechanism is at the upper position, i.e. clear of both tables.

Each process state is reset automatically as the operation advances to the succeeding process. In Fig. 5.20, the stepladder state number and process step number are shown in the same function block; however, any state S from the range supported by the chosen PLC may be used. Remember that an out-of-sequence state may be used in place of consecutively numbered states. For example, in Fig. 5.20 state S602 may be replaced by S642, which still resets state S601 automatically when set.

When a graphic programmer or printer is used, the function chart description of the process is expressed in ladder-diagram format as shown in Fig. 5.21.

Comparison of the ladder program in Fig. 5.21 with the function chart in Fig. 5.20 will show the close similarity between the 'plan' and its stepladder implementation. In practice, the automatic sequence program can be directly programmed from the function chart, so close is this relationship. This is a major advantage of stepladder sequence control.

Notice that the same output instructions are used in both the single-operation

```
          Single-operation program        Automatic program →                    ← Automatic program
LDI  X  407      LDI  X  410      OUT  Y  433
CJP     700      ANI  X  411      LD   X  403
LD   X  413      ANI  X  412      S    S  604
AND  X  402      CJP     701      STL  S  604
OUT  M  100      LD   X  410      LD   X  405
MC   M  100      ANI  X  400      OUT  Y  430
LD   X  400      OUT  M  574      LD   X  401
ANI  Y  434      LD   X  412      S    S  605
OUT  Y  433      AND  X  400      STL  S  605
LD   X  406      S    M  200      R    Y  431
ANI  Y  433      LD   X  411      OUT  T  450
OUT  Y  434      OR   X  406           K  1.5
MCR  M  100      R    M  200      LD   T  450
LD   X  500      LD   X  400      S    S  606
OUT  M  101      AND  X  402      STL  S  606
MC   M  101      AND  X  404      OUT  Y  432
LD   X  400      S    S  600      LD   X  402
S    Y  431      STL  S  600      S    S  607
LD   X  406      OUT  Y  430      STL  S  607
R    Y  431      LD   X  401      OUT  Y  434
MCR  M  101      S    S  601      LD   X  404
LD   X  501      STL  S  601      AND  M  200
OUT  M  102      S    Y  431      S    S  600
MC   M  102      OUT  T  450      RET
LD   X  400           K  1.7      EJP     701
ANI  Y  432      LD   T  450      LD   M   71
OUT  Y  430      S    S  602      OUT  F  671
LD   X  406      STL  S  602           K  600
ANI  Y  430      OUT  Y  432      OUT  F  672
OUT  Y  432      LD   X  402           K  647
MCR  M  102      S    S  603      OUT  F  670
EJP     700      STL  S  603           K   26
```

Fig. 5.22   Coding of the overall program (Mitsubishi F Series PC).

and automatic programs. This approach is normally undesirable, since 'double coils' can cause unpredictable switching of the PLC outputs. In this situation however, any possible contention is avoided by the use of conditional jumps 700 and 701, which ensure that only one of the two programs has access to the outputs in any one cycle.

Finally, the complete program is entered into the target programmable controller, either in ladder format or using the resident instruction set. Fig. 5.22 lists the program code for a Mitsubishi F series PC, one of a range of controllers that support stepladder instructions.

It was stated previously that ladder diagrams and relay instructions can be difficult to interpret for complex programs. The above example, although fairly simple, illustrates the need for full documentation and comments on all PLC programs.

Exercise 5.3   Design a stepladder solution to the chemical-process example used in earlier chapters, incorporating branching paths as necessary. Comment on the solution as compared to the other methods used, in terms of simplicity, readability and structure.

5.5   SEQUENCERS AND DRUM TIMERS

The final sequential control method to be examined is the drum timer or sequencer. All but the smallest programmable controllers normally provide this useful function, either in a dedicated form or as an extension feature on a specific group of counters and timers within the PLC. Fig. 5.23 describes the conceptual operation of a typical sequencer. This consists of a master counter or sequencer that has a range

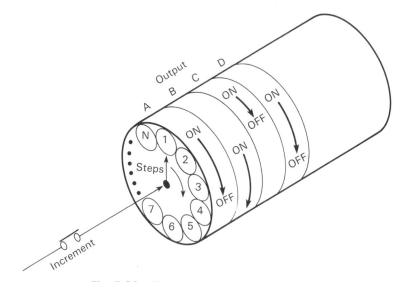

Fig. 5.23   Illustration of sequencer operation.

of steps corresponding to the count value. The sequencer will move (increment) from one step to the next in numerical order, as directed by a controlling signal such as a push button.

Operation

Each step in the sequencer relates to a certain output or set of outputs, which may be programmed to turn on and off in any of the sequencer steps as necessary. These outputs are formed from internal marker relays, and are used in turn to activate output points to process devices (in Fig. 5.23, relays A, B, C, D). Typically a PLC sequencer supports 64 or more such outputs − more than enough for most applications. As usual, each marker can support a large number of related contacts. The number of possible sequencer steps varies from 64 to over 10 000, depending on the PLC manufacturer and model. Each sequencer starts at the zero or home position when reset, and moves through the successive steps until reaching the last, when it restarts the cycle. Alternatively, if not all the steps are required (e.g. 6 out of 64), then a reset facility is used to recommence the sequence after step 6.

Thus the sequencer/drum timer can be considered as a step/output matrix that describes the status of each output at any particular step in the process, or with respect to time. This approach is also the method used in practice to specify sequence requirements. In Fig. 5.24 the sequence map or timing diagram is given for an example problem that has an overall cycle time of 45 s. The requirement is to divide the process into increments of 5 s, controlling the six listed output devices. A timer is therefore to be incorporated in the program, resetting every 5 s. This function is often provided within the sequencer itself, resulting in the alternative name − drum timer.

In this example, six outputs are to be controlled in the sequence given, each with specified ON periods as shown by the arrows in Fig. 5.24. The timing diagram also lists the step numbers and assigned output points for a General Electric Series 1 PLC. An alternative way of displaying this information is by a matrix chart, as in Fig. 5.25, where the desired step times are entered below the step number. ON states are signified by a 1 in the matrix, OFF states by 0.

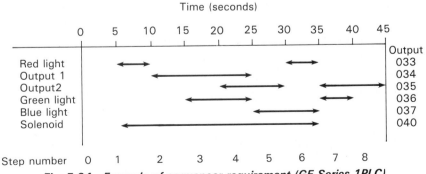

Fig. 5.24 Example of sequencer requirement (GE Series 1 PLC).

Drum 01 Reset = 01		Outputs assigned								
Step	Time	33	34	35	36	37	40	41	42	etc.
0	0	0	0	0	0	0	0	0	0	
1	05	1	0	0	0	0	1	0	0	
2	05	0	1	0	0	0	1	0	0	
3	05	0	1	0	1	0	1	0	0	
4	05	0	1	1	1	0	1	0	0	
5	05	0	0	1	0	1	1	0	0	
6	05	1	0	0	0	1	1	0	0	
7	05	0	0	1	1	0	0	0	0	
8	05	0	0	1	0	0	0	0	0	
9		0	0	0	0	0	0	0	0	
⋮										

etc.

NOTE: Time = length of time outputs are in the specified state for each step
1 = output will be on; 0 = output off during step.

Fig. 5.25   Matrix chart of sequence requirements (equivalent to Fig. 5.24).

Programming on two different controllers

This example application will be described for two popular small PLCs – a GE series 1 and a Mitsubishi F2 machine, since they each have a slightly different approach to the problem.

General Electric series 1 PC

This PLC can have up to 64 sequencers, each with up to 9999 steps. Timers have to be programmed separately from the drum sequencer, as there is no integrated facility.

Referring to Fig. 5.26, a timer T600 is programmed to self-reset every 5 s, if activated by input 030 being held closed. This produces a pulse that is offered to the sequencer/counter C601, causing it to step on one increment. Input 031 being closed resets the sequencer to zero, irrespective of the current position. Normally the counter C601 increments to 8 before resetting (see the program listing). The sequencer outputs are programmed in ladder logic, by specifying the counter used, followed by the step number/count number that is to cause output operation. For example,

Output 037 (blue light) is to be operated in steps 5 and 6:
...STR   601;   counter identifier
...   5;   sequence step for activation
OR   601;
...   6;   other step that is to activate this output
OUT   037;   operate blue light

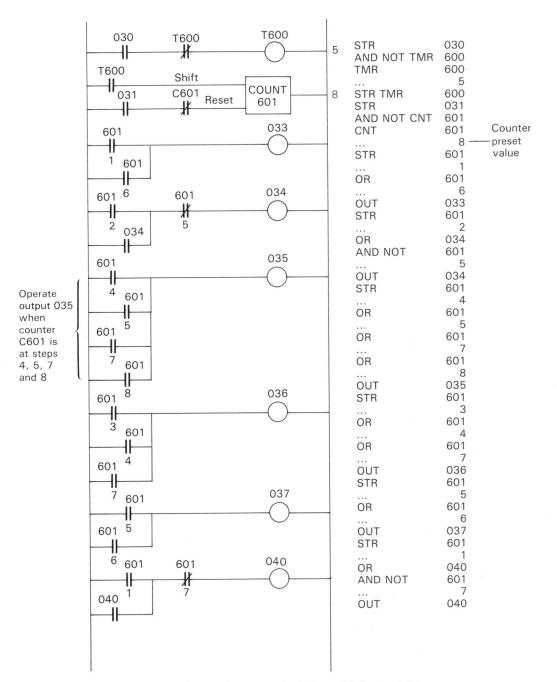

Fig. 5.26   Sequence logic for a GE Series 1 PC.

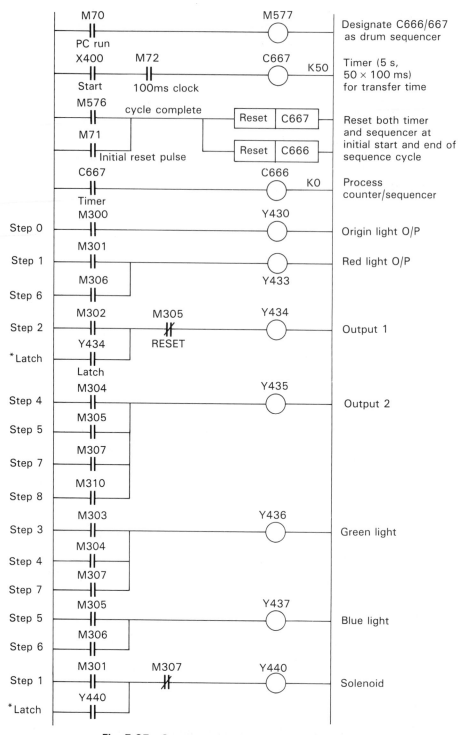

Fig. 5.27   Sequence logic for a Mitsubishi F2 PC.

If an output is ON for more steps than it is off, the logic solution may be simplified by using normally closed contacts. Also, notice the ladder logic for outputs 034 and 040: in both cases the output is activated early in the sequence, then latched to hold continuously for several steps, finally being reset by the normally closed contacts related to the first non-active step.

Mitsubishi F2 PC

The drum sequencers in this popular range of small PLCs are implemented using internal counters and certain special function relays. In this example, counter C666 is used as the sequence counter, driving a set of up to 64 internal marker relays, M300–377. The associated counter C667 is used to set up the process transfer times, either individually or as a standard. Special relay M577 is used to designate the above counters as a drum sequencer, with M70 operating to initiate this action when the PLC is switched to RUN mode. M71 provides a fleeting reset pulse to the counter reset rung, which also has an M576 contact in parallel. This special relay is activated when all steps in the drum sequencer are completed, being reset by the completion of the resulting counter reset. For this example, outputs Y430–440 have been used to correspond to the GE number range.

The equivalent ladder program is shown in Fig. 5.27. The timer C667 is driven by an X400 start contact ANDed with an M72 contact that pulses at 100 ms intervals. The timer is loaded with a target of 50 pulses, which equates to a time-out of 5 s with the 100 ms pulsing of M72. On time-out C667 increments the sequencer C666 via an associated contact, causing it to step on to the next process stage and change the output status in response, via the M relay block. This continues until all assigned stages have been completed, when M576 operates to reset the process back to step 0 for the next cycle.

Exercise 5.4 Using a drum timer or sequencer, design a program for the programmable controller of your choice, to control the chemical process plant described in Chapter 4.

5.6 SUMMARY

In this chapter we have investigated several techniques and facilities for the solution of sequential process control problems. The choice of a particular method depends not only on the particular application, but also on the chosen PLC, together with the user's experience and preference.

Sequence control forms the majority of programmable controller applications, stemming from the times when they were equipped only with logic functions. Today, a growing number of PLCs provide many facilities that are powerful additions to basic logic control, allowing their use for continuous control, analog processing and communications. These topics are discussed in the next chapter.

5.7 FURTHER EXERCISES

(Exercises 5.1–5.4 are included within the body of the chapter.)

5.5   Explain how a sequencer operates.

5.6   List the different methods that may be used to control a sequential control application.

5.7   Design a sequencer solution to provide a four-step sequence with delays of 5, 10 and 15 s between the respective steps.

5.8   What is a matrix chart used for in connection with sequencer design?

5.9   Describe the use of function charts in designing a sequence-control program. Refer to the annotation of state-change information and desired outputs on to the chart.

5.10   With the aid of a diagram, explain how to represent a control situation that has several concurrent actions occurring at one point in the sequence.

ematical calculation and conversion, they also require corresponding amounts of processor time. Depending on system demands and program length, this may result in slow response/scan times for the PLC, which may be unacceptable in high-speed applications. In this situation, the function module with dedicated processor has obvious advantages, including the release of the main CPU from these processor-intensive tasks. This leaves it free to handle the conventional logic operations, I/O processing, etc., requiring only an overall 'management' or supervisory involvement with additional function modules. The main CPU will pass operating parameters (from the controlling program) to the function module, which then performs the task or tasks using this data. In turn, the module may pass status information or process data to the main CPU, where the data can be used in the main program as part of the overall control process.

Input/Output modules either have terminal block or plug and socket type connectors.

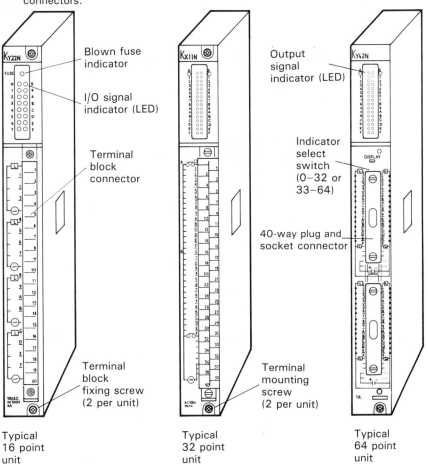

Typical
16 point
unit

(a)

Typical
32 point
unit

Typical
64 point
unit

Fig. 6.2 (Continued overleaf)

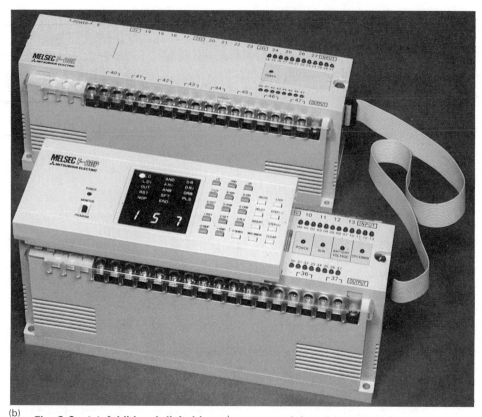

(b)
Fig. 6.2　(a) Additional digital input/output modules; (b) small PLC expansion unit (courtesy Mitsubishi Electric UK Ltd).

Range of functions

The usual approach is for manufacturers to provide a range of optional function modules, allowing the selection of only those modules necessary for a particular application, as in Fig. 6.1. Since the performance and sophistication of these modules generally results in a high unit cost per function, this approach is financially more attractive than a system that incorporates these special features as standard. Modularity of the PLC system is intended to produce a cost-effective control system, tailored to the requirements of each application. In this manner, it is possible to use a mid-range PLC and add special functions that would only have been available on a much more powerful and expensive system. Also, a system that is initially used for conventional switched control can have the advanced features fitted at a later date.

The more common special function modules and their applications are examined in the remainder of this chapter.

6.3 ANALOG INPUT/OUTPUT UNITS

The nature of analog signals was introduced in Chapter 1, defining them as signals where the level or amplitude may vary continuously with time, assuming any of an infinite number of values between minimum and maximum limits. Many sensors generate analog signals, for example to detect pressure, flow, speed, vibration, etc., and are easily connected to a controller that can react to the varying amplitude and/or frequency of the signal. Similarly, many actuators require to be driven by an analog signal, for instance a d.c. servomotor or a variable heating element (see Fig. 6.3).

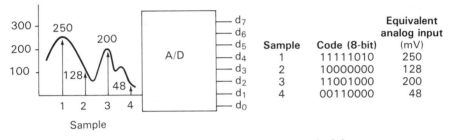

Fig. 6.3 Analog-to-digital conversion principles.

Analog-to-digital conversion

Traditionally the control system dealing with these continuous signals is an analog control system, employing op-amp circuitry to process the signals. However, the use of analog-to-digital (A/D) and digital-to-analog (D/A) converters enables a digital system to process the information contained in these signals, as was discussed in Chapter 1. This also applies to programmable controllers, where A/D converters are fitted to analog input modules that may be added to the PLC rack. The A/D module converts analog values from input sensors into binary data at specific intervals (sampling) which the digital PLC can interpret and then use in mathematical operations as directed by the program.

Conversion precision

Depending upon the manufacturer and model, A/D modules provide various levels of *resolution*, an indication of the conversion precision, and hence accuracy. The resolution refers to the number of bits in the binary word used to represent a sampled value of an analog signal. Since the number of bits has a direct relationship to the maximum number and size of 'steps' in the range, the specification of a particular A/D resolution may be an important factor in the system choice.

For example,

> 8-bit converter: resolution = 256 different digital values between minimum and maximum ($2^{-8} = 0.004$)
> 10-bit: resolution = 1024 digital levels ($2^{-10} = 0.001$)
> 12-bit = 4096 levels
> ($2^8 = 256$; $2^{10} = 1024$; $2^{12} = 4096$)

If we take, for example, an input signal that may vary between $+10$ and -10 V (20 V range) then the step size differs for 8- or 10-bit A/D converters:

> 8-bit: $10/256 = 0.0391 = 39.1$ mV per step, with 25.6 steps per volt.
> 10-bit: $10/1024 = 0.00976 = 9.76$ mV per step, with 102.4 steps per volt.

Thus for two additional bits, the conversion error has dropped from ± 18 mV to ± 5 mV – an improvement of almost a factor of 4.

   Choice of a particular A/D module depends on additional factors such as conversion speed, number of channels supported, and signal range. Fig. 6.4 gives the specification for an A/D unit for the Mitsubishi A series of programmable controllers.

A/D input points

As with the majority of industrial control instruments, the analog inputs to the module are 4–20 mA or ± 10 V, since most sensor outputs fall into one of these categories. In most cases, the module input range is selected by means of DIP switches located on the card.

   The module in Fig. 6.4 supports 4 analog input channels per unit, passing the converted data to the controller as 16-bit words via 16 input points, as shown in Fig. 6.5.

   Since each A/D conversion can take 10–50 μs, a single converter IC serving all four inputs can carry out conversions on all inputs within one millisecond, fast enough for most applications and PC scan times. At the opposite extreme, the

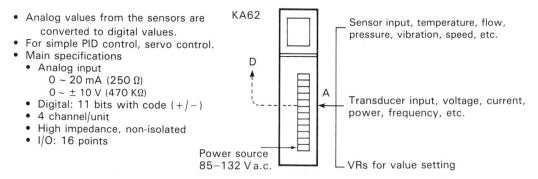

* Analog values from the sensors are converted to digital values.
* For simple PID control, servo control.
* Main specifications
  * Analog input
    * 0 ~ 20 mA (250 Ω)
    * 0 ~ ± 10 V (470 KΩ)
  * Digital: 11 bits with code (+/−)
  * 4 channel/unit
  * High impedance, non-isolated
  * I/O: 16 points

KA62

Sensor input, temperature, flow, pressure, vibration, speed, etc.

Transducer input, voltage, current, power, frequency, etc.

Power source 85–132 V a.c.

VRs for value setting

Fig. 6.4   Analog-to-digital module for Mitsubishi A series PLC.

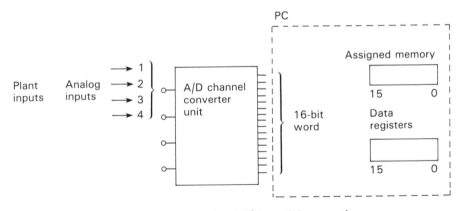

Fig. 6.5 Schematic of A/D module operation.

provision of one converter IC per input point would give maximum performance, but at a correspondingly greater cost. The compromise solution of one A/D converter IC per four inputs is therefore fairly common. Obviously consideration must be given to the frequency and number of analog inputs when an A/D module is being selected. Typically, a worst-case conversion rate of five conversions per channel per second can be expected, with all channels being updated every 200 mS.

Programming analog inputs

As described above, a PLC converts an analog quantity into a digital representation. Analog input signals are usually stored as 16-bit digital words. This digital value is placed in a register located in the memory, often as part of a dedicated data table, allowing this data to be accessed by the standard PLC arithmetic and logical instructions. The size and identification of these registers depends on the programmable controller being used, but they are used by including the identifier in a ladder rung or section of code (see Fig. 6.6(a)). Depending on the PLC, you may be able to use the same input addresses for temporary storage of results if they are not being used for analog inputs. For example, registers AI100–115 relate to the 16 analog input signals connected to an A/D card.

    If it was necessary to compare the measured value in register AI100 with a programmed preset in register R200 and trip an output when the measured value drops below the preset, this could be obtained as follows:

GET AI100	fetch the contents of register 100
CMP R200	compare this data with the contents of R200
*	
A 071	
OUT 040	use I/O address 071 to control output 40

Further examples are given in Fig. 6.6(b).

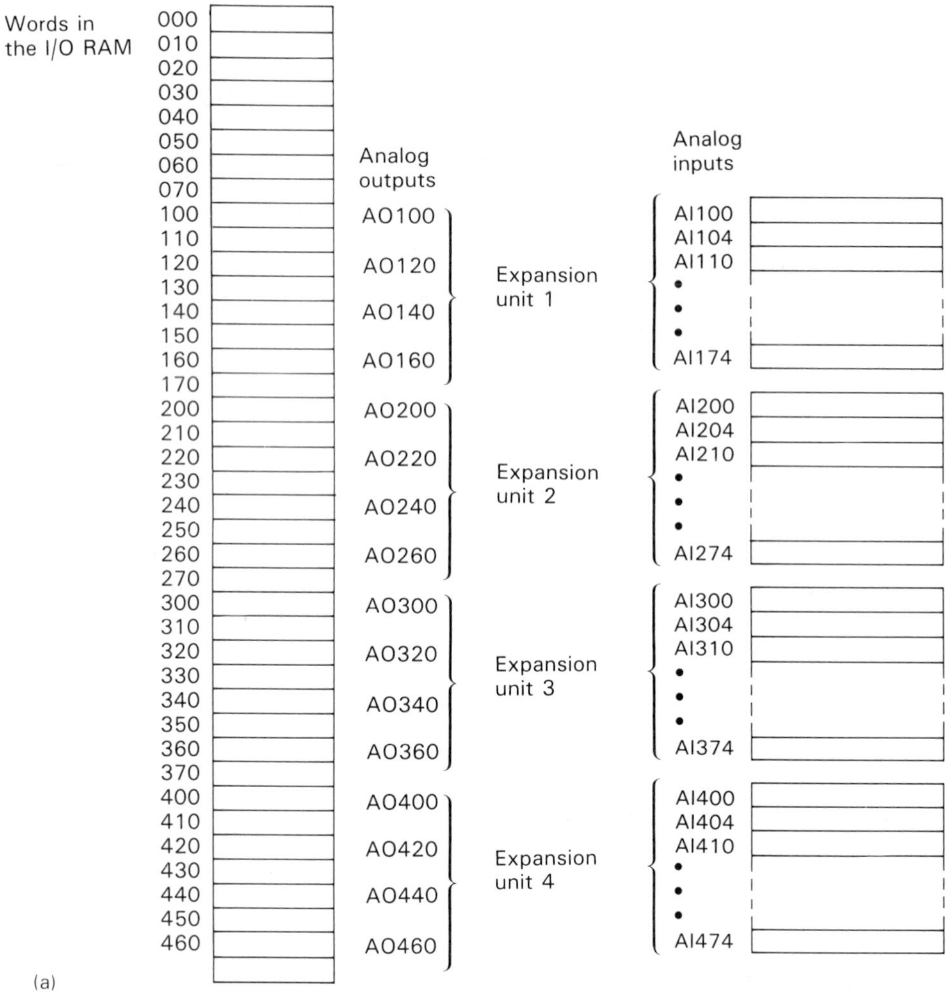

Words in the I/O RAM				

Words in the I/O RAM

```
000
010
020
030
040
050        Analog                          Analog
060        outputs                         inputs
070
100        AO100                           AI100
110                                         AI104
120        AO120                           AI110
130                   Expansion            •
140        AO140      unit 1               •
150                                         •
160        AO160                           AI174
170
200        AO200                           AI200
210                                         AI204
220        AO220                           AI210
230                   Expansion            •
240        AO240      unit 2               •
250                                         •
260        AO260                           AI274
270
300        AO300                           AI300
310                                         AI304
320        AO320                           AI310
330                   Expansion            •
340        AO340      unit 3               •
350                                         •
360        AO360                           AI374
370
400        AO400                           AI400
410                                         AI404
420        AO420                           AI410
430                   Expansion            •
440        AO440      unit 4               •
450                                         •
460        AO460                           AI474
```

(a)

Fig. 6.6 *(a) Typical layout of analog I/O memory; (b) analog I/O using a SattControl 05–20 PC.*

Trip points

Trip points can be set anywhere between 0 and 100% of the maximum value of the analog input. The output of a trip circuit is normally to a switched device. Depending on whether the trip preset level has been reached, the output is 1 or 0. This form of output may be used to display status information, switch alarms, etc., and as a condition in the remainder of the program. Both high- and low-level trip points can be used, together with 'hysteresis' operation where the trip output changes from 1 to 0 at a different level than from a 0 to a 1. This prevents the signal from repeatedly changing states when the analog input value fluctuates around the

Low-level trip

We want to detect when the analog signal AI144 drops below 50% of its max. value (32 788 is 50% of 65 536). We use I/O address 71 to control output 40

```
GET    IA144
CMP    K32768
*
  A         071
-( )- =     040
*
```

Digital output 040

ON 1
OFF 0

0 20 40 60 80 100   AI144 (%) Analog input

High-level trip with hysteresis

When the analog input signal AI144 reaches or exceeds 85% (K55705), the trip signal is set to 1 (output 250)

```
GET    AI144
CMP    K55705
*
-||-A        073
-||-O        072
-(S)- = S    250
*
GET    AI144
CMP    K49151
*
-||-A        071
-(R)- = R    250
```

The trip signal turns off when the input signal drops below 75% (K49151)

250

1

0

0 20 40 60 80 100   AI144 (%)

Ramp function

```
Step 0          Step 3
0%              45%
-||-A   075     -||-A   S3.02
-||-O   S3.05   -||-A   403
-||-A   400     -( )- = S3.03
-( )- = S3.00   GET     K29490
GET    K00000   *
*
                Step 4
Step 1          60%
15%             -||-A   S3.03
-||-A   S3.00   -||-A   404
-||-A   401     -( )- = S3.04
-( )-   S3.01   GET     K39320
GET    K09830   *
*
                Step 5
Step 2          75%
30%             -||-A   S3.04
-||-A   S3.01   -||-A   405
-||-A   402     -( )- = S3.05
-( )- = S3.03   GET     K49150
GET    K19660   *
*
                STO     AO100
                *
```

Correct value set at output

Analog O/P AO100

75%
60%
45%
30%
15%
0%

0 1 2 3 4 5   Step

(b)

trip point. When driving analog outputs, another common feature is to program a 'ramp' function, where the output will change in specific steps, say, as a result of series of steps in a control sequence. Fig. 6.6(b) gives examples of these programs.

Analog-to-digital applications

In many industrial processes it is necessary for action to be taken when an analog input sensor returns a signal of a certain value, for instance a heat sensor circuit providing a voltage output that is linear with respect to temperature, changing 0.5 mV per $^{\circ}$C change in temperature. At 0°C, the sensor voltage is 0.0 mV. To activate an output device when the process temperature $= 800^{\circ}$C, the appropriate sensor utput is 800×0.5 mV $= 400$ mV. (See Fig. 6.7.)

Using an A/D module with 10-bit resolution (10 data + 1 sign bit), we have a range of 1024 binary values covering 0 to $+10$ volts, with approximately 10 mV divisions, so the conversion error will be a maximum of ± 5 mV. The following table shows the relationship between the analog and digital values.

	Signal voltage	*Bits*	
		9 - - - - - - - - 0	
	0.0 V	0000000000	base 2
	1.0 V	0001100110	$= 102$ steps $(64 + 32 + 4 + 2)$
	10.0 V	1111111111	
thus	400 mV	0000101000	$= 40$ steps $(32 + 8)$

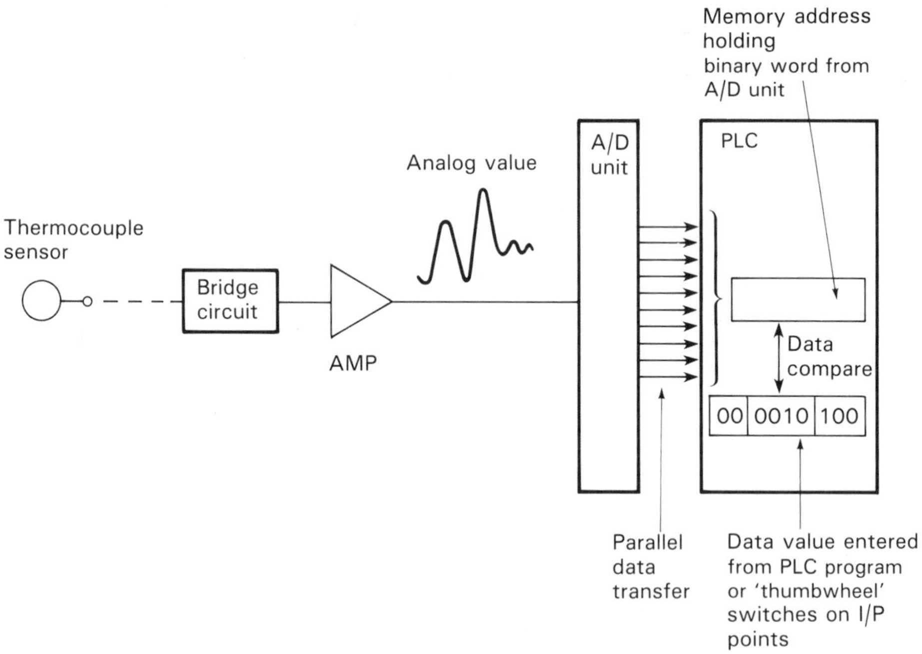

Fig. 6.7   Analog-to-digital application schematic.

In operation, the A/D module samples and converts the sensor input once every time the host PLC passes through the control program, passing the 10-bit binary value to the main CPU at a specified point in the cycle. The CPU has been programmed to examine this input for a value equal to (or greater than) 101000 binary, using conventional PLC mathematical instructions to compare the A/D data with a preset or programmed binary word held in a data register. When this value is detected, the control logic is enabled and operates any assigned outputs. This constitutes an analog in/switched output form of control, a common requirement in industry (continuous/discontinuous o/p).

Alternatively, the PLC may perform mathematical scaling operations on the input values prior to the comparison, or may even pass the results directly to a D/A module to directly drive continuous actuators in the process. This forms the basis for full continuous control, which is discussed in Section 6.6.

Examples of scaling

In many cases the user will require to scale incoming signals to represent the actual voltage or current values being received at the input points. The examples in Table 6.1 indicate the resolution provided by the various signal ranges, and how the values contained in a register can be scaled to their actual values.

Table 6.1

Range	Program operation	Result	Resolution
4–20 mA	R100 = 109/5*8 + 4000	Read reg 100 in mA	8 mA per bit
0–5 V d.c.	= R109/2	Read reg 100 in mV	2.5 mV per bit
0–10 V d.c.	= R109	Read reg 100 in mV	5 mV per bit
± 10 V d.c.	= R109*2 – 10000	Read reg 100 in mV	10 mV per bit

6.4 DIGITAL-TO-ANALOG MODULES

PLC analog outputs may be provided on the same function module that performs the conversion of analog inputs, but are more commonly supplied as a dedicated output unit carrying out digital-to-analog conversion on a number of channels. A typical D/A module is the Mitsubishi analog output unit described in Fig. 6.8. In common with the complementary input module described in the previous section, this unit uses 10-bit data words as the digital input from the PC (plus one sign bit), with two analog output channels per unit.

(Although 16 digital output points are used to connect the D/A unit to the PLC, only 11 of these are used to pass convertible data – the others are for timing purposes)

These channels provide a 20 mA current loop or ± 10 V outputs to drive

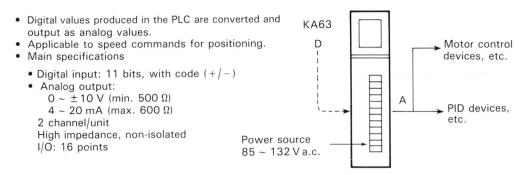

- Digital values produced in the PLC are converted and output as analog values.
- Applicable to speed commands for positioning.
- Main specifications
  - Digital input: 11 bits, with code $(+/-)$
  - Analog output:
    $0 \sim \pm 10$ V (min. 500 Ω)
    $4 \sim 20$ mA (max. 600 Ω)
    2 channel/unit
    High impedance, non-isolated
    I/O: 16 points

KA63
D

Motor control devices, etc.

A

PID devices, etc.

Power source
$85 \sim 132$ V a.c.

Fig. 6.8 Digital-to-analog converter for a Mitsubishi A Series PC.

continuous actuators. The word length used to represent an analog signal is often 16 bits in size and, in order to allow rapid handling of these signals, the words are temporarily stored in the PLC I/O RAM, where the locations are read by the D/A units. The memory cells used for this function cannot simultaneously be employed as general working memory.

Operation

When the parallel binary code is offered to the D/A module, the converter outputs a corresponding voltage level, as illustrated in Fig. 6.9. Again, the range of output values is discrete (1024 for the 10 bit word), resulting in a stepped output when the digital input changes. To more closely represent a true analog signal, the converter output is passed through amplifier and filter stages that smooth out the stepped signal. The accuracy of the output from a D/A function module also depends on the frequency of updating the digital input word. This is often related to the PLC scan time, since the processing of data that results in an input to the D/A unit occurs only once per program cycle.

Analog output signals are provided in the same range as for analog inputs, i.e. 4 to 20 mA, 0 to $+5$ V d.c., 0 to $+10$ V d.c., etc., selected by DIP switches on the module.

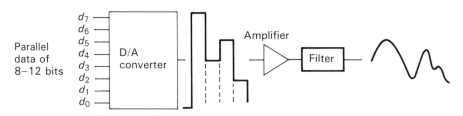

Parallel data of 8–12 bits

d_7
d_6
d_5
d_4
d_3
d_2
d_1
d_0

D/A converter

Amplifier

Filter

Fig. 6.9 Digital-to-analog signal conversion principles.

Applications

Where it is necessary to drive continuous output devices such as d.c. servomotors, variable speed pumps, etc., this type of module can be used by the PLC in conjunction with either switched input signals or, more commonly, with analog inputs.

In the rare cases when a D/A unit is used on a programmable controller with simple switched inputs, the status of several input points must be used together to form a parallel data word that can be interpreted to select which of a restricted range of output levels is to be output. Fig. 6.10 illustrates this operation, with four digital inputs being used to determine which of 16 possible speeds from a variable drive output is selected.

When all input switches are open, the drive is at rest (off). Half speed is selected by switch D being closed, and full speed by all switches closed. Each switch combination results in a different voltage level being output from the D/A module to the drive motor.

However, analog output modules are most commonly used in conjunction with analog input modules, as part of a continuous control system. This is the topic of Section 6.6.

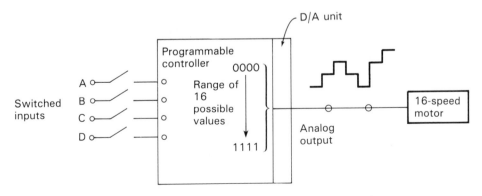

Fig. 6.10   Switched input, analog output example.

Analog status and fault information

In many programmable controller systems, the special function modules provide a certain amount of status and fault information, often as the contents of a specific memory register, as in Fig. 6.11. This status byte is for an A/D converter, but the format applies to all types of function card. As with all other registers, various bits can be tested using the PLC instruction set within applications programs, allowing action to be taken if a tested condition is met.

AS00

X	X	X	X	exp 4	exp 3	exp 2	exp 1	exp 4	exp 3	exp 2	exp 1	exp 4	exp 3	exp 2	exp 1

bit no.:   15 14 13 12   11 10   9   8   7   6   5   4   3   2   1   0

N/A   A/D converter fault   Reference voltage   Analog expansion
                                            fault   unit connected

                   Set to 1 if the A/D   Set to 1 if an   Set to 1 if there is a
                   converter in the   analog expansion   fault on the internal
                   expansion unit has not   unit is connected in   voltage in the
                   responded for a   the corresponding   corresponding
                   preset number of   position   expansion unit
                   milliseconds.   (units 1–4).   (units 1–4).

Fig. 6.11   Typical status byte for a special function module.

6.5 SIGNAL-CONDITIONING MODULES

In the event that plant signals do not correspond to any of the standard analog ranges, it may be necessary to use a signal-conditioning module in the PLC rack.

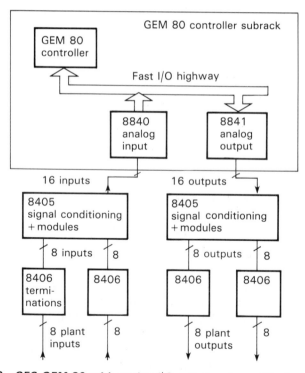

Fig. 6.12   GEC GEM 80 with analog I/O and signal-conditioning modules.

Table 6.2

Application	Input span
Thermocouple modules	0–5 mA to 0–120 mA (μA/$^\circ$C)
Resistance thermometer	2 Ω to 1 kΩ (mV/$^\circ$C)
Slidewire input	0–100 Ω to 0–5 kΩ
Strain gage module	10 mV to 10 V
Process control I/P	0–400 mV to 0–100 V
Process control O/P	0–4 mA to 0–20 mA
etc.	

(This is intended to provide buffering and scaling of plant signals to a standard 5 V, allowing 5 V A/D and D/A cards to process the signals. A wide range of signal-conditioning modules is available from most manufacturers, for example from GEC (Table 6.2).

The overall accuracy of any channel will depend on the accuracy of both the signal-conditioning module used, and the analog input or output module in the programmable controller rack. Errors and offsets in the various process transducers, etc., should be compensated for in the controlling program by using available functions (such as linear conversion) or by mathematical algorithms.

Fig. 6.12 shows the layout of a control system with signal conditioning.

6.6 CONTINUOUS CONTROL FUNCTIONS

As mentioned earlier, a PLC equipped with analog input channels may perform mathematical operations on the input values and pass the results directly to an analog output module to directly drive continuous actuators in a process. The level and sophistication of possible control algorithms depends on the speed and mathematical capability of the programmable controller, but may include proportional, integral and derivative control terms. For example, proportional control can be performed by PLCs with basic mathematical facilities (see Fig. 6.13(a)), by programming the following operations:

(a) Input the converted analog sensor value; compare this actual output (measured value MV) with the desired output (setpoint SP), producing the error value.

$$E = SP - MV$$

(b) Multiply the error value by a preset proportional constant Kp, which is the system gain.

(c) Pass the result of this calculation to the D/A converter module, which will output this new correction signal to the actuator.

Return to (a)

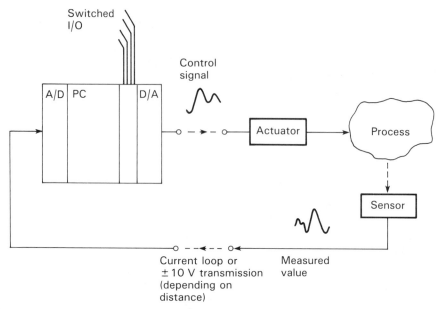

(a)

(b)

Fig. 6.13 (a) Simple proportional feedback control using PLC function cards and mathematical programming; (b) GE series 6 PLC with 'Proloop' control modules (courtesy General Electric).

In practice, the output of the controller will include an initial bias (I), plus or minus the amount proportional to the error:

$$MV = I + Kp \cdot E$$

The reader is referred to Chapter 1 for further details of continuous control plans. As discussed there, proportional control alone is rarely adequate for the majority of control tasks, requiring the inclusion of integral or derivative terms to deal with factors such as rapid response, oscillation, permanent disturbances, etc.

In many cases where a PLC provides integral PID control as standard, the control algorithm is supplied as a subroutine resident in memory. The user simply passes the desired parameters and I/O locations to the subroutine via the main program, and calls the routine as required. These terms may also be approximated by a PLC with a reasonable mathematical capability, but at a cost of processor time, resulting in longer program cycle times. In a worst-case situation, the response of a programmable controller may be slowed to the extent that proper machine control is not possible.

PID function modules

For this reason, a preferable alternative is the use of add-on intelligent PID function modules, produced by most PLC manufacturers for their particular range of modular systems. These modules contain both input and output analog channels, together with a dedicated microprocessor to carry out the necessary control calculations. This processor operates in parallel with the main CPU, relieving it of all the PID mathematical calculations, only requiring the passing of initial parameters to configure the software control loops already installed in ROM memory within the module. When the main CPU requires status data from the PID module, it reads the relevant locations in I/O memory where the PID processor places this information each time it completes a control cycle. The photograph in Fig. 6.13(b) shows a GE Series 6 PC with two such modules – 'proloop' control units.

In addition to the formal control functions, it may be necessary to carry out some form of pre-processing of the data coming from the process, for example to smooth fluctuating numeric values. Specific PLCs may have specialist functions to provide this type of processing. The GEC GEM 80 series has a function that provides exponential (first-order) smoothing of input changes, termed *ANALAG* (*ANA*log *LAG*). This function has a user-programmable time constant and may be switched into or out of the system within the control program. Fig. 6.14 shows an example of the programming and effect of this function.

Programming PID function modules

Depending on the PLC system, PID modules may be programmed either by conventional ladder programming panels/terminals, or by special programming units. In most cases the process is very straightforward. With ladder diagrams, the

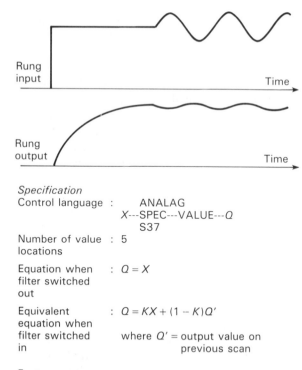

Specification

Control language :   ANALAG
                         X---SPEC---VALUE---Q
                         S37

Number of value : 5
locations

Equation when   : $Q = X$
filter switched
out

Equivalent   : $Q = KX + (1 - K)Q'$
equation when
filter switched   where Q' = output value on
in   previous scan

Features

* Equivalent to a single time constant filter
  in analog systems

* User-programmable time constant

* Provision for switching and filtering action
  into or out of the system, and for
  initializing the output

Fig. 6.14 Smoothing function for GEM 80 (ANALAG).

PID loop will be called as a special function with user-defined parameters entered into the function as, say, the preset time is entered into a simple timer circuit.

Figure 6.15 shows elements of a PID control rung for a GEC GEM 80 controller. Further details of this are contained in Appendix A.

Special programming panels usually provide a 'menu' of questions and options relating to the set-up of the internal control loops (see Fig. 6.16). Typical options include: one-, two- or three-term control (P, PI, or PID); addresses of I/O points to act as sensor input and actuator; gain parameters; integral (reset) times; differential (rate) times; sampling rate and engineering units to be used in the calculations.

Many intelligent PID units contain buffer memory to permit the storage of process status/performance data. This data can be accessed by the main CPU for use in sequencing and alarm circuits, or for passing to other control loops that are part of the same process.

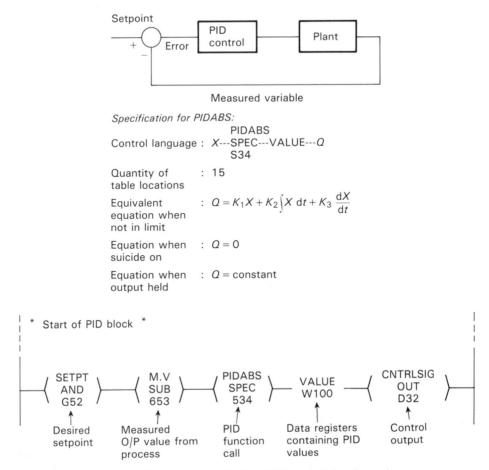

Specification for PIDABS:

PIDABS

Control language : X---SPEC---VALUE---Q
S34

Quantity of   : 15
table locations

Equivalent   : $Q = K_1 X + K_2 \int X \, dt + K_3 \dfrac{dX}{dt}$
equation when
not in limit

Equation when   : $Q = 0$
suicide on

Equation when   : $Q = $ constant
output held

Fig. 6.15 Example of ladder PID control configuration.

It should be understood that a modular PLC will perform discontinuous control using the conventional digital I/O points and ladder programming, together with continuous control via the optional analog or PID-type of modules.

PID control modules – applications

Direct digital control algorithms as used in these PLC modules are extremely versatile, being fast, accurate and simple to configure for most applications. Common areas of use include motor-speed control, hydraulic-pressure regulation, temperature control and energy management.

In many control situations there is a requirement for the sensing and control of several variable quantities. Certain applications may involve sampling tens or even hundreds of such variables to carry out the desired control. Additionally, the control

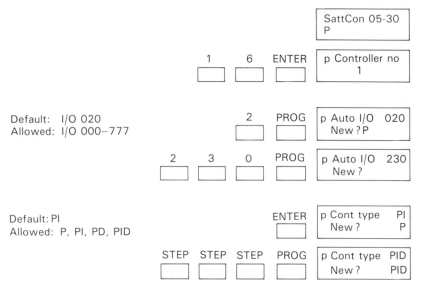

			SattCon 05-30 P
1	6	ENTER	p Controller no 1

Default: I/O 020 Allowed: I/O 000−777		2	PROG	p Auto I/O 020 New ? P

| 2 | 3 | 0 | PROG | p Auto I/O 230
New ? |

| Default: PI
Allowed: P, PI, PD, PID | | | ENTER | p Cont type   PI
New ?   P |

| STEP | STEP | STEP | PROG | p Cont type   PID
New ?   PID |

Fig. 6.16   Example of a Menu-driven PID control program on a Sattcontrol 05−30 PC (see also Fig. 2.21)

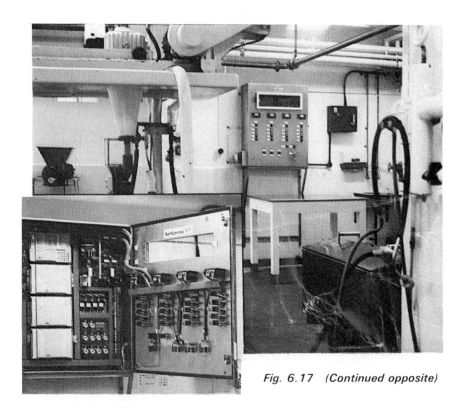

Fig. 6.17   (Continued opposite)

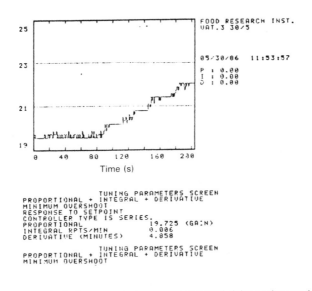

Time (s)

Temperature control of cheese vats in an experimental dairy makes an interesting application for a SattCon 05-30 controller. The systems department recently supplied a complete control package to the AFRC Institute of Research at Shinfield, near Reading.

The system controls the water jacket temperatures of four vats used for the production of various types of cheese for experimental purposes. As well as holding each vat's temperature to a preset value, the system can increase the temperature to a new value at a programmable ramp rate. Since the SattCon 05-30 can handle up to four PID loops, it made an ideal choice for this application.

Jacket temperatures are measured by resistance thermometer probes and displayed on digital meters located on the front of the stainless steel wall-mounted panel. Temperature signals are connected to the SattCon 05-30 via its XACV analog expansion unit which also provides the control signals to thyristor units driving the vat heaters. Digital expansion units are used for the operator push buttons and indicators, as well as the three sets of thumbwheel switches which enable the start/stop temperatures and ramp rate to be programmed for each vat in turn.

The system also monitors the operation of the circulating pumps, checks the water levels in the jackets, and detects operation of circuit breakers and motor overloads. Should a fault occur, the operator's attention is drawn to it by an alarm lamp on the panel, and a text message on the SattCon 05-30's integral liquid crystal display, visible through the window in the door of the panel, describes the nature of the fault.

To speed up commissioning of the system, the PID loops were tuned using the Protuner 1100 process analyzer which was connected to the input and output of each loop. A step change was made to the setpoint and, within a few seconds, the Protuner calculated suitable P, I and D parameters which were then stored in the SattCon 05-30's memory. Because of the system's slow thermal response to changing output signals, conventional loop tuning would have taken considerably longer to achieve satisfactory control.

Fig. 6.17 Use of a SattControl 05–30 PC for continuous control.

system frequently consists of more than a single control loop, possibly with one loop interacting with another. These requirements can all be met by PLC closed-loop controllers, which can easily be programmed to sample and process data from one control loop and then pass the resulting data to a further loop. This ease of programming extends to the alteration of control-loop parameters by process technicians. In order to 'fine tune' a control loop or system, internal loop parameters such as the setpoint, gain and integral time can be repeatedly altered and tested until the ideal 'model' is achieved. System responses to each modified control plan can be stored for later analysis, if so desired. In these ways the programmable controller can become an extremely effective tool and diagnostic aid in the continuous control field.

Case study – Continuous control using a Sattcon 05 PLC

Fig. 6.17 is from a Sattcontrol publication, describing the application of an 05–30 programmable controller to the temperature control of cheese vats.

6.7 GRAPHICS AND MIMIC DIAGRAMS – OPERATOR INFORMATION

In all industrial control situations, there is a need to provide the operating personnel with process status information, including alarms, prompts for manual inputs, product quantity, times, etc. Virtually all programmable controllers provide I/O point status as standard, using LED indicators on the PLC body or I/O cards. Several output points are frequently used to control alarm lamps, bells, etc. to draw the attention of maintenance operators when necessary. In addition, many PLCs are equipped with small, integral liquid crystal display panels that can display short messages referring to process status, operator prompts, etc. (see Fig. 6.18).

However, the amount of meaningful information that can be gained from these facilities is limited, particularly if complex or high-speed processes are involved. The requirement is for easy-to-use operator interfaces that can deal with

For example:
You can instruct SattCon 05 to display the dynamic
status of analog signals with 7 ENTER. AI100 and AI104 are displayed first.

Fig. 6.18   LCD display screen on a SattControl 05–30 PC.

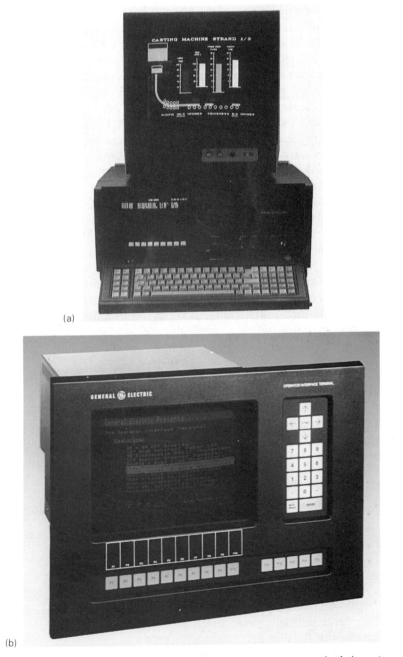

Fig. 6.19 (a) Color graphics displays; (b) operator terminal (courtesy General Electric).

large amounts of data, presented in a readily understood form. The use of VDU terminals or microcomputer-based color monitors as operators' consoles fulfills this requirement, and the majority of PLCs now provide standard communication links for this purpose. This link commonly takes the form of a function card that may be added to the PLC rack, handling all programmed communication to and from the operator terminal, including system status information, production statistics, diagnostics, process mimics with 'zoom' facilities, data entry and alarm logging.

To provide permanent records of process performance, a printer is often connected to the PLC or microcomputer terminal. Data can be logged for various events in each process cycle, or purely for alarm conditions.

The display screen will show general schematics of a complete plant or process, and also detailed diagrams and tables for the subsections of the process. Dynamic mimic displays can be produced which directly show changes in the status of process transducers and actuators, either by color change or flashing symbols. The operator can normally select which mimic screen to view by keyboard entry. The inclusion of color graphics monitors allows the most effective use of the display capabilities, allowing different aspects of process status to be represented in suitable colors. Figs. 6.19 and 20 provide a selection of typical color graphics displays. CRT-based color monitors are less costly and more flexible than conventional engraved ceramic mimic/alarm lamp panels, and have the great advantage of small size and portability, allowing their installation in a variety of locations.

Many specialist display terminals are available which offer very high-resolution graphics and bulk storage facilities for process data.

Influence over the process

The information flow is not always unidirectional from the process to the operator, but may be configured to allow direct operator input to certain aspects of the process:

- acknowledge alarms;
- inhibit signals;
- control actuators in the process;
- select from a range of control programs (e.g. for different recipes or parts);
- alter parameters in a control program;
- halt the process or parts of a process.

Operator input is carried out using the standard terminal keyboard or a custom panel with keys related to a particular application. Other operator input devices may be used, including light pens, touch screens, mouse, etc., which provide direct interaction with the display.

The display screen usually has specific areas or lines reserved for operator communication. For example, the bottom four lines:

- alarm line, displaying the last unacknowledged alarm;

Colour display screen

Function keyboard

Fig. 6.20   *Color display screen and function keyboard.*

- message line, indicating the identity of the current diagram;
- operator's line, repeating the characters keyed in by the operator on the terminal;
- response line, showing any replies to operator questions, and also error messages.

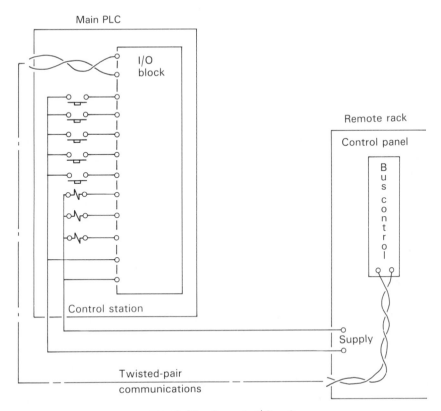

Fig. 6.21   *Remote I/O unit.*

Programming process mimics

Mimic diagrams and status tables for a particular process are usually created by the system supplier using graphic construction languages (often a version of BASIC – see the case study in Chapter 8 on a chemical treatment plant).

This involves programming the various mimics and tables, then relating each element or value to the corresponding memory location or I/O point in the controlling PLC. Once these relationships are established, the mimic will reflect changes in these elements in the PLC, presenting this information to the operator. Each mimic display can be stored on disk in the microcomputer terminal and can be selected in its turn by the operator. Facilities that allow the user to alter or expand the graphics mimics may be provided, enabling the system to be developed in line with process changes.

6.8 REMOTE INPUT/OUTPUT AND COMMUNICATIONS

The input/output structure of a programmable controller is designed to provide the largest number of I/O points in the smallest physical area. Large PLCs are often equipped with several hundred (or thousand) I/O points, however, and so input/output racks can still occupy large amounts of floor space.

Remote input/output

When large numbers of input/output points are located a considerable distance away from the programmable controller, it would be uneconomic (and bulky) to run connecting cables to every point. The adoption of communications technology into the manufacturing industry contributed to the development of *remote input/output systems*, i.e. locating I/O racks several hundred or thousand meters away from the controlling PLC, communicating over a twisted-wire pair or a fiber-optic cable (Fig. 6.21).

This acts as a concentrator to monitor all inputs and transmit their status over a single serial communications link to the programmable controller. Once output signals have been produced by the PLC they are fed back along the communications cable to the remote I/O unit, which converts the serial data into the individual output signals to drive the process. External cabling costs are substantially reduced by the use of remote I/O techniques.

In many cases a group of small PLCs can be linked together, with a master unit sending and receiving I/O data from the other local or slave units. Typically, this involves small PLCs being 'daisy-chained' together from a larger master PLC. The slaves are equipped with data link hardware and software, but do not contain control programs, since all control processing is carried out by the master PLC. (Fig. 6.22.)

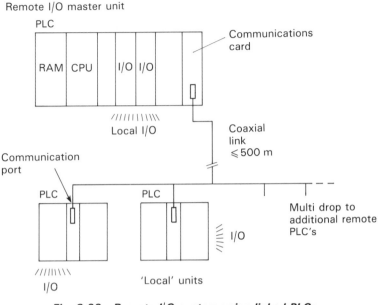

Fig. 6.22 Remote I/O system using linked PLCs.

I/O Diagnostics

It is commonplace for programmable controllers to provide certain self diagnostic features, such as the detection and reporting of CPU failures, module failures, etc. These often result in the triggering of fail-safe operations, including a system halt in cases of critical failures. Other actions may be specified by the PLC manufacturer, including the provision of output-disabling relays which may be specified in rungs of a control program. When enabled, this type of coil disables/disconnects *all* PLC outputs, rendering the system as halted. (Note that care must be taken when using this form of error/fault treatment with continuous and/or time-critical processes, to ensure the part-finished product is not abandoned unless it is unavoidable. Additionally, failure and error conditions may, to a certain extent, be monitored and isolated by the use of appropriate program sections. This is accomplished by including regular (often timed) data transfers and checks between constituent units in a distributed control system, or even between modules in a single, centralized system.

However, these forms of testing tend to be confined to the 'main' PLC hardware. Associated I/O systems can make up 75% of a complete PLC installation, and also account for the same proportion of total system faults. (The pie chart in Fig. 6.23 shows the distribution of control-system faults among the constituent parts.) It is therefore highly desirable that PLC fault diagnostics be extended to cover this important area, in order to further improve reliability and maintainability of the total control system.

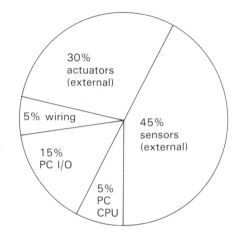

Fig. 6.23 *Appropriate distribution of faults in a PLC system.*

This facility can be provided by the inclusion of a PLC input/output system that can automatically detect, isolate and report faults occurring both within and outside the physical I/O units. An example of such a system is the 'Genius' I/O from General Electric (USA).

Genius is designed to interface with larger programmable controllers and the CIMSTAR industrial microcomputer from the GE range, either as the sole I/O system or in combination with conventional I/O. The system consists of a bus controller module mounted in the main PLC rack, plus associated remote I/O blocks connected onto a common twisted-pair bus. Each bus can have up to 480 I/O

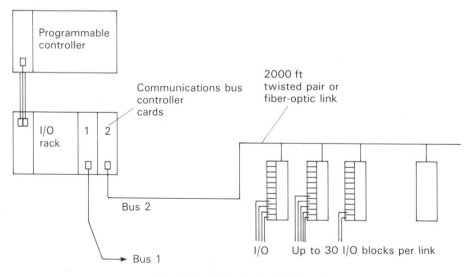

Fig. 6.24 *GE Genius I/O layout.*

Fig. 6.25   The Genius I/O system from General Electric (courtesy General
Electric).

Fig. 6.26 The CIMSTAR Industrial Microcomputer from General Electric. It combines the advantages of a personal microcomputer and a programmable controller. It supports the Genius I/O system via plug-in bus controller modules in its back plane (courtesy General Electric).

points distributed over a single twisted pair, running up to 2000 ft from the bus controller (see Figs. 6.24–6.26).

This communications bus is designed to allow the transmission of both I/O data and diagnostic information back to the CPU with minimum delay, using a data rate of 150 kbits/s. It is a multidrop design, allowing the connecting of I/O blocks at any point. This permits block failures to occur without affecting the operation of the rest of the system.

Diagnostics on the Genius system are very comprehensive, reporting errors such as: bus errors; failed switch; no load; short circuit; over-current and over-range; under-range. The system uses high-speed power electronics to sense over current conditions, disabling the output point within 5 μs when tripped. This provides improved protection of wiring and circuitry, eliminating the need for

Fig. 6.27 Centraflex 11600 flexographic press at Kidder Stacy, fitted with a GE Series 6 controller and Genius I/O system.

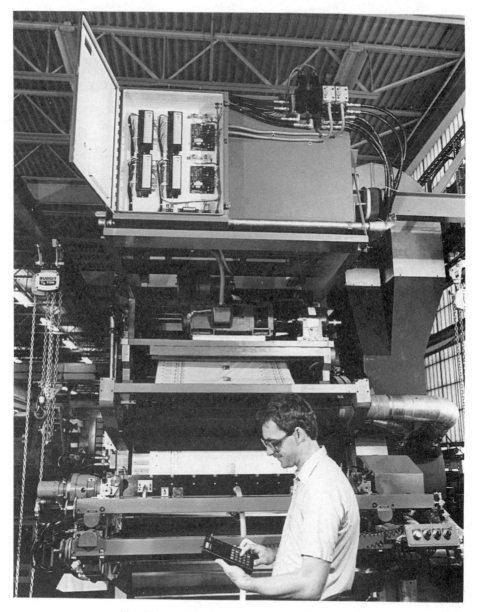

Fig. 6.28   Remote station on the Genius bus.

conventional fuses. The I/O blocks contain intelligence that allows other valuable test functions to be carried out, including the checking for faulty wiring and output actuator failure. This is based on a 1 ms pulse test that toggles the state of each output in turn. During this time the voltage and current drawn from the output is checked for possible errors that would indicate a fault. This ability to detect and localize faults is of immense value to maintenance staff trying to find faults on a complete control system.

Application example of the GE Genius system
(Original source: General Electric)

The Genius input/output system was installed in an advanced-technology printing works for the Kidder Stacy Company in Massachusetts, USA. It was required to remotely control a new multistation, 6-color printing press developed by the same company. This control was to be performed by a Series 6 programmable controller operating over a Genius remote I/O system to the various press stations in the plant.

In Fig. 6.27 you can see the Series 6 PC housed at the bottom left of the photograph. Bus controller cards for communications and control of the Genius system are contained in the Series 6 rack. At the middle/right of the picture is one of the remote stations equipped with three Genius blocks. Fig. 6.28 shows a second remote station, together with a view of the communications cabling to and from the rack.

The Genius system proved to be extremely useful during the debugging and testing stages of the new press, with the distributed nature of the system requiring less than half the wiring of a conventional PLC installation. It was possible to place the remote I/O blocks directly at the point of control. This also reduced the need for cabinets, trunking and labor for the installation, with obvious financial savings.

The integral diagnostic capabilities of the system detected various faults during the development period, with maintenance having access to the Genius system via a hand-held monitor or the larger programming terminal.

As a 'total' package, the Centraflex press/Series 6 PC/Genius I/O allows easy alterations to the press functions, together with comprehensive field testing and troubleshooting aids.

6.9 SUMMARY

This chapter has examined a selection of the more common function cards available for programmable controllers, together with several typical applications. This has involved mainly the driving and sensing of process outputs and inputs, or the direct presentation of information via graphics screens, etc., and as such forms the 'front line' between the PLC and the real world. Chapter 8 contains a further selection of industrial applications of PLCs, including the use of various special functions.

When it becomes necessary to control larger processes, or several smaller processes working in synchronization, the need arises for communications between programmable controllers and any other intelligent devices found within the process.

Communications and the requirements for plant automation are discussed in Chapter 7.

6.10 EXERCISES

6.1 Explain the need for different input/output modules to handle digital and analog quantities.

6.2 What are the most common signal levels (a) for digital input points and (b), for analog input points?

6.3 Describe the process of converting an analog signal into an equivalent digital representation.

6.4 What is meant by converter resolution, and how does it relate to the precision of the converted value?

6.5 Using an 8-bit converter, what will be the step size when sampling the analog voltage input from a position potentiometer which varies between +5 and 0 V?

6.6 What PLC functions would you use to examine data obtained from an A/D module? Describe a typical application of this technique.

6.7 What is meant by the term 'scaling'?

6.8 When may a signal conditioning be required in a control situation? Give two examples of conditioning modules that may be used on a programmable controller.

6.9 Describe the basis of a closed-loop control system. What is PID control?

6.10 Describe two different ways that PID control may be implemented on a PLC. In each case give the advantages and disadvantages.

6.11 Why is a dedicated microprocess often used within a dedicated PID module?

6.12 State the benefits of graphic mimic displays in process control systems.

6.13 What is a remote I/O unit? Describe where such a system might be used, and for what reasons.

6.14 Describe the facilities offered by a Genius remote I/O system. What implications does this type of system have for maintenance and troubleshooting of the complete control system?

## SEVEN

PLC Communications and Automation

7.1 COMMUNICATIONS

The importance of inter-machine communications has greatly increased over the last few years, and this trend will continue as more industrial companies strive for improved production and efficiency through linked automation systems. This has far-reaching implications for plant design and operation, where discrete elements such as CNC machines, component transfer systems, etc., will be integrated into a complete manufacturing system. Existing flexible and dedicated manufacturing systems will also be subject to further developments in the areas of management information and computer integrated manufacturing (CIM). These topics are discussed later in this chapter.

All these aspects of advanced manufacturing and control are dependent on communications systems, from simple machine-to-machine serial links through to local and wide area networking where tens or hundreds of intelligent machines communicate over a common data highway. For this reason, this chapter deals exclusively with data communications and the applications in industrial automation.

PLC communications

The need to pass information between PLCs and other devices within an automated plant has resulted in the provision of a communications facility on all but the most elementary controllers. In the case of small PLCs, the necessary communications hardware and software is incorporated in the programmable controller body, whilst larger PLCs have a range of communications modules available to suit different applications (see Fig. 7.1).

Common uses of PLC communications ports

1. Presenting operating data and alarms, etc. via printers or VDUs.
2. Data logging into archive files or records (normally on a micro/minicomputer): to be used for process performance analysis and management information (via software applications packages).

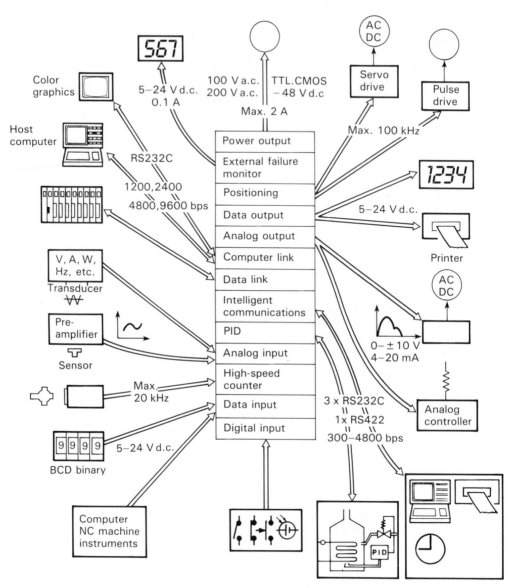

Fig. 7.1 PLC communications needs.

3. Passing values/parameters into existing PLC programs from operator terminals or supervisory controllers.
4. Changing resident PLC programs – uploading/downloading from a supervisory controller (PLC or microcomputer).
5. Forcing I/O points and memory elements from a remote terminal.
6. Linking a PLC into a control hierarchy containing several sizes of PLC and computer.

Serial communications – RS232 and derivatives

PLC communications facilities normally provide serial transmission of information, for example sending logged data to a printer or display screen (VDU) over a standard serial link, where the binary words (bytes) are transmitted one after the other.

There are, of course, standards for just how data is to be transmitted between devices. The most common standards are RS232 and a later derivative, RS422/423. These are serial communication standards issued by the US Electronics Industry Association.

RS232C communication is virtually *the* standard in short-distance computer communications, with the majority of computer hardware and peripherals, such as printers and VDUs, having RS232 interfaces and connections. RS232C ('C' indicates current revision) simply defines physical and electrical connections, the interrelationship between the signals, and the procedures to be used for exchanging information. The 'D'-type 25-way connector has become universally associated with RS232 communications, and is provided on most programmable controllers, computers and peripheral devices.

CCITT recommendation V24 is almost identical to RS232, being common in Europe. However, V24 defines only the procedures for information exchange, requiring another recommendation, V28, to define the pin allocations and signal voltages. In general, however, RS232 and V24 can be regarded as being equivalent. RS232 and 422 can also be considered as similar, except for the differences in voltage level of the outputs, and in the potential transmission rates.

Standard communications requirements

There are three aspects that must be considered in serial communication. Firstly, the rate or speed of the transmission, that is the number of bits per second that are to be sent over the communications link, and the duration of each of these bits. Secondly, we have to consider the logic levels, in other words what signal represents logic 1 and 0, together with the order in which we are going to transmit information. Finally, we have to consider a method of synchronizing the data to enable the receiving device to understand the transmission. The arrangement used in RS232 is described in Fig. 7.2. Here a start bit is used – a logic 1 pulse for one bit duration – to tell the receiving device that data is following. The data is then transmitted, followed by the stop bit. RS232 allows the inclusion of date-checking characters, called *parity bits*. Parity is a method of checking the number of 1s or 0s in a data word and adding a 'parity' 1 or 0 to make this number odd or even, depending on the type of parity checking in use – odd or even parity.

RS232 allows the selection of six, seven or eight data bits, at transmission speeds ranging between 75 to 19 000 bps. Common standard rates in baud (bps) include 300, 600, 1200, 2400, 4800, 9600.

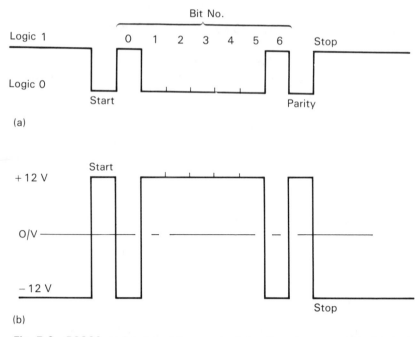

Fig. 7.2 RS232 serial data bits required for the character 'A': (a) logic levels; (b) signal levels. The 10-bit package contains 1 start bit, 7 data bits, 1 parity bit and 1 stop bit.

Obviously the sending and transmitting devices must be operating at the same data rate, with the same data format, otherwise the transmitted data will not be received correctly.

Transmission distances

RS232 has a maximum effective distance of approximately 30 m at 9600 baud, due to the amount of stray capacitance in the cable, and its effect on signal transition time between the recognized limits. An RS232 transmitter produces a voltage between + 5 and + 25 V for one of the two possible signal states (space), and a voltage of between − 5 and − 25 V for the other (mark) signal (normally ± 12 V). Since these voltages are not usually provided by a PC or computer which operate at the standard + 5 V TTL/CMOS levels, an additional power supply may be necessary. Operating distance can be extended up to 100 m when using screened cables and a lower data rate.

Referring to Fig. 7.3, not all RS232 pins are used in every application, but the minimum requirements for interfacing are: transmitted data, received data, signal ground and frame ground. Different manufacturers often use only a minimum of the standard RS232 signals, jumbling and interchanging others to obtain a certain

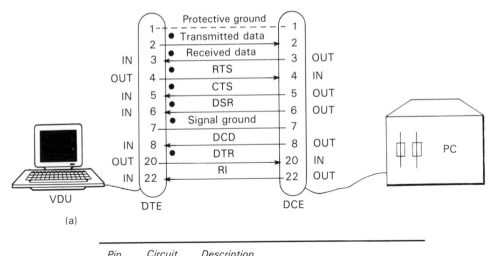

(a)

Pin	Circuit	Description
1	AA	Protective ground
2	BA	● Transmitted data
3	BB	● Received data
4	CA	● Request to send
5	CB	● Clear to send
6	CC	● Data set ready
7	AB	● Signal ground
8	CF	Received line signal detector
9/10	—	(Reserved for data set testing)
11	—	Unassigned
12	SCF	Secondary received line signal detector
13	SCB	Secondary clear to send
14	SBA	Secondary transmitted data
15	DB	Transmit signal element timing (DCE source)
16	SBB	Secondary received data
17	DD	Receive signal element timing
18	—	Unassigned
19	SCA	Secondary request to send
20	CD	● Data terminal ready
21	CG	Signal quality detector
22	CE	Ring indicator
23	CH/CI	Data signal rate select (DTE/DCE source)
24	DA	Transmit signal element timing (DTE source)
25	—	Unassigned

(b)

Fig. 7.3 *(a) Typical RS232C female connector (25-pin 'D' type) with pin assignments; (b) minimum configuration (shown dotted).*

configuration for their equipment – RS232-compatible may not mean universal compatibility!

Long distance two-way transmission is often required between intelligent devices or computers, and may be provided using RS422 or another common method of serial communication, the 20 mA current loop.

Table 7.1.

Interface	Distance (m)	Max. transmission rate (bits/s)
RS232C	10	10 000
	100	1 000
	1000	—
RS422	10	1 000 000
	100	100 000
	1000	10 000
20 mA current loop	100	10 000
	300	10 000
	2000	1 000

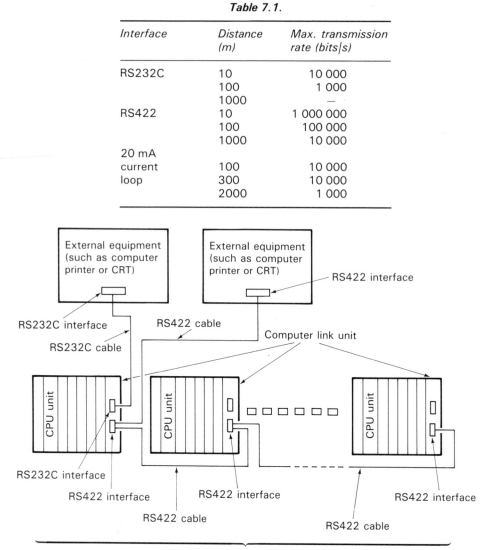

Fig. 7.4   Example of RS232/422 links to interconnect several PCs.

The 20 mA current loop

The 20 mA loop is more suitable for long-distance use than RS232, and is used in many industrial systems in which the communication path is in a noisy electrical environment, because of its high noise immunity. A switched 20 mA circuit is used to generate the serial data. In comparison, at a data transmission rate of 9600 baud

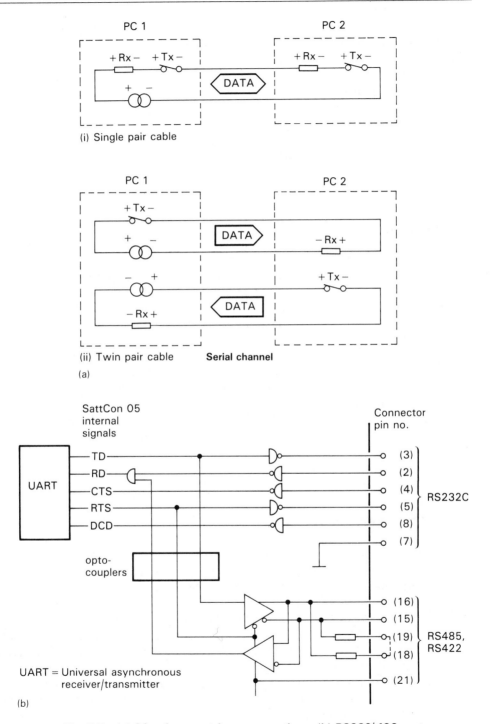

(i) Single pair cable

(ii) Twin pair cable   **Serial channel**

(a)

(b)

UART = Universal asynchronous
         receiver/transmitter

Fig. 7.5   (a) 20 mA current-loop connections; (b) RS232/422 ports.

RS232 signals can be used up to only 15 m, whereas 20 mA signals can be used up to 300 m at the same data rate. The current-loop system requires a pair of wires each for send and receive, and can easily be fitted with opto-isolators to give electrical isolation between the transmit and receive nodes (Fig. 7.5(a)). Unfortunately, current loop is not directly compatible with RS232C and requires hardware interfacing if the two media are to communicate.

Disadvantages of current-loop communication include: no standardization of circuitry; lack of control (handshake) signals provided with some RS232 channels. (Handshake lines allow either node to signal various conditions to the other, e.g. data ready for transfer/not ready; terminal busy, etc.)

RS422/423

However, the world is not moving totally away from RS232-type communication, having developed an improved specification, RS422, that overcomes some of the defects of RS232, as well as combining some of the advantages of current-loop systems. RS422 uses two wires for each signal – a balanced electrical interface with differential input and output lines to provide a greater distance and higher data rate than RS232. RS422 has a much narrower signal transition region (0.4 V against 6 V in RS232), allowing the interface to be implemented using the 5 V supply from most microprocessor-based equipment.

Virtually all PLCs with a communications facility have RS232/V24 ports, often with an additional RS422 port derived from the RS232 lines (Fig. 7.5(b)). The RS232 port is used for short-distance links, say to a VDU, printer or computer in close proximity. The RS422 or current-loop ports are used for longer-distance links, often between several PCs in a distributed system, as in Fig. 7.4.

7.2 FLOW CONTROL

In communications between any types of microcomputer, mainframe or programmable controller, we have to consider flow control – the direction of data flow, and whether this is in one or both directions at one time.

Simplex operation is where data flows in only one direction at a time, one device always being the transmitter and the other device the receiver: for example, a PLC or computer talking to a printer. The printer never needs to talk to the PC, therefore the PLC is always the transmitter and the printer a receiver. Where two intelligent devices such as programmable controllers or computers are involved, it may be necessary for two-way communication to occur. We may have communication in which one device talks and then the other (*half-duplex*) or in which both talk simultaneously on the same channel (*full-duplex*).

Programmable controllers may require to use any of these types of data flow, depending on the application.

Protocol for transmission

To provide complete and correct communications between two devices, we also have to be able to control the flow of data, so that one device can signal the other to start or stop sending data. For example, if a microcomputer is sending data to a PLC at a faster rate than the PLC can manage, then the microcomputer must be signaled to stop transmission or pause until the PLC is ready to receive more data. Flow control or protocol is handled either by using additional signal lines or by sending control characters on the communications channel.

It is common to use two extra signal (or handshake) wires connecting the transmitting and receiving devices – one to signal the receiver that the transmitter is ready to send (RTS), and the other to tell the transmitter that the receiver is ready for the data – clear to send (CTS). One device requests transmission by driving RTS low (active). The other device replies by driving CTS low, then the data is transmitted until CTS is turned off (high) by the receiver. RTS/CTS lines are provided in RS232 links, and are commonly used in PC intercommunication (see Fig. 7.6).

Two very common forms of protocol using additional control characters on the transmit/receive wires are known as XON/XOFF and ENQ/ACK. The basic operation of these flow-control methods are illustrated in Fig. 7.7(a, b). With XON/XOFF, when a receiving device wishes to stop receiving data, it transmits an XOFF command (03 hex) to the sending device, which then stops transmitting and waits to receive an XON command (01 hex) before resuming transmission. The ENQ/ACK protocol on the other hand, provides data 'packets' that are sent to a receiver together with a query character – ENQ (03 hex). This signals the end of a data packet, and once the receiver has processed the data, it can request another datablock from the transmitter by sending back an ACK (acknowledge) command (06 hex).

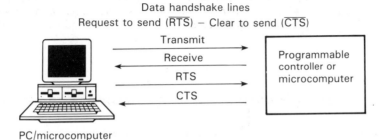

Fig. 7.6 *RTS and CTS control transmission of data from one device to another.*

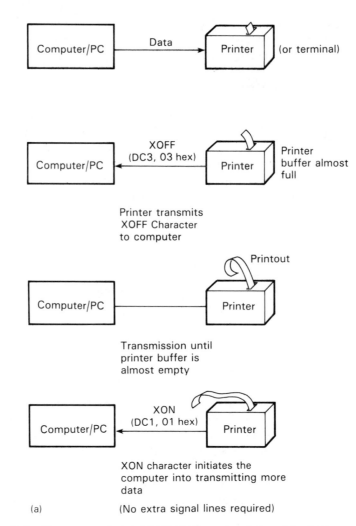

Printer transmits
XOFF Character
to computer

Transmission until
printer buffer is
almost empty

XON character initiates the
computer into transmitting more
data

(a)   (No extra signal lines required)

Fig. 7.7 Flow control: (a) XON/XOFF protocol (used in full/asymmetric duplex since XOFF must be sent against the main data stream); (b) ENQ/ACK protocol (can be used in half-duplex since sending device transmits ENQ byte).

7.3 PLC COMMUNICATIONS – MODULES AND PROGRAMMING

Whether the communications facility is provided internally or by an additional module, programmable controllers require to have the communications options configured to suit each application. This may include:

* baud rate/transmission speed;

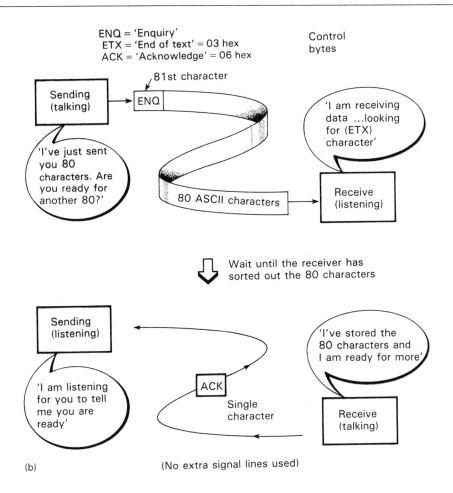

ENQ = 'Enquiry'
ETX = 'End of text' = 03 hex
ACK = 'Acknowledge' = 06 hex

Control bytes

81st character

Sending (talking) → ENQ

'I am receiving data ...looking for (ETX) character'

'I've just sent you 80 characters. Are you ready for another 80?'

80 ASCII characters

Receive (listening)

Wait until the receiver has sorted out the 80 characters

Sending (listening)

'I've stored the 80 characters and I am ready for more'

'I am listening for you to tell me you are ready'

ACK

Single character

Receive (talking)

(b)   (No extra signal lines used)

- data format – start, stop, parity and number of data bits;
- protocol used – e.g. RTS/CTS or XON/XOFF.

The selection of each option may be either by hardware switches on the communications module itself as in Fig. 7.8, or by software from within the PLC (Fig. 7.9).

Data transmitted over PLC links

The nature and type of data required to be sent out from a PLC can vary from a single status byte, reflecting a register value, to a complete program transfer. The list in Table 7.2 for a Sattcon 05-30 is typical of the message requests supported by PLC communication modules.

A programmable controller has to be programmed to respond to requests for data output, and to handle incoming data from external devices. Again, it depends

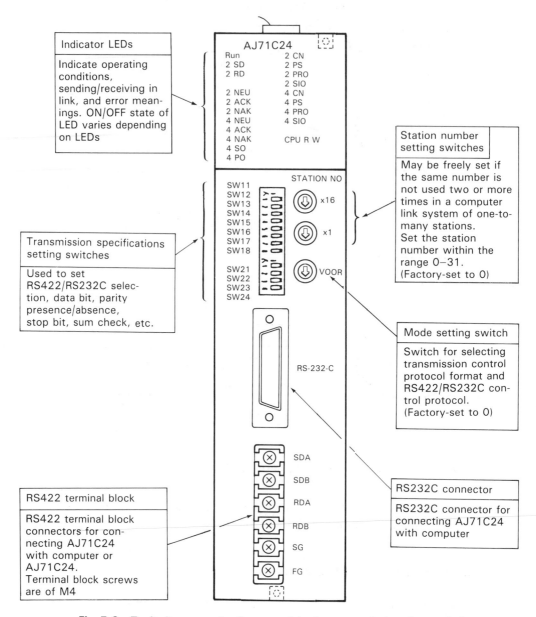

Fig. 7.8 Typical communications module front panel showing switches and RS232/422 connectors (Mitsubishi).

> Programming COMLI parameters

COMLI 05 – SATTCONTROL
SERIAL COMMUNICATIONS ON AN 05 PC

General

SattCon 05 can communicate with other systems via the serial channel using the communication protocol COMLI (communication link). This section only briefly presents the method used to program Satt-Con 05 for COMLI.

The communication is point-to-point (RS232C) or multipoint (RS485). The SattCon 05 always has the following setting.

- binary data transmission
- 8 data bits
- 1 stop bit
- Odd parity

COMLI parameters
(channel setting)

A SattCon 05 always communicates as the SLAVE, subordinate to a MASTER, which controls the transmission. Therefore you need only program the following parameters on the SattCon 05:

- COMLI on/off
- Baud rate (300–19 200 baud)
- Identify the SattCon 05 (1-247)
- Signal delay (for modem communication)

The COMLI parameters are programmed at command level 5 ENTER. In a question/answer program you give the parameter the correct value before proceeding to the next using ENTER.

'COMLI:off' means that the serial communication has been switched off, and that the serial channel can then be used for listing on a printer.
The COMLI channel is enabled, and SattCon 05 can receive orders from the master.

The COMLI baud rate, which is now displayed, need not be the same as for the printer, and is therefore set separately. The baud rate must be identical to that of the master. The correct value is selected using STEP.

The identity of the SattCon 05 is displayed. SattCon 05 only accepts COMLI messages which have this identity as their address. A new identity (1–247) can be entered with n PROG.

The signal delay (modem delay) need only be set for modem communication. To adapt the SattCon 05 to different types of modems, the handshaking signal RST may be delayed in steps of five milliseconds. The delay should normally be 0.

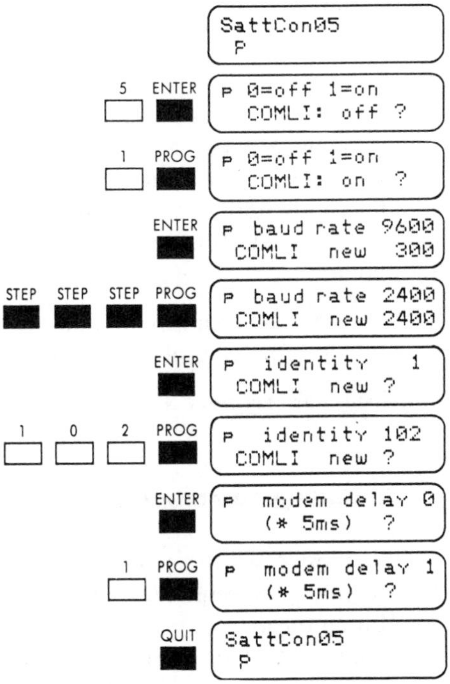

```
                                            SattCon05
                                            P

                          5   ENTER     P  0=off 1=on
                          [ ]  [█]          COMLI:  off  ?

                          1   PROG      P  0=off 1=on
                          [ ]  [█]          COMLI:  on   ?

                              ENTER     P  baud rate  9600
                               [█]         COMLI   new   300

          STEP  STEP  STEP  PROG         P  baud rate  2400
          [█]   [█]   [█]   [█]             COMLI   new  2400

                              ENTER     P   identity      1
                               [█]          COMLI   new  ?

                          1   0   2   PROG  P   identity 102
                         [ ] [ ] [ ]  [█]      COMLI   new  ?

                              ENTER     P   modem delay  0
                               [█]          ( * 5ms )    ?

                          1   PROG      P   modem delay  1
                          [ ]  [█]          ( * 5ms )    ?

                              QUIT      SattCon05
                               [█]      P
```

Fig. 7.9   Setting communications options via software (SattControl and GEC systems).

Table 7.2   SattCon communications messages

Transmitted control code	Meaning	Slave receive	Master receive	Master transmit
0	Transmit I/O RAM status or register value	*	*	*
1	Acknowledge		*	
2	Request I/O RAM status or register value	*		*
3	Transmit single I/O bit status	*	*	*
4	Request status of single I/O bit	*		*
9	Request analog input values	*		*
A	Transmit analog input values		*	
D	Enter terminal mode	*		
I	Request date and time	*		*
J	Transmit date and time	*	*	*
W	Request for DUMP data	*		
X	Transmit LOAD data	*		
Y	Transmit VERIFY data	*		
	Transmit time-labeled event			*
	Request time-labeled event	*		*
^	Request system information	*		
—	Transmit system information	*		

on the PLC system whether this programming is done by conventional relay ladder symbols, a symbolic communications language, or by a high-level language similar to BASIC. The actual communications processing may be done by a dedicated microprocessor in the communications card or by the main CPU, depending on the system used. Whatever method is used, it will involve specifying the following points:

- the condition(s) that are to initiate data transfer;
- the type, location and amount of data that is to be transferred.

These are required for both the transmission and reception of data.

Initiating communications

In most cases there is no requirement to perform communications continuously, requiring a specific event or events to trigger transmission or reception of serial data. Initiation may be by an internal event, say a contact closing, or by an external request for data from another PLC or computer.

The majority of PLC communication facilities are programmed to provide specific responses to remote requests for data. These requests are normally signaled by the transmission of one or more control characters from the initiating device. Table 7.2 lists typical codes and their meaning. The characters are normally transmitted in ASCII format over the link, together with any protocol characters used. (ASCII – American Standard Code for Information Interchange. This is the most common method of representing alphanumeric characters for digital transmission.

These methods of initiating communication are used to obtain and transfer PC/plant status or fault information. When status data is required only at specific time intervals – once a second, hour, day, etc. – then internal timer circuits may be used to control the contacts that initiate communications from a remote PC to its supervisor/master. Likewise, timer circuits in the supervisory controller can be programmed to send the aforementioned control code requests at preset intervals, simply by storing a message request code/ASCII character in an output register and then transmitting this to the remote station.

The data that is to be transferred is usually stored in one or more PLC registers, often as part of a larger data table. By this means, information on the I/O status, process events and/or text messages may be transferred from one device to another.

The exact details of communications programming and internal organization differ between programmable controller systems, but the main concepts are similar. The following example deals with a ladder format of communications programming, illustrating several important points in the process.

Example – communications rung operation

This example concerns the SY/MAX Square D system. The communications rung is made up of three parts:

1. the PLC registers involved,
2. the status register, and
3. the contact that initiates the transfer.

The information within a rung describes which registers will be involved in transfer (transmission/reception) between PLCs, and the number of registers to be transferred in the same rung. In Fig. 7.10 the communications rung consists of three information areas: local, remote and count.

The 'local' entry refers to registers within the PLC that will initiate the ladder rung. The term 'remote' identifies registers in the PLC that are to receive the incoming data from the serial link, whilst 'count' defines how many registers will be involved in the transfer. The word 'read' or 'write' at the start of the rung specifies the direction of the data transfer. In the above example register 15 is the source location for data that will be sent to register 20 in the remote programmable controller. A count of 1 means that only one register per PLC is involved in the transfer, rather than an array or table of source and destination registers.

The status register is used to monitor the communications operations, providing information on message progress, communications errors, and can be examined (read) by a remote device to check for correct transfer.

The contact that initiates the rung can consist of one or more normally open or closed contacts. When the enabling path is complete (contacts closed), the

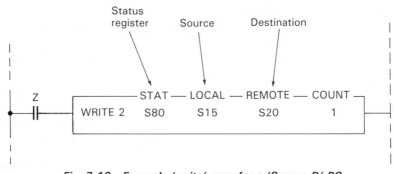

Fig. 7.10   Example 'write' rung for a 'Square D' PC.

communication rung is activated and the data transfer is carried out. If the count
value was, say, 4 instead of 1, then the contents of the next three registers would also
be transferred. That is,

Local			*Remote*
15	⎫		20
16	⎬ Write ⎰	21
17			22
18	⎭		23

In some PLC systems, data tables are designated for communications use, with
one table for data to be received from a remote device, and another to hold data to
be transmitted to a remote device. In this case the programmer may only have to
specify a read or write operation, without the necessity for exact register definitions.

7.4   COMMUNICATIONS BETWEEN SEVERAL PLCs

When several PLCs or machines have to send or receive data to and from a single
source, it is possible to merge these signals using a concentrator unit. This unit
switches between channels under software control from the supervising controller,
or in response to a request signal from one of the linked PLCs. (Fig. 7.11).

However, when it becomes desirable or necessary to interconnect a large
number of terminals or intelligent machines, a communications network is often
used. A *local area network* (LAN) provides a physical link between all devices plus
providing overall data exchange management or protocol, ensuring that each device
can 'talk to' other machines and understand data received from them.

7.5   LOCAL AREA NETWORKS (LANs)

Local area networks (LANs) constitute a relatively recent innovation in microcomputer development. In the 1970s the use of versatile and inexpensive microcomputers

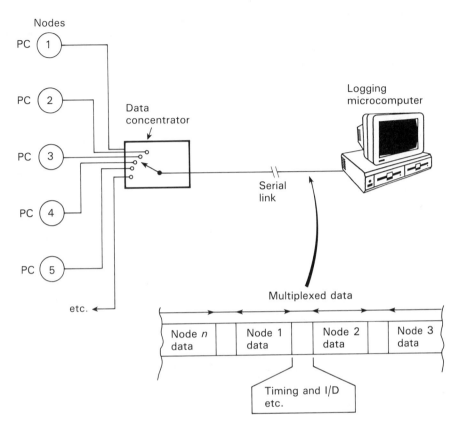

*Fig. 7.11 **Linking several PLCs for data logging into a single supervisor.***

became firmly established as people moved away from the constraints imposed by large mainframe installations.

The advent of the local area network offered a solution to this problem, providing a data transmission system linking computers and associated devices within a restricted geographical area (up to 10 km, although 1 km is more typical).

Reasons for installing a network instead of point-to-point links are:

- All devices can access and share data and programs.
- Dispersion of equipment makes cabling for point-to-point impractical and prohibitively expensive.
- A network provides a flexible base for contributing communications architectures.

A network provides a compromise between long-distance and high data rates of up to 10 Mbps. LANs provide the common, high-speed data communications bus which interconnects any or all devices within the 'local' area.

LANs are commonly used in business applications to allow several users to share costly software packages and peripheral equipment such as printers and hard disk storage. There are many different networks available, for example

ETHERNET, IBM Token Ring, etc., all having different physical, electrical and protocol standards. One particular microcomputer 'family' may be compatible with a certain network but no others, preventing future expansion of the system to include different computers or intelligence devices.

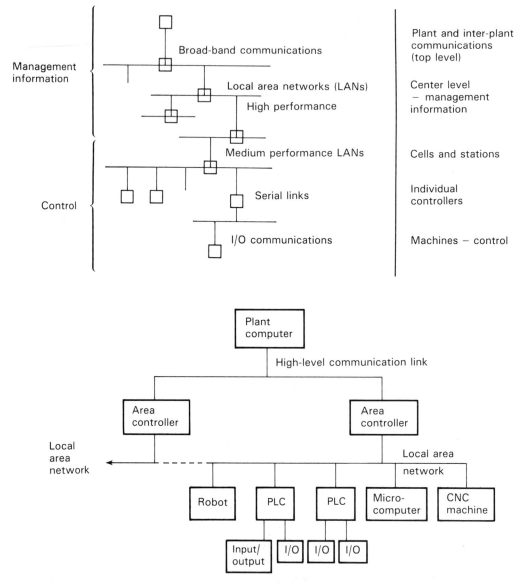

Fig. 7.12 Communications networks—a hierarchy.

Programmable controllers and networks

In industry, with increasing automation of plants and processes to raise efficiency and output, the need for overall communications exists. This is necessary to allow communications and control on a plant-wide basis, interconnecting microcomputers, minicomputers, robots, CNC machines and programmable controllers (often on a real-time basis).

Most manufacturers provide a dedicated network system that may be used for communications between controllers from their own product range. Some examples are given in Table 7.3.

A PLC may be linked into its manufacturer's communications network by using a proprietary network interface module, if one is available for the particular PLC. It is rarely possible to use PCs from different sources on a single vendor's networking system.

The illustration in Fig. 7.13(c) shows a Texas Instruments TIWAY data network. This is used to transport data from any suitably equipped Texas PLC to any other, or to a central mini/mainframe computer.

The GE NET factory LAN in Fig. 7.13(b) is the General Electric proprietary network, which also conforms to recent worldwide 'open systems' standards under MAP – discussed in a later section.

Before looking at PLC communications networks in more detail, it is worthwhile considering the role such communications must serve.

Table 7.3.

Manufacturer	Network
Allan-Bradley	Data Highway
Gould	Modbus
General Electric	GE Net Factory LAN
Mitsubishi	Melsec-NET
SattControl	Comli
Square D	SY/NET
Texas Instruments	TIWAY
etc.	

7.6 DISTRIBUTED CONTROL

Communication facilities also allow the programmable controller to act not just as a dedicated controller on one particular machine, but also as a controller of multiple stations within a large manufacturing area. Thus the PLC can become part of a *hierarchical control structure*, in which a coordinating programmable controller supervises several dedicated PLCs or other intelligent devices such as robots or CNC machines (Table 7.4).

Table 7.4.

Level	Function/control medium
1. Plant	Planning and main database central mainframe computer
2. Sections	Production and scheduling data minicomputer or powerful micro
3. Cells	Coordination of multiple stations large programmable controller or microcomputer
4. Stations	Controlling plant and machinery programmable controller or microcomputer
5. Machinery	Plant and equipment input/output interfacing

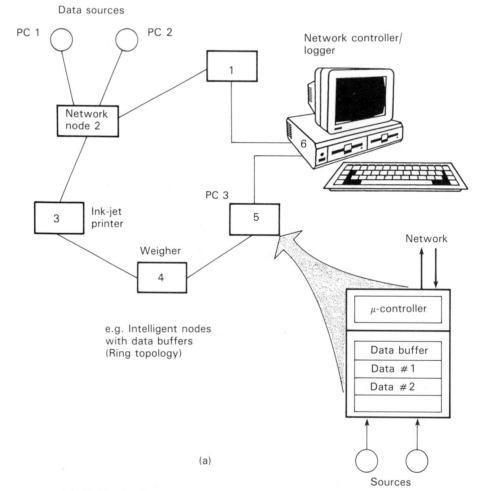

(a)

Fig. 7.13 *(a) Networks; (b) GE Net; (c) Texas Instruments' TIWAY.*

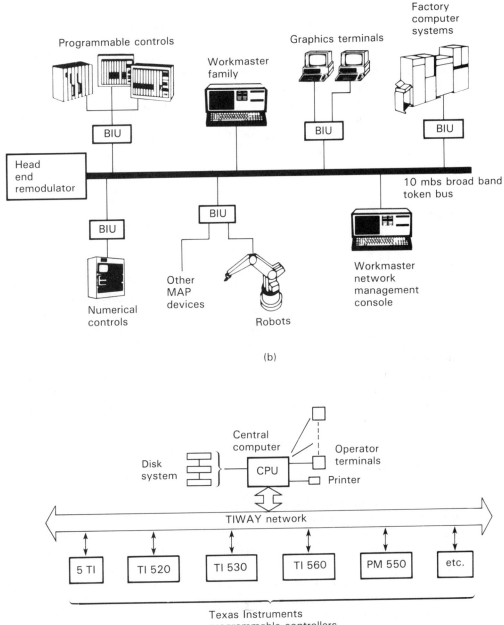

(b)

(c)

This illustrates the concept of 'top-down' control currently being adopted in the automation of complete factories. It shows the use of mainframe, mini- and microcomputers for coordination of all levels of plant operation from top management to shop-floor machinery.

To develop this automated factory, we must establish communications and control on a plant-wide basis. This may involve communication from one PLC to another, or to micro- and minicomputers, as well as to other data-processing equipment (Fig. 7.12). This includes the integration of process data into management databases, allowing immediate presentation of information to those involved in planning and production scheduling. Thus the creation of acceptable communications standards (for both hardware connections and data format) is highly desirable.

Range of requirements

The communicating needs of this control hierarchy vary from level to level (see Fig. 7.12). At the bottom of the pyramid, real-time control may be necessary, with the LAN carrying the control and data signals to and from programmable controllers, robots, CNC machines, etc., linking with an *area controller* (which may be a larger PLC or computer). At this level, communication is mainly small amounts of high-speed control data. From the area controller upwards to the plant computer, data is mainly information on plant performance rather than control. The data-rate capabilities of the constituent elements may also vary greatly.

This implies a hierarchical network design, with stepped levels of capacity and performance. For example, twisted-pair connections for low and medium data rate devices, linking into a 'backbone' high-speed (high-bandwidth) system based on coaxial or fiber-optic cable.

Transmission medium

The physical environment that the network will have to operate in also affects the choice of medium. Electrical 'noise' has always been a problem in manufacturing industry, where electrical plant, welding and cutting machines produce electromagnetic radiation. When communications cabling passes close to these noise sources, it may be difficult to obtain reliable, high-speed data transmission.

Twisted-pair cabling is commonly used on the factory floor, but often has to be routed through grounded steel conduit to obtain satisfactory communications.

Coaxial cable can operate at higher data rates than twisted pair, and does not require additional shielding. Transmissions may be simple *base band*, without the use of a carrier, and with only one channel defined in the system. Alternatively, a *broad-band* system may be used, having several channels multiplexed in frequency across the wide bandwidth of the coaxial cable. Broad band is relatively unaffected by noise, and is therefore ideally suited to the factory environment. However, it is many times more expensive than a base band system, due mainly to the need for frequency modulations/demodulations at each node.

   Fiber-optic cabling will eventually replace the above media, due to its greater bandwidth, noise immunity, small size and flexibility. Point-to-point links are relatively simple, but other connections to the optical fiber cable cannot be made without regenerating the signal.

Network configurations

Although there are several hundred different network systems on the business and industrial markets, they all possess certain common features. Each device on a network, referred to as a station or node, has to have a suitable interface. All stations are linked into the system by lengths of cable, which may be single twisted pair, coaxial, or fiber-optic cable, but its function remains identical – to carry data from one network station to another.

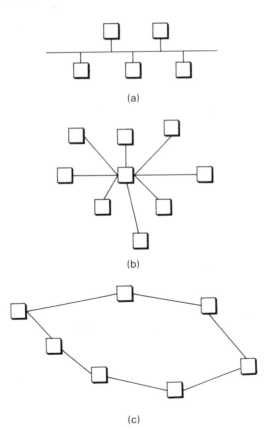

Fig. 7.14   Network topologies: (a) bus topology; (b) star topology; (c) ring topology.

The network requires controlling software to correctly handle all file transfers within the system, dealing with station access, data validity, etc.

The topology of a network is the physical arrangement of the stations and their inter-connections (Fig. 7.14). There are three main patterns in use:

1. a bus topology (Fig. 7.14(a)) consisting of a central cable with all stations connected to it by spurs;
2. a star topology (Fig. 7.14(b)) with stations clustered around a single, central device that acts as a 'file server';
3. a ring topology (Fig. 7.14(c)) consisting of stations connected together in a complete circle or loop.

A 'file server' is normally used on all types of network, the term referring to a station that is used to store shared files or software and may also provide access to shared peripherals such as printers, hard disk storage, etc.

However, a network is not fully defined by its topology, since each configuration can operate in several different ways.

Channel access and control

With several stations on a network, there must be a mechanism for deciding which station gains 'access' to the common channel to transmit or receive information. Under heavy traffic conditions, there is more than one station trying to access the network at one time, causing response times to deteriorate (often dramatically). It is therefore important that the traffic is controlled, to allow a smooth, efficient operation and to reduce the chance of data corruption caused by the collision of two data streams on the network.

Protocols

The use of communications protocols can ease problems of congestion on a network. There are two main protocols currently in use: (a) CSMA/CD and (b) token passing.

(a) *CSMA/CD* In a carrier sense, multiple access/collisions detection (CSMA/CD) system, stations gain access to the network on a 'first come, first served' basis. The *carrier sense* feature means that a station 'listens' on the network to check for other traffic. If no carrier signal is present (no traffic) the station accesses the network. If a carrier is detected, the station defers transmission, waiting for a certain time before trying again. This 'looking before crossing the road' strategy reduces, but does not eliminate, the risk of data collisions.

Multiple access means that any station may transmit data as soon as it senses the channel is free.

Collision detection indicates that a station can listen on the network as it transmits data, allowing it to detect additional, contending data on the channel. The station then breaks off transmission, waiting for a short, random interval before trying to retransmit. The 'random back-off' reduces the possibility of further collisions.

CSMA/CD techniques offer fast response at low traffic rates, but as the load increases, so does the waiting time.

(b) *Token passing* This type of protocol uses a special token or data packet that passes control from one station to another. Any station wishing to transmit information must wait until it has received the token. Having completed transmission, it passes the token to the next node.

Token passing is used in both ring and bus topologies, providing a relatively slow response at low traffic rates compared with CMSA/CD, but with little deterioration of response time as the load increases.

Industrial and business LANs

A wide variety of LAN standards exist in both the industrial and business markets, due to the competitive and 'lock-in' nature of the suppliers, resulting in continually changing specifications. Once a firm has purchased a system of this type, it is normally constrained into expanding the system using equipment from the same source for reasons of compatibility.

In a business situation, it may often be possible to specify equipment from a single manufacturer. However, in industry the type and diversity of equipment and plant often precludes single-sourcing, since the PLCs, automatic tools, CNC, etc., inevitably come from different manufacturers.

A PLC may be linked into its manufacturer's communications network by using a proprietary network interface module, if one is available for the particular PC. Most manufacturers provide a networking system that allows the connection of most of their products, but very rarely will it be possible to use PLCs from different sources on the same system.

If no standard LAN were adopted, each manufacturer would fit their own choice of interface, resulting in obvious difficulties when attempting to link different devices into the same network.

At levels 3 and 4 of the control pyramid shown in Table 7.4 (cell and station), the specialized nature of PLCs, robots, etc., dictates the level of communications required, normally involving real-time control and high functionality. In most situations, a company using a single manufacturer of, say, programmable controllers, finds it convenient and economic to use a proprietary network for communications at this level.

7.7 NETWORK STANDARDS – ISO, IEEE AND MAP

ISO – open systems interconnection

Several manufacturers have followed the International Standards Organization (ISO) model for open systems interconnection (OSI). This model was defined in 1979 to expedite communications between dissimilar computer systems, and is now the standard that computer and equipment manufacturers have adopted or will provide as an option to their in-house specification.

The ISO/OSI model provides a framework for network communications standards, but not an actual specification for communications protocols.

A communications link between items of digital equipment is defined in terms

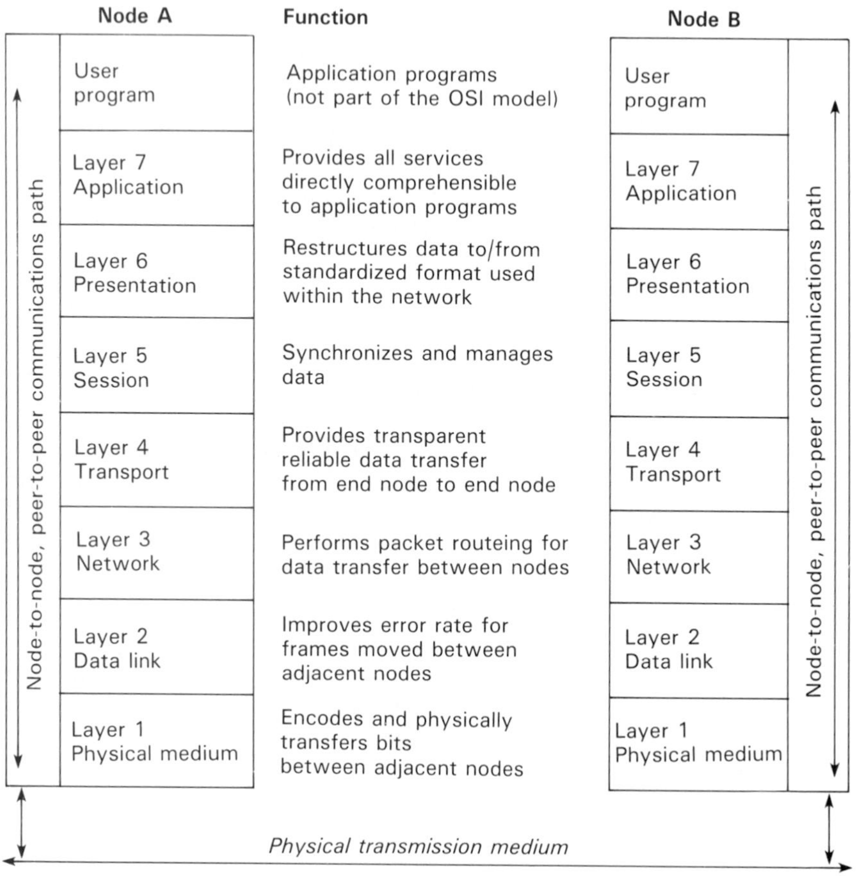
Node A
Function
Node B
User program
Application programs (not part of the OSI model)
User program
Layer 7 Application
Provides all services directly comprehensible to application programs
Layer 7 Application
Layer 6 Presentation
Restructures data to/from standardized format used within the network
Layer 6 Presentation
Layer 5 Session
Synchronizes and manages data
Layer 5 Session
Layer 4 Transport
Provides transparent reliable data transfer from end node to end node
Layer 4 Transport
Layer 3 Network
Performs packet routeing for data transfer between nodes
Layer 3 Network
Layer 2 Data link
Improves error rate for frames moved between adjacent nodes
Layer 2 Data link
Layer 1 Physical medium
Encodes and physically transfers bits between adjacent nodes
Layer 1 Physical medium
Node-to-node, peer-to-peer communications path
Node-to-node, peer-to-peer communications path
Physical transmission medium

Fig. 7.15 ISO/OSI seven layer model (from **Through MAP to CIM**, *DTI, 1987).*

of physical, electrical, protocol and user standards. The OSI model defines all aspects of communications from user to user with a seven-layer breakdown shown in Fig. 7.15. Each layer must be able to stand alone, performing its task and transferring its results on the next layer (up or down depending on the direction of transmission).

In essence, layers 1–3 are common, providing the transmissions media and lower-level interfaces necessary for data exchange between compatible components (e.g. similar PLCs). The higher levels (4–7) are necessary to allow intelligent communication between dissimilar devices using individual software applications. Without these layers, the LAN would allow messages to be exchanged between the dissimilar devices, but the message would not be understood and could not be accessed by the user.

IEEE

In America, the IEEE project 802 has developed LAN standards using a layered approach similar to the ISO reference model, and many of the IEEE standards are used to fully define networks that are based on the OSI model, particularly at the lower levels.

GM MAP – the industrial standard?

In the US, General Motors (GM) were faced with the task of equipping their factories with around 20 000 systems by 1990, all of which are required to talk to each other. They made firm commitments to six of the layers in the OSI model, defining their manufacturing automation protocol (MAP). This resulted in all major control systems and information systems suppliers also committing themselves to MAP compatibility. The specification was available free to anyone.

General Motors specified a *manufacturing automation protocol* (GM MAP) for a LAN to integrate all levels of control systems, including the microcomputer-based products listed above. The GM MAP has been adopted by many manufacturers of control equipment and semiconductor manufacturers who now produce interface ICs to this specification.

GM's aim was for all the systems on the shop floor – programmable controllers, robot systems, welding systems, vision systems, etc. – whoever their manufacturers are, to be able to communicate with each other by 1988.

GM MAP – outline of the specification

General Motors chose the token bus, as defined by the IEEE 802-4 standard, for MAP's physical layer. The reason for this choice was the requirement for 'deterministic access time' – a declared and operational maximum access time – necessary for the network to efficiently cope with heavy traffic conditions, and to avoid data

OSI layer		MAP protocol	Ethernet	
Number	Definition			
7	Application	ISO file transfer MMFS, FTAM, CASE	e.g. MS net software	Implemented in software (various for Ethernet)
6	Presentation	Null		
5	Session	ISO session kernel		
4	Transport	ISO transport class 4	ISO class 4	
3	Network	ISO Internet	Null	
2	Data link	IEEE 802.2 class 1 IEEE 802.4 token bus	IEEE 802.3 CSMA/CD bus	Implemented in hardware/firmware
1	Physical	IEEE 802.4 broad band	IEEE 802.3 base band	
	Link	10 mbps coaxial cable with RF modulators	10 mbps coaxial cable with top connectors	

Fig. 7.16   MAP layer model and standards.

collisions. (CSMA/CD) has probabilistic access time – i.e. the operational access time can be estimated for different traffic conditions, but not specified – and theoretically has an inferior response than token systems under heavy traffic – for example IEEE 802.3 – Ethernet type systems.)

MAP uses a broadband coaxial medium, at a data rate of 10 mbps, with the cable bandwidth divided into many separate channels with different positions in the frequency spectrum. Two channels are used for transmit and receive paths for MAP network data, with several other channels available for other purposes, such as video and voice communication. A head-end frequency converter is required to terminate the cable. (Some plants already have broadband cabling installed, which may be used for a MAP network link.) Fig. 7.16 gives a summary of the other MAP layers, together with the 'equivalent' Ethernet standards for comparison.

Comparison with Ethernet

Ethernet is compared here because it was the first major network product offered using non-proprietary protocols and interfaces. Ethernet was created jointly by Xerox, DEC and Intel, who defined products based on published communications standards. Ethernet systems are widely used in both industrial and office environments, with compatible products available from a wide range of suppliers.

Physical layer

Ethernet has bus topology, with a data rate of 10 mbps over coaxial cable (10 mbps) using *baseband transmission*. (Baseband transmission means the sending of data without a carrier frequency, using only one channel within the system.) Access to the coaxial cable is obtained by a 'vampire tap' that clamps into the cable, thus allowing the connection or removal of stations without disrupting system operation. Then data is transmitted using 'Manchester encoding', which provides a timing component for synchronization, because a logic inversion always occurs in the middle of each bit.

Data link layer

Brief summary of layer:

Network control: Multi-access – fairly distributed to all nodes.
Access control: CSMA/CD.

Layers 4–7: The Ethernet specification did not include any standards for these levels, but suitable software is available from several sources to provide the necessary data management functions.

Implementing the MAP standard

Layers 1 and 2 are implemented in hardware electronics, whilst for layers 3–7 the standards are implemented in software.

The application layer serves as an access point for a user's application programs to the system, supporting three elements: common, application-specific and user-specific.

(a) The common element allows communications irrespective of the nature of the package.
   • CASE – common applications service
(b) The application-specific aspect provides information (file) transfer facilities.
   • FTAM – file transfer
(c) The user-specific element applies to functions that are specific to an area of application, for example PLC's or robots.
   • GM MMFS – manufacturing message format standard.

Each of these service elements provides a set of commands that will be interpreted and executed in a fully defined manner, by the specific target device or software package.

For example, Fig. 7.17 shows the subset of commands for programmable controllers and robots (1986). These instructions may be sent from any suitable, authorized terminal on the network, and allow both data transfers and control instructions.

MAP in plant-wide communications and open systems

The GM-MAP is becoming the standard factory backbone network, providing communications between all production units within the factory. Subnetworks often exist which provide communications within each production unit and, as previously mentioned, many proprietary LANs exist which fulfil the requirements of these

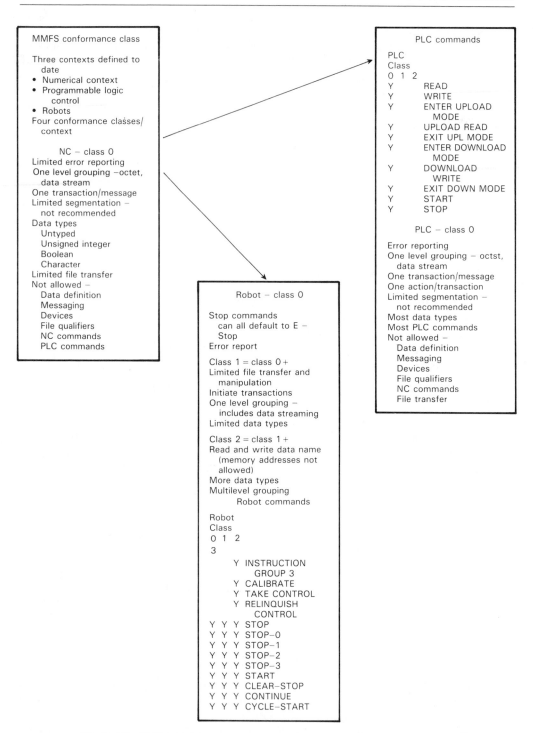

MMFS conformance class

Three contexts defined to
  date
• Numerical context
• Programmable logic
  control
• Robots
Four conformance classes/
  context

          NC – class 0
Limited error reporting
One level grouping –octet,
  data stream
One transaction/message
Limited segmentation –
  not recommended
Data types
  Untyped
  Unsigned integer
  Boolean
  Character
Limited file transfer
Not allowed –
  Data definition
  Messaging
  Devices
  File qualifiers
  NC commands
  PLC commands

PLC commands

PLC
Class
0 1 2
Y   READ
Y   WRITE
Y   ENTER UPLOAD
            MODE
Y   UPLOAD READ
Y   EXIT UPL MODE
Y   ENTER DOWNLOAD
            MODE
Y   DOWNLOAD
            WRITE
Y   EXIT DOWN MODE
Y   START
Y   STOP

          PLC – class 0

Error reporting
One level grouping – octst,
  data stream
One transaction/message
One action/transaction
Limited segmentation –
  not recommended
Most data types
Most PLC commands
Not allowed –
  Data definition
  Messaging
  Devices
  File qualifiers
  NC commands
  File transfer

Robot – class 0

Stop commands
  can all default to E –
  Stop
Error report

Class 1 = class 0 +
Limited file transfer and
  manipulation
Initiate transactions
One level grouping –
  includes data streaming
Limited data types

Class 2 = class 1 +
Read and write data name
  (memory addresses not
  allowed)
More data types
Multilevel grouping
          Robot commands

Robot
Class
0 1 2
3
        Y INSTRUCTION
            GROUP 3
        Y CALIBRATE
        Y TAKE CONTROL
        Y RELINQUISH
            CONTROL
Y Y Y STOP
Y Y Y STOP–0
Y Y Y STOP–1
Y Y Y STOP–2
Y Y Y STOP–3
Y Y Y START
Y Y Y CLEAR–STOP
Y Y Y CONTINUE
Y Y Y CYCLE–START

Fig. 7.17   MAP MMFS (applications layer) commands for PLCs and robots.

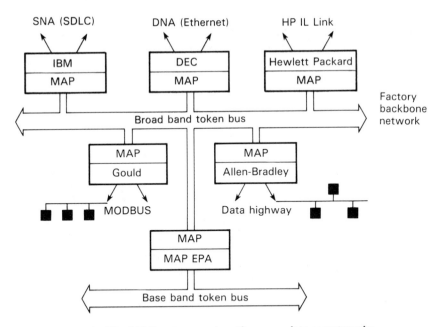

Fig. 7.18   MAP gateways to other proprietary networks.

units. These continue to be used, with the provision of 'gateways' to the MAP backbone from the majority of nonstandard industrial networks, e.g. Gould Modbus, AB Data Highway, DEC DNA, etc. Gateways allow the integration of segmented manufacturing systems into MAP, without having to replace existing equipment (see Fig. 7.18).

MAP gateways

Depending on the equipment to be used, a gateway may be either a self-contained unit, or consist of a MAP interface board that fits into the backplane of a programmable device or computer, plus memory software to support the MAP data management layers. A gateway acts as a nontransparent message store, forwarding messages from one node on a non-OSI network to another node on a MAP-OSI network, automatically performing protocol conversions.

Costs

Due to the hardware and software specifications of MAP, unit connection costs are relatively high. The cost of MAP interfaces and gateways has fallen with the production of VLSI MAP chip sets, but the need for broadband interface RF modems keeps this cost high, precluding the provision of MAP level communications on very small, low-cost programmable controllers and similar equipment.

There are several reasons for this performance and cost shortfall, including:

- high software overhead to implement all upper layers of OSI model;
- expensive broadband transmission with RF modems;
- token-passing access proving relatively slow when used in practice with MAP software.

MAP performance

As the work of specification progressed, it became increasingly clear that MAP, in the shape of its original model, did not cover the needs of networks that were more local in scope. These may entail, for example, fast real-time communication between robots and welding systems and between systems and PLC systems, or a vision need relating entirely to local communications within an FMS cell. The requirement in this case is that communication can operate at high speed with small amounts of data. Besides this, the communication units must be as simple and cheap as possible.

To this end, the MAP specification will be complemented by a variant of the original concept taking the above factors (among others) into account. This selective MAP is expected to cover only OSI levels 1, 2 and 7. Signal transfer will not be broadband, which will markedly reduce the costs for the physical construction of the network. On the debit side, considerably fewer units will be connectable to each communications interface.

There have also been numerous suggested names: the most common are Mini-MAP, Collapsed MAP, Real Time MAP, PROWAY and MAP Enhanced Performance Architecture. Irrespective of which of these names eventually comes to be accepted, this is a highly interesting development in communications, and will certainly be at least equal in importance in the future with the original MAP, particularly in the context of programmable controllers.

Full MAP architecture will coexist with enhanced (real-time) performance architectures via bridge or gateway interfaces.

Network communications in the factory

As described in the previous sections, the route exists from manufacturers' proprietary networks into a factory backbone (MAP) network. This allows companies to specify and install small-scale communications as necessary, with the future option of expanding into a factory-wide system using MAP. Once experience is gained with the application of networks with PLCs and other systems, it will become apparent what can be achieved in a particular plant in terms of productivity improvement.

7.8 PROPRIETARY PLC NETWORKS

As mentioned, a programmable controller can normally be equipped with a proprietary network interface card to allow connection to the manufacturer's communications network. In many ways this is similar to a serial communications module, providing all the necessary signal conversion and protocol facilities for data transfer. However, since there may be more than two devices on the network (up to 200 nodes in many systems), provision must be made to uniquely identify each node.

Network interface modules (NIM)

A device number or identifier is used to specify the source and destination of messages placed on the network. When first configuring the network, device numbers are selected either by thumbwheel-style switches on the interface module (Fig. 7.19), or by software programming. In many cases this also selects the time slot the module will have to 'talk' on the network – since all communications occur over a common cable, access to the network must be controlled to avoid data collisions that could happen if any node could access the channel at any time.

Network identifiers may also determine the *priority* of a node on the network – its ability to gain access to the channel before another node with a lower priority number. Generally, the lower the identifying number, the higher its priority. This

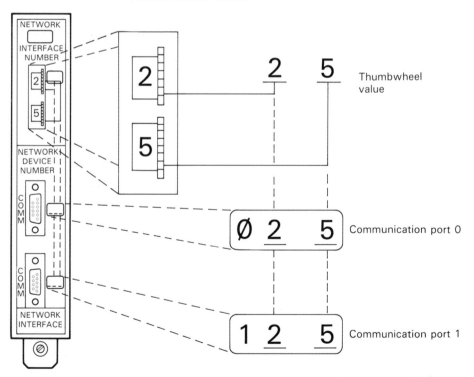

**Fig. 7.19 *Selecting a PLC address number on a network interface module
(Square D PC).***

facility is useful where devices with differing roles are connected to the network, allowing, say, a supervisory controller to have priority over several subordinate plant controllers.

Once this is done, device numbers can be used to route messages from any node on the network to any other node, by adding the identifiers to the message data.

Example   Ladder rung to pass message between nodes on a PLC network (Square D)

In Fig. 7.20, PLC X is connected to a serial port (1) of a NIM with identifier 7. It therefore has a network node number 107. PLC Y is connected to the serial port 1 of NIM number 15, so the node number is 115. This information is inserted in a ladder program rung as shown, resulting in the eight register contents being sent over the network from PLC X to PLC Y.

The type and quantity of data to be transferred over a network is similar to that for serial communications, as are the conditions used to initiate the transfer – ladder rungs, control characters, etc.

Messages are placed on the channel with destination and source data. When the destination module receives a message from the network, it waits until the complete message is received and then sends an acknowledgement back to the source node. If errors are detected in the message, it will send back an error message. The NIM will have on-board RAM to hold messages that have been received, or that are to be transmitted.

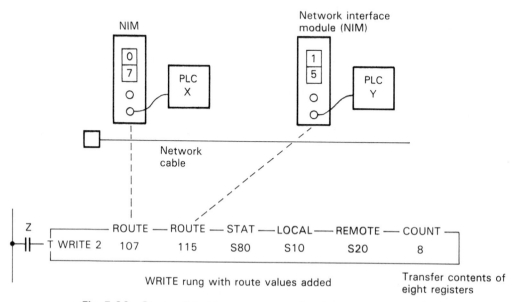

Fig. 7.20   *Square D ladder rung to transfer data over a network route.*

Protocols

The protocol used on proprietary PLC networks varies between manufacturers, although some are adopting 'standard' methods of message handling. For example, several PLC networks use HDLC protocol (high-level data link control) as the data structure. This provides data transfer over the channel, message checking and error control, and overall channel timing/synchronization. Other protocols are also in common use. For information on HDLC and other protocols, the reader should refer to one of the many texts covering data transmission (see p. 441).

Adaptable LANs

Other forms of local area network exist, and are often employed to bridge the gap between the restrictions of a manufacturer's proprietary network and the cost and overheads of a MAP or MAP-derivative. One such family uses networked intelligent nodes that are connected to virtually any programmable controller or other device that has serial RS232 or current-loop links. Each node may be programmed to deal with the particular protocol and bit-rate requirements of the attached device, providing great flexibilty of application. Network-management software is tailored to each application, dealing with message transfer, error handling and node priority, etc. This type of LAN is usually very cost-effective in any situation involving different PCs and intelligent units, providing secure, high-speed data transfer within an adaptable environment.

Two examples of this type of network are the Nine-Tiles 'Multilink' and the 'Clearway Ring'. A case study of an application using such a network is given in Chapter 8.

Levels of network utilization

Initially a LAN need only be used for data collection, introducing the technology and providing information on how the process can be controlled in a more efficient manner. This can be expanded in the future to form a partial or complete distributed control system, possibly culminating in a full computer integrated manufacturing (CIM) system.

7.9 CIM – FULL INTEGRATION

CIM involves the integration of all the activities of a business enterprise – direct production control, operational planning functions, management functions, design, marketing, etc. Fig. 7.21 shows the ESPRIT model of such a system, from Yeomans *et al.*, (1985)

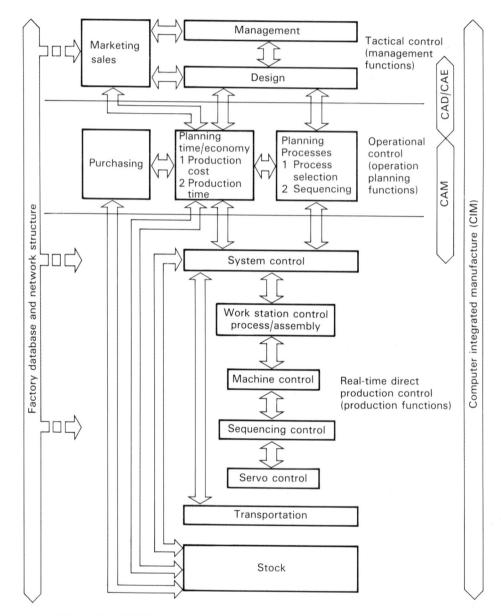

Fig. 7.21   ESPRIT CIM model, showing integration of all constituent areas.

CIM and management information

Computer integrated manufacture (CIM) is the highest level of automation, and is likely to affect every department within a business. CIM includes:

- computer control
- CAD
- management and planning functions
- stock control
- data handling on a plant-wide basis.

CIM is created using a plant-wide communications network that links these consistuent areas, allowing information to be passed to and from all relevant sub-units. The ESPRIT model of an integrated manufacturing system shows the linking element described as a 'factory database and network structure'.

This allows supervisory and management levels to examine shop-floor production data, status, etc., purely by accessing the database. Management information is presented using various handling packages, including spreadsheets, graphical analysis and database management systems. The use of these 'decision support' aids allows data to be condensed, summarized and presented in a form that promotes the recognition of production trends, areas for concern/action, etc., as well as simple report generation. The integration of data into a factory database from all areas of a manufacturing plant allows production and resource planning based on current levels of demand, sales, stock and work in progress. Fig. 7.22 illustrates the type of options available to a supervisory/management-level station.

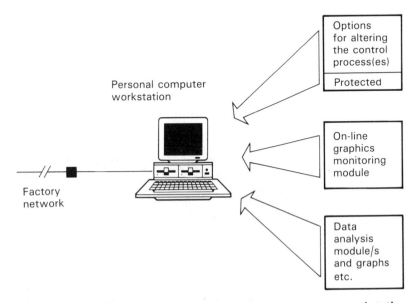

Fig. 7.22 Typical options available at a CIM management workstation.

The role of programmable controllers in an integrated manufacturing system is threefold:

1. (obviously) to control a process or item(s) of plant via a resident control program;
2. to allow the selection of control options for the process, by way of
    (a)   selection signals received from an authorized station on the network,

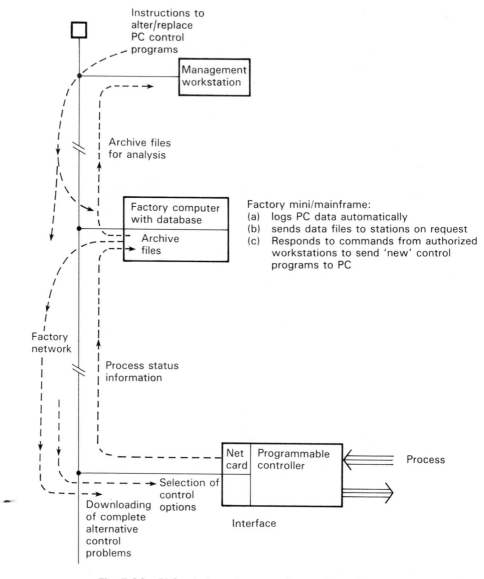

Fig. 7.23   PLC role in an integrated manufacturing system.

(b) downloading complete control programs from the network (stored in a computer system higher in the control/CIM hierarchy);

3. Transmitting process/plant status information to archive databases via the communications network (later utilized in plant analysis, etc).

Fig. 7.23 summarizes these operations.

It should be realized that items 2 and 3 above can also be implemented using straightforward serial communications or proprietary LAN systems. See Chapter 8 for an example case study of small-scale networking of PLCs for data collection.

7.10 SUMMARY

This chapter has briefly examined the communications facilities available on programmable controllers, together with their implications for industrial automation and communications between all intelligent machines on the factory floor. Programmable controllers constitute an important element on the route towards full automation, and their ability to communicate over proprietary and open networking systems is fundamental to the achievement of this goal.

7.11 EXERCISES

7.1 Explain why communications facilities are being provided on virtually all new programmable controllers.

7.2 List six uses of communications ports on a PLC.

7.3 Describe the differences between serial and parallel data transmission.

7.4 What is RS232 communication? State four of the data rates that are supported by RS232.

7.5 Compare RS232, RS422 and 20 mA current-loop communications systems in terms of transmission distance and data rate.

7.6 What is meant by communications protocol? Give an example to support your answer.

7.7 Describe how communications modules may be programmed to pass data using ladder instructions.

7.8 What benefits are achieved by using a local area network (LAN)? How do LANs contribute to the development of hierarchical, distributed control systems?

7.9 Explain the need for international standards in network communications. What part do the IEEE and ISO organisations play in this scenario?

7.10 State the names of proprietary data networks for Allen-Bradley, Gould and DEC. What are the advantages and disadvantages of this type of network when linking machines together?

7.11 What is MAP? Describe how the concept of MAP was initiated, and the facilities it offers for factory-floor automation.

7.12 What is computer integrated manufacture and what part do programmable controllers play in achieving it?

## EIGHT

Applications of Programmable Controllers

This chapter comprises several case studies and examples of the application of PLCs, highlighting various features such as special functions, communications and advanced manufacturing systems.

8.1 APPLICATIONS TO ROBOTICS

Programmable controllers are mainly used for two specific tasks in robotics:

1. as the 'controller' or reprogrammable part of a robot;
2. as an overall 'system controller' (executive) in a manufacturing cell containing one or more robots.

The above applications will now be investigated.

The PLC as a robot controller

Today, the most common robot controller is a microcomputer. In fact, microcomputers are the 'logical' choice for controlling continuous-path and language-driven robots, where large amounts of memory are required for storage of path information or control programs respectively. In these applications the 'conventional' microcomputer is ideal for processing data in complex control programs. Often, 16-bit microcomputers are necessary to provide adequate processing speed for control of industrial actuators.

Programmable controllers are designed to process larger numbers of input/output signals, often performing only simple logic functions on these inputs. This makes small PLCs a reasonable choice for (a) controlling limited sequence/point-to-point robots, and also (b) for providing interlock control between a robot or process and associated sensors.

Limited sequence control

In this type of application the programmable controller replaces the relays and

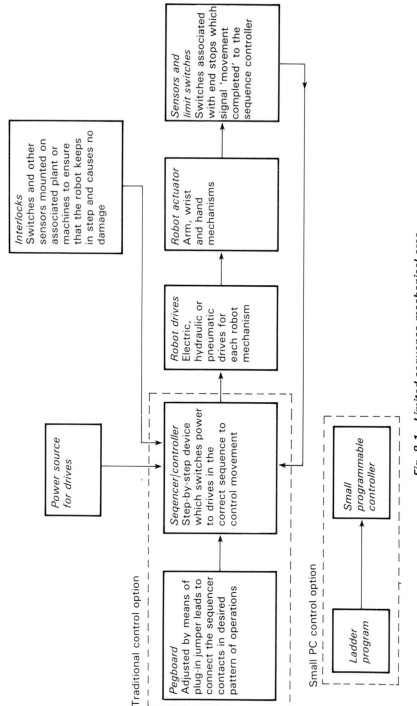

Fig. 8.1 *Limited sequence mechanical area.*

electromechanical stepping switches normally used to control the actuator's movements in response to limit switches that detect and signal when the actuator has reached a specific point in its sequence. Ladder contacts (X) are used to represent limit switches, with output solenoids driven directly from the controller output coils (Y) (Fig. 8.1).

The example given in Chapter 5 illustrates a typical ladder logic solution for controlling a limited sequence robot arm, using a small PLC.

Flexibility

The small PLC used in the above example is capable of executing programs up to 320 steps in length, which if necessary can be amended quickly *in situ* to alter the robot arm sequence.

Interlock and sequence control

Robots are often required to load and unload articles to and from other machines, for example conveyors and CNC tools. It is a straightforward matter to link all the constituent machines together using a PLC, allowing one machine to signal the next when a certain action is complete (or about to start).

Example A conveyor system carries pistons towards a pick-and-place robot. When a piston reaches a certain position, sensors signal this fact to a PLC, which tells the conveyor to stop and the robot to pick up the item (once the conveyor has stopped – interlock) and place it in a certain bin.

It is equally straightforward to design the PLC program so that the sequence can be varied automatically, depending on the information received from associated equipment. Fig. 8.2 illustrates how this might be achieved. Thus, if sensors (or a machine loading the conveyor) transmitted signals to indicate that (say) a larger piston was now on the conveyor, the PLC/robot could react by 'enabling' a different

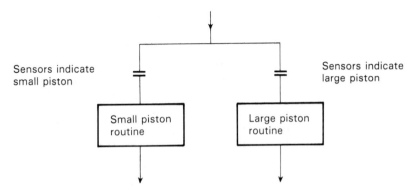

Fig. 8.2 Sequence selection in a PLC program.

section of program to deal with this larger item. For example, a wider gripper opening, and then placing the item in a different bin.

If we require information dealing with, for example, the number of cycles performed, or the number of large and small pistons moved, the programmable controller can easily be configured to output pulses at the end of each batch or cycle. These signals would normally be routed to a computer or PLC that was supervising the work cell.

8.2 FLEXIBLE MANUFACTURING SYSTEMS (FMS)

The logical development from linking machines in this manner is to group programmable machines into 'flexible manufacturing cells', each capable of machining a variety of products under fully automatic control. A typical cell (Fig. 8.3) might contain:

- 2 CNC machine tools e.g., lathe and mill, with tool racks
- 2 robots
- Conveyor system
- Host microcomputer and PLC

All tooling on the machines can be changed automatically, and programs for the CNC machines and robots downloaded from the host computer whenever a product change is signaled. A programmable controller is used to provide overall synchronization between all the machines. In addition, a larger PLC with remote I/O links is often used to supervise and control a group of cells within a distributed, hierarchical control structure as described in Chapter 7.

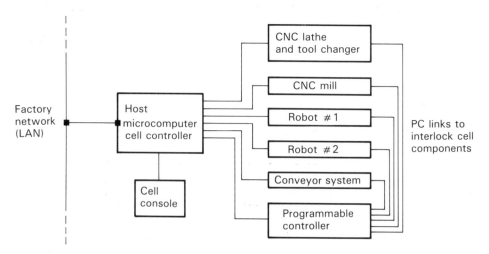

Fig. 8.3 FMS cell.

The FMS approach is making substantial improvements to levels of productivity within the factory system.

Application of programmable controllers to factory automation: the Austin-Rover car-body assembly plant at Cowley

This section describes a working installation of PLCs, robots and advanced manufacturing techniques. See Hill & Jones, (1985).

Process overview

The plant produces multi-style car body shells (Maestro/Montego) from individual body panels. The process consists of the following activities:

1. make up sub-assemblies from panels – e.g. doors, underframe;
2. tag sub-assemblies together (to permit handling by machine);
3. pass tagged bodies to a main 'jig' for automatic alignment and framing, then spot weld (done by several workstations);
4. conduct material transfer, in which sub-assemblies are selected, transported and delivered to workstations by conveyor systems;
5. maintain quality control by automatic monitoring and manual inspection of each process.

A distributed, hierarchical control system is used, as shown previously in Fig. 7.13(a) and is now shown as a 'pyramid' structure in Fig. 8.4.

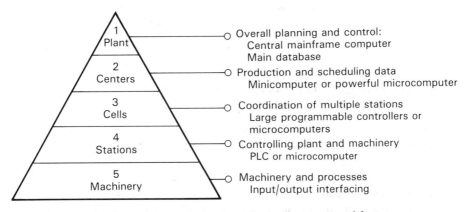

Fig. 8.4   Pyramid control hierarchy in an automated factory.

Structure

Level 1 is the mainframe computer, handling the main database for production scheduling, sales and other management needs.

Level 2 is the body-shop supervisor, a DEC PDP-11 minicomputer with a duplicate backup system. This supervisor is instructed from level 1 as regards production requirements, and monitors/coordinates all programmable controllers at level 3 to achieve the required body assemblies. A local area network (LAN) provides communications between levels 2 and 3, allowing data to pass up or down the hierarchy, as well as from PLC to PLC.

Level 3 Each PLC is totally responsible for the monitoring and control of its particular plant and equipment. Seventy-six Gould Modicon programmable controllers of various sizes are installed in the area, connecting to 50 000 I/O points using remote and local I/O units. The PLCs vary in size from small units with 40 I/O, up to large multiprocessor PLCs with rack-mounted function cards.

Level 4 The individual process units include robots, conveyors, transfer machines (such as overhead conveyors) and automatic welders. In this area 68 robots are used for welding applications, with a further four used for sealant application (Table 8.1).

The ASEA robots use continuous-path control for seamwelding, compared to the spotweld robots, which are point-to-point. The robots and automatic welders together carry out 60% of total welds.

Table 8.1.

Robot	Number	Job
Unimate 4000/2000	36	Spotwelding
KUKA (floorbased)	20	Spotwelding
KUKA (overhead)	4	Spotwelding
ASEA	8	Seamwelding (CO_2)
Unimate Puma	4	Application of sealant

Control of robots and material transfer devices

Within a particular production area, a Gould programmable controller supervises all robots and controls the associated transfer devices, together with safety interlocks. The PLC tracks each component as it moves through the production area, communicating this information to each appropriate robot as necessary. Data sent between PLC and robot includes handshaking signals to indicate robot busy, parked, action complete, etc. Data in binary-coded-decimal (BCD) form is used to send component information and weld sequences from the PLC to the robot, which must acknowledge receipt of the correct data before the PLC will allow it to commence operation.

The application of programmable controllers and robots by Austin Rover has obtained the following benefits:

1. consistent, high-quality production;

2.   increased manufacturing efficiency;
3.   flexible manufacturing capability;
4.   increased share of the market;
5.   profitable operations.

8.3   KOBRA PRESS UNLOADER WITH GENERAL ELECTRIC SERIES 6 CONTROLLER

(Presented here with the permission of General Electric)

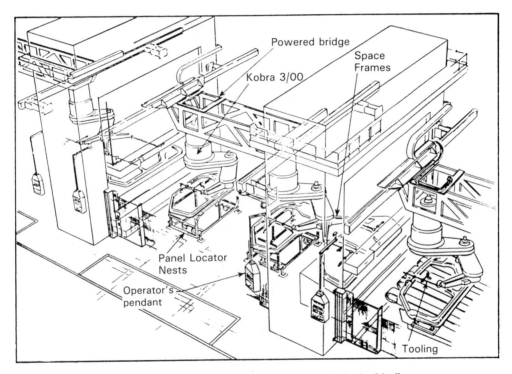

Fig. 8.5   Part of KOBRA Automatic Bodyside line.

Sahlin Automation Ltd have been making press shop equipment in the USA since 1947 and in the UK since 1958. By the late 1970s Sahlin realized their old products were looking dated with the car industry's demand for hi-tech automation and robots.

Many of Sahlin's products could and have been updated by the use of digital readouts, thyristor d.c. drives and programmable controllers, etc., but one of their products, the iron hand (a press unloader) was not such a simple matter to update to high technology.

Requirements

The press shop industry demanded a pick-and-place mechanism faster than any other industry, more flexible in operating envelope and with a capability of handling a cantilevered payload which would fatigue most existing robots in a few hours.

Many previous machines to fulfill this specification had a horizontal axis on a carriage-and-rail principle, but to achieve the maintenance-free speeds necessary, Sahlin's first parameter was that the machine must be a link-arm construction.

The simplicity of the KOBRA design is that the straight-line horizontal carriage-less axis generated by three pivots is driven by only one motor whereas other robots would have three computerized drive motors to achieve the same configuration.

The resultant KOBRA mechanism provides a fast effortless travel from end to end, achieving a peak velocity of 225 ins/s (325 metres/min).

Control system

Having solved the basic mechanical engineering design, Sahlin then needed a control system to harness the power and speed of the KOBRA. Their knowledge of the controls market led to their insistence on a system that would provide:

1. acceptance in world markets;
2. flexibility to different press lines and components;
3. accurate high-speed axis control,

and yet be readily understood by maintenance personnel.

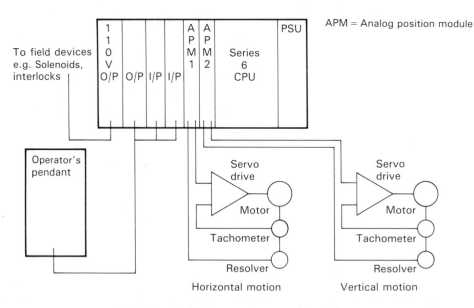

Fig. 8.6 KOBRA control system block diagram.

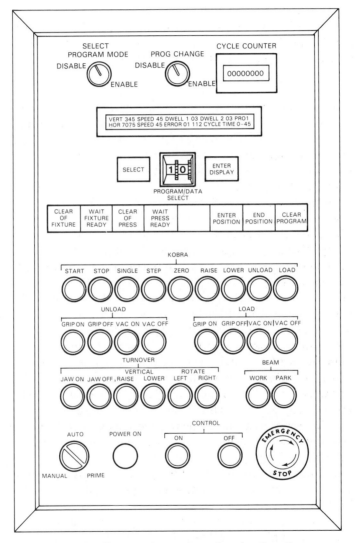

Fig. 8.7 Operator's pendant showing functions.

Sahlin decided that the answer to these requirements was a programmable controller with integrated axis control (Fig. 8.6).

A General Electric Series 6 PLC was put forward by Amptronik Ltd, a West Midlands (UK) systems house, who have been applying General Electric PCs since 1978.

A Model 60 4K CPU with extended instruction set was used to provide the heart of the system. The Model 60 CPU has six I/O slots in its 19 inch frame and into these were fitted two axis-positioning modules (APMs), two 5–50 V d.c. HD (32-way) input modules, a 10–50 V d.c. (32-way) output module and an 8-way 110 V a.c. output module. This provides a very compact solution to the KOBRA control.

The axis-positioning module is an intelligent, programmable, single-axis positioning controller that provides real-time interface between the Series 6 PC and a servo-controlled axis. The APM I module uses resolver feedback to provide high resolution and a high slew rate along with high noise immunity. The module has over 80 instructions and programmable parameters and over 60 diagnostic error codes and status points reported. It communicates to the PC via the CPU backplane and with the high scan rate of the Series 6, this ensures a high-speed interface. The APMs feed into thyristor d.c. drive amplifiers to power the motors.

To set the KOBRA to work is a relatively simple task requiring the minimum of training. There is no requirement, as with other systems, for the complexity of a CRT and alphanumeric keyboard. All operations are performed via the operator's control pendant (Fig. 8.7).

The software for the KOBRA provides the operator with teach-and-learn profile programming, the 1024 16-bit data storage registers enabling up to 12 such profile programs to be taught and stored. In some applications it has also been possible to add software to control d.c. motor-driven inter-press orientation units, all within the 4K of user logic. The operator, whilst teaching a profile, enters in the points at which grippers and vacuum cups operate, when press interlocks come into play and when turnover units operate. He can also enter in dwell times in each press and a percentage feed rate. This feed rate and the dwell times can be changed when running. The Series 6 calculates the relative velocities of the two axes to achieve the required profile and sends them to the axis-positioning modules. The operator can check the file by jogging through, and change any point in the program.

For future installations the family nature of the General Electric Series 6 enables operator interface via intelligent I/O modules with BASIC compilers and NEMA 4 CRT terminals.

Also of prime importance to the automotive industry is communications and here the GE commitment to MAP is paramount. General Electric set up a joint venture company with Ungermann Bass. The resultant GE net factory local area network (LAN) is a fully MAP-compliant 10 Mbit/s broadband token bus network.

Bus Interface Units (BIU) for the General Electric Series 6, 3 and 1 PCs and for the Mark Century 2000 mean complete communications between all GE automation control products and any MAP-compliant device.

At present there are press lines fitted with KOBRA robots in the Ford Dagenham and General Motors, Ellesmere Port Press Shops (Fig. 8.8).

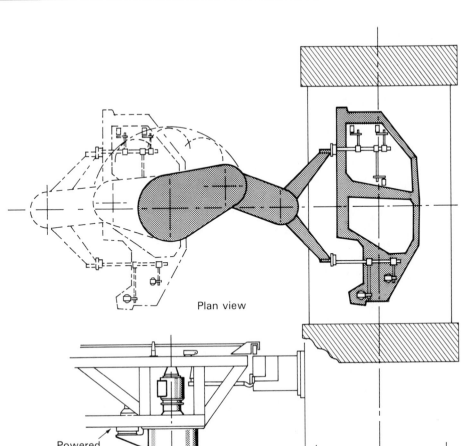

Plan view

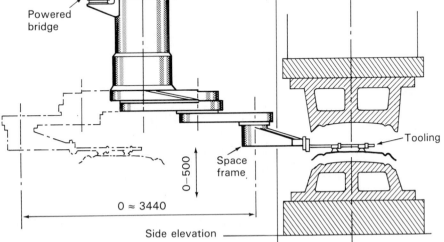

Powered
bridge

Tooling

Space
frame

0 – 500

0 ≈ 3440

Side elevation

Fig. 8.8   KOBRA 3/00.

8.4 PROGRAMMABLE CONTROLLERS IN PROCESS CONTROL

Chemical plating of alloy components (Courtesy of ISL Ltd)

A large manufacturing company was using an electroplating line to treat a range of high-precision alloy components requiring different treatment cycles. Components were usually presented in small batch quantities, with a high individual component value. For example, 50 items each of value $70 000. The existing process line consisted of 24 tanks, containing various chemical solutions for etching, coating or washing treatments. An overhead transport system was used to move components from tank to tank, lifting and lowering a loaded 'flight bar' under manual control (See Fig. 8.9).

Flight bars are loaded with the part for treatment, and each bar/part is moved through the necessary bath sequence. There could be several flight bars in the system at one time. This line was manually controlled, with the operator having to carry out several tasks:

- move flight bars from tank to tank as required for a loaded component;
- leave components in each tank for a particular length of time – often a critical period for correct treatment of the part;
- correctly track and process more than one part at any one time.

Typically, a sequence involves the raising of a loaded flight bar; lowering into tank A for n seconds; transfer to tank C for m seconds at a control current of 10 A; move to a wash tank B, etc. An additional flight bar was to be interleaved with the sequence of the first where possible. The combination of these tasks meant that incorrect cycling and timing of different parts could and did occur.

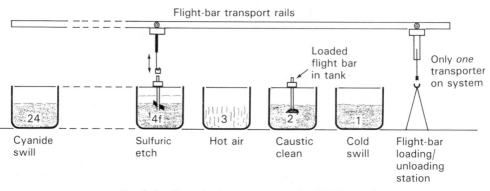

Fig. 8.9 Chemical treatment tanks (24 in total).

Automation

The company required the line to be refurbished and automated, to provide increased productivity and component quality. The specification included requirements to:

- allow up to 12 different cycles;
- process 3 or more parts concurrently;
- provide a comprehensive operator interface and manual switch mode;
- allow simple inclusion of additional cycles at a future date.

The specification contained details of all required part cycles and tank conditions, etc., and by its nature required a sophisticated software solution running on a large programmable controller.

Problem solution: GE Series 6 and Series 1 PLCs running custom control software, developed and installed by ISL

The main Series 6 PLC was chosen to provide the necessary control facilities and operator interface, together with more than adequate input/output capacity. This was specified with both digital and analog I/O cards. Additionally, a Series 1 GE controller was used to control the transport system, with a communications link to the Series 6 machine for overall control. Fig. 2.18(b) and (c) illustrate the GE Series 6 range of PLCs and the GE CIMSTAR.

The solution to this application is unusual in this type of industry, in its ability to handle several different cycles asynchronously – most plating lines do not have this facility. (This requirement also required an experienced and very able software design team.) Problems that had to be overcome included the very different cycle timings and tank sequences, plus the need to avoid 'log jams' when a target tank is occupied with another bar. (This problem would obviously be more likely as more components were entered into the system.)

Part of the PLC system specification for this application is given in Fig. 8.10. This is typical of the documentation detail that would be offered on tender for a particular job.

Software development

The overall control task was broken down into constituent sections (device handlers) that could be individually designed and developed, for later linking into a complete software solution. Modules include:

1. device handlers for each tank, containing timers and flags to indicate loaded or unloaded status. Certain tanks required a specific open-circuit voltage to be set before a part was immersed, or a specific current to be passed through the liquid.
2. device handlers for each flight bar (i.e. process sequence for a loaded component).
3. a transporter handler, providing speed control ramping and positioning at any station in the line. This module had also to store requests for service from any flight bar handler module.

Inspection Etch Plant

2.3

⋮

2.4   An interlock will be fitted to prevent lowering into an occupied station. If this should be attempted in auto mode, an alarm will be raised. In manual mode the operation will simply be disabled without raising an alarm.

2.5   Brackets will be fitted to the transporter to allow it to locate and lift the flight bars in exactly the same manner as at present. The design of the transporter will be such that no oil will be able to drip into the vats.

3.0 Control System

3.1   The control of the system will be based on the General Electric Series 6 Programmable Logic Controller (PLC). All inputs and outputs will be represented by LEDs. This equipment will be 19 inch rack mounting and will be mounted inside the main control enclosure behind a glazed door.

3.2   The system will have two modes of operation, viz.: manual and auto. This will be selectable by a rotary selector on the front panel. An illuminated indicator will confirm the system's status (i.e. auto or manual) at any time. It will be a principal feature of the design that the manual facility will be separately wired and will not operate through the PLC. This will ensure that, in the unlikely event of a failure of the PLC, movement of the transporter can still be manually controlled with the PLC disabled or even totally removed.

3.3   Similarly, the emergency stop will not be controlled through the PLC. Its operation will directly inhibit all rectifiers and motor starters and will be a normally energized fail-safe system.

3.4   The transporter position will be detected by inductive proximity switches located on the transporter. These will sense binary-coded flags placed alongside the vats. This will be an absolute system and will allow the transporter to know each station number even after a power failure. Two extra proximity switches will be located to identify the approach to a station and hence allow the transporter to decelerate.

3.5   In manual control, the transporter's traverse and hoist movements will be controlled from either the local pendant or from the buttons on the main control panel. In either case all manual traverse movements of the transporter, once initiated, will continue to the next vat, where the transporter will decelerate and stop on station.

3.6   If the system is running in auto mode, it may be switched to manual by the selector switch. Manual control will then be made available at the pendant and control panel following the completion of the automatic movement currently being executed. All rectifiers will be switched off but the elapsed time in the vats will still be monitored. Having returned the transporter to the position that it last occupied in auto mode, the operator may reselect auto and the system will continue from where it stopped. If the transporter has not been returned to its correct restart position, attempting to reselect auto will cause an alarm. Clearly some flight bar times in vats may be exceeded if excess time is spent in manual but this will become the responsibility of the operator.

3.7   In auto mode, the operator will select the desired program number by thumbwheel. On pressing the 'Sequence Start' button, which will then be illuminated to acknowledge that it has been pressed, the transporter will come and remove the flight bar situated at the load station and progress it through the required sequence of vats. This request will be ignored if no flight bar is detected at the load station.

3.8   On lowering of a flight bar into an electroplating vat, the system will switch on the

Fig. 8.10 Sample sections of PLC specification.

rectifier and regulate the voltage or current as required. The system will ensure that the rectifiers are switched off before a flight bar is inserted into or removed from a vat.

3.9 From an inspection of the six plating sequences we have determined that it is necessary to provide control of plating parameters as follows:

85	Caustic/cyanide clean	Voltage
86	Caustic/cyanide clean	Voltage
88	Sulfuric etch	Voltage and current
89	Sulfuric etch	Voltage and current
91	Blue etch anodize	Voltage

These will occupy 7 channels of an 8-channel analog input module on the PLC. Each of these channels will have an input range of ± 10 V.

Since the plant signals vary from 0–75 mV at the smallest to 30 V, at the largest, it will be necessary to condition these signals. We shall supply a signal-conditioning amplifier for each channel which will condition each signal to the ± 10 V range. These amplifiers will be modular and will be fitted into their own 19 inch rack situated below the PLC racks.

These will be isolating amplifiers with preset gains and offsets. This will allow accurate calibration of the signals to the required span. The isolation feature of these amplifiers is necessary to ensure that common-mode voltages will not introduce measurement errors into the system. This will also ensure that if there is any accidental connection of a mains voltage to the signal inputs, only one module will be damaged and the PLC equipment will be protected.

Please note that no instrumentation has been included to measure the current flowing in vats 85, 86 and 91.

3.10 The auto system will cater for any random combination of programs which may be selected by the operator. The movements of the transporter will be determined by the schedule of priorities as defined in your specification.

3.11 All alarms will cause an audible annunciator to sound. The appropriate fault lamp will also flash. This annunciator will be muted and the lamp will be continuously lit when the 'Accept' button is pressed. A 'Cancel' button will clear the fault lamp. Any attempts to cancel faults which have not been rectified will restart the alarm sequence.

3.12 Six digital readouts of two digits each will be provided to indicate information on the status of flight bars within their program sequences. This will require further discussions to determine what information can most usefully be displayed.

3.13 Panel-mounted LEDs will indicate transporter movements, transporter position and which vats are occupied. N.B. Since the vat occupancy is based on the PLC remembering that it has loaded or unloaded each vat and no instrumentation exists to indicate otherwise, these latter facilities cannot be provided in manual mode.

3.14 We have noted that the time spent in the hydrofluoric vats appears to be a process variable which is likely to be adjusted by the operator to suit differing batches of components. We have, therefore, included a set of thumbwheels for each of programs 2–6 inclusive. These will indicate to the system the time required in this vat when that program is executed.

4.0 Main Control Panel

4.1 The main control panel will be a free-standing robust steel enclosure constructed in high-quality zintec sheet affording damp and dust protection to IP54. Paint finish inside will be white enamel and on the outside will be orange to BS 06E51. The panel will have rear access doors with a fixed front plate on which all instrumentation will be mounted.

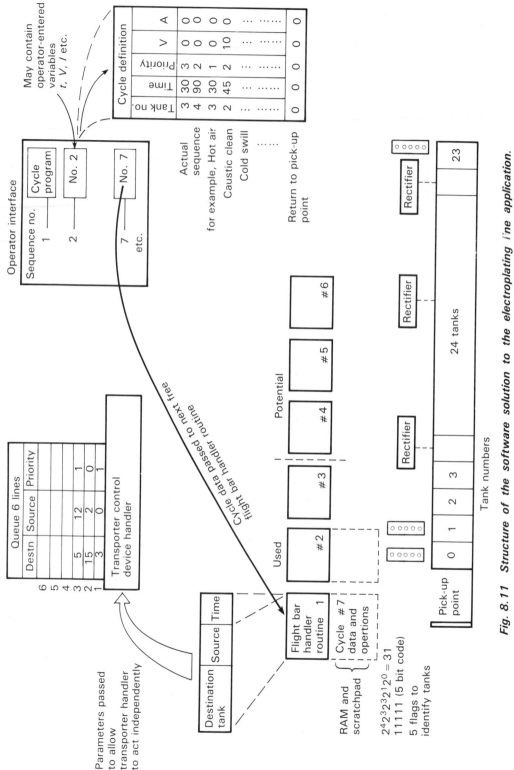

Fig. 8.11 Structure of the software solution to the electroplating line application.

```
07/01/87     19:27:11                 GE SERIES SIX DOCUMENTATION
                           Plating Line Automatic Control Ladde. Listing
                                  Instrumentation Systems Limited

 << RUNG    74 >>

   $$$$$$$$$$$$$$$$$$$$$$$$$$$$$$$$$$$$$$$$$$$$$$$$$$$$$$$$$$$$$$$$$$$$$$$$$$$
   This timer is used to inhibit the hoist untill the transporter has been
   stopped at the station for the time set which is 2 seconds.
   $$$$$$$$$$$$$$$$$$$$$$$$$$$$$$$$$$$$$$$$$$$$$$$$$$$$$$$$$$$$$$$$$$$$$$$$$$$

  On                                                              On
Station                                                         Station
 Prox.                                                          Timer.      Transporter is stopped on station and has
                                                                            been so for at least 2 seconds.  This time
 I0038                                                  Const   00800       will be modified on commisioning the line
--] [---------------------------------------------------[PRESC]---(TS)      to allow the flightbar to become stable
                                                          002     (  )      before a raise or lower command is issued.
  On                                                              (  )
Station                                                           (  )
 Prox.                                                            (  )
 I0038                                                 R00400     (  )
--]/[-----------+---------------------------------------[ACCRG]---( R)
                |
     Trans.     |
      Dest.     |
     Loaded     |
 00005   00425  |
--] [------] [--+

 << RUNG    75 >>

   $$$$$$$$$$$$$$$$$$$$$$$$$$$$$$$$$$$$$$$$$$$$$$$$$$$$$$$$$$$$$$$$$$$$$$$$$$$$
   The next twenety seven rungs are used to decode the station proximity
   switches and set a flag to indicate the station number that the
   transporter is presently at.
   $$$$$$$$$$$$$$$$$$$$$$$$$$$$$$$$$$$$$$$$$$$$$$$$$$$$$$$$$$$$$$$$$$$$$$$$$$$

 Prox.   Prox.   Prox.   Prox.   Prox.                                    At
 2^4.    2^3.    2^2.    2^1.    2^0.                                    Load
                                                                        Station
 I0037   I0036   I0035   I0034   I0033                                   00257
--]/[-----]/[-----]/[-----]/[-----] [-------------------------------------(  )

 << RUNG    76 >>
 Prox.   Prox.   Prox.   Prox.   Prox.                                    At
 2^4.    2^3.    2^2.    2^1.    2^0.                                   Station
                                                                          F95
 I0037   I0036   I0035   I0034   I0033                                   00258
--]/[-----]/[-----]/[-----] [-----]/[-------------------------------------(  )
```

```
07/01/87     19:29:09                 GE SERIES SIX DOCUMENTATION
                           Plating Line Automatic Control Ladder Listing
                                  Instrumentation Systems Limited

 << RUNG   129 >>

   The next         rungs of logic are used to control the transporter
   movement.  The first      rungs read in the station number as the
   transporter passes the flags, if the destination station is the next one
   then control is passed to the transporter PC to decelerate the
   transporter and bring it to a halt on station.

  On       Prox.     Raw
Station    2^0.    station
 Prox.             flags.
 I0038    I0033    R0302
--] [----[ I/O   TO REG ]-                                         (  )

 << RUNG   130 >>
  Raw     Station  Present
station    flag    Station
flags.     mask.
 R00302   R00303   R00300    Const
-[  A   AND  B   =    C       LEN ]-                                (  )
                              001
 << RUNG   131 >>
$$$$$$$$$$$$$$$$$$$$$$$$$$$$$$$$$$$$$$$$$$$$$$$$$$$$$$$$$$$$$$$$$$$$$$$$$$$$$$$$
$                                                                              $
$ NB                                                                           $
$ The next four rungs are only applicable when in computer manual mode.        $
$                                                                              $
$ $$$$$$$$$$$$$$$$$$$$$$$$$$$$$$$$$$$$$$$$$$$$$$$$$$$$$$$$$$$$$$$$$$$$$$$$$$$ $
$ $$$$$$$$$$$$$$$$$$$$$$$$$$$$$$$$$$$$$$$$$$$$$$$$$$$$$$$$$$$$$$$$$$$$$$$$$$$ $
$                                                                              $
$  This rung is used to cause the transporter to decelerate and stop at the    $
$ load station even if the operator does not remove his finger from the        $
$ button.                                                                      $
$                                                                              $
$$$$$$$$$$$$$$$$$$$$$$$$$$$$$$$$$$$$$$$$$$$$$$$$$$$$$$$$$$$$$$$$$$$$$$$$$$$$$$$$
Present   Test    Moving                                             At 2
Station   Reg =    Right                                            Let go
           2 ?                                                      Button
 R0300    R0803   01017                                              00380
-[  A   :    B ]---] [-------------------------------------------------(  )
```

```
07/01/87     19:30:17                 GE SERIES SIX DOCUMENTATION
                           Plating Line Automatic Control Ladder Listing
                                  Instrumentation Systems Limited

 << RUNG   132 >>
$$$$$$$$$$$$$$$$$$$$$$$$$$$$$$$$$$$$$$$$$$$$$$$$$$$$$$$$$$$$$$$$$$$$$$$$$$$$$$$$
$                                                                              $
$ This rung is used to cause the transporter to decelerate and stop at         $
$ station twenty seven even if the operator does not remove his finger         $
$ from the traverse button.                                                    $
$                                                                              $
$ It has been modified to suit the constraints of the line at present, see     $
$ the explanation at rung 134 for the changes required to restore the          $
$ availability of the full line.                                               $
$                                                                              $
$$$$$$$$$$$$$$$$$$$$$$$$$$$$$$$$$$$$$$$$$$$$$$$$$$$$$$$$$$$$$$$$$$$$$$$$$$$$$$$$
Present   Test    Moveing
Station   Reg =    Left.
           23 ?
 R0300    R0824   01016                                             A00015
-[  A   :    B ]---] [-------------------------------------------------(  )
```

Fig. 8.12 Example program sections from ISL chemical plating control system.

The 'log jam' problem mentioned above was dealt with by assigning a priority to each tank session. Level 1 priority indicated an exact dip-time required. Level 2 service requests would be taken if no level 1 requests were current, and level 3 priority requests were noncritical, being serviced after any level 2 calls. In operation, this was realized by the controlling software causing the transporter to remain with a level 1 priority part until it entered a lower-priority section of the cycle. If no level 1 operations are present, the transporter then deals with the next level 2 (or level 3) demand in the queue, and is allowed to move between such parts to service other calls. The final software solution potentially allows up to six flight bars to be present in the system at any one time, although in practice the demand has only been for three bars concurrently.

Fig. 8.11 provides a conceptual model of the overall software solution, showing each module and the links to other modules in the system.

The control software was developed in ladder logic for a GE Series 6 PC, with an operator interface written in BASIC for the graphics and process mimics running on a special function card and monitor. This interface allows operators to alter several process variables by key control.

The priority system mentioned above requires the software to store a priority code for each product on particular tank dips. Referring to Fig. 8.11, we can see a priority column in the 'cycle definition' block at the right of the illustration. This data block contains all relevant information on the sequence a part is to follow, including tank numbers, immersion time and/or electrical settings. From this file, data is passed to the flight-bar handler routine which then determines source and destination tanks, together with immersion time. The flight bar handler routine passes parameters to the transporter control device handler (running on a Series 1 PLC), which can then act independently of the main PLC. A request queuing facility is provided in the Series 1, allowing up to six lines of data as shown. This queue also supports priority labeling, with a logic 1 signifying a level 1 part and 0 for level 2 or 3 parts. This ensures priority 1 requests are serviced before the others, which are treated as non-urgent. The transporter will leave a priority 0 item to deal with a more important request if one occurs.

Example sections of ladder programming from this application are given in Fig. 8.12 and in Appendix C.2. Note the use of labels and text comments to aid program readability.

8.5 CASE STUDY OF SMALL-SCALE NETWORKING OF PLCS FOR DATA COLLECTION – BP CHEMICALS LTD HIGH DENSITY POLYETHYLENE PLANT, GRANGEMOUTH

The background to this case study is given in Fig. 8.13. See also Fig. 8.14.

BP Chemicals, the largest European producer of polyethylene, have a high-density polyethylene line which was installed at their Grangemouth plant in 1986. The line produces polyethylene pellets, which have to be packaged into 25 kg

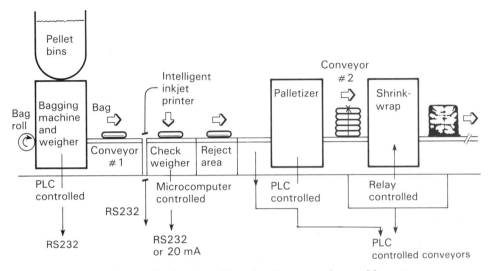

Fig. 8.13   Outline of bagging line operation and layout.

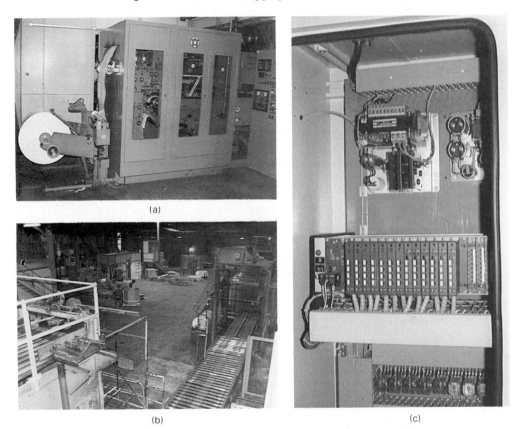

(a)

(b)   (c)

Fig. 8.14   Plant machines and controllers: (a) bagging machine; (b) BP chemicals weigher line at Grangemouth; (c) PC controlling palletizer.

bags. As part of a drive to improve customer service, BPCL commissioned a feasibility study of this line to investigate the potential for computer-based data acquisition.

The bagging line consists of pellet hoppers with flapper valves to allow pellets to drop into a bagging machine/weigher. This machine weighs 25 kg of product, then drops the batch into a bag which has been taken from a continuous roll and held open to receive the batch. The bag is next sealed and then carried by conveyor to a checkweigher and inkjet printer which prints product data on each bag.

Plant control

The line equipment is individually controlled as follows:

Bagging machine	Programmable controller A*
Inkjet printer	Microcomputer based
Checkweigher	Microcomputer based
Palletizer/conveyor 1	Programmable controller B*
Conveyor 2	Programmable controller C*
Shrink wrapper	Relays

PLCs B and C are from the same manufacturer, PLC A is from a different source.

Constraints

The data communications/acquisition systems must operate without slowing the bagging line speed, in particular the bagging machine, which produces one bag every 3 seconds.

Possible future extensions to the specifications would be to provide a certain amount of control over the line from the supervisory microcomputer, an IBM XT with 20 Mbytes of hard disk storage.

All PLCs require the fitting of specific communications cards to allow serial data transfer. The other microprocessor-based devices have hardware communications facilities, but require suitable programming of the device.

Requirement

The company wishes to collect data from all controllers in the line, in order to create a management information system based on a personal computer. The information is to be used locally for:

- production monitoring/analysis
- productivity analysis;
- fault finding;

- fault monitoring and reliability;
- packaged material stock control;
- input to factory production records systems.

This last item requires that all data files produced are of compatible format with the factory mainframe computer, and that provision be made to link the local microcomputer with the factory system. In addition, there is likely to be a future need to integrate a further two packaging lines into this system.

TASK   To identify and install a communications system capable of meeting present and likely future needs (which may include plant control from management terminals).

The plant and machines are controlled by a wide variety of programmable controllers and integral microcomputers with differing communications protocols, although all support RS232C or current-loop facilities. The personal computer (data store) is located over 150 m from the product line, in a different building.

Several options were considered.

(a) *Proprietary programmable controller network*   Only two of the five intelligent devices were from the same manufacturer, and no station was capable of controlling the communications needs of the complete system. Any solution based on a standard PC network, therefore, would additionally require a supervisory PLC. This would communicate with each plant device via network nodes.

DECISION   The need for a fairly powerful supervisory PLC made this option prohibitively expensive.

(b) *Concentrated serial links* (Fig. 8.15)   Serial channel concentration units are available, mainly supporting RS232C communications. Typically, six or eight data sources may be multiplexed onto a single channel to the destination terminal, which has to separate the data stream back into individual sections, for passing into RAM or disk storage files.

DECISION   this approach would be an adequate solution to the present requirements, but could not be easily expanded beyond the capacity of the concentration, should this be necessary. Also, the passing of control data down to shop-floor equipment may be complicated by the need to multiplex handshaking signals down the common channel, especially if current loop transmission is adopted. Finally, the tasks of the supervisory microcomputer had to include the demultiplexing of the concentrated data stream, in addition to necessary housekeeping duties such as controlling the concentrations, creating RAM caches for each logged channel, spooling channel data into disk files, and running applications packages to process these files.

These reasons meant that the 'concentrated channels' scheme was rejected in favor of the following scheme.

(c) *Self-contained local area network with intelligent nodes*   By selecting an industrial network based on intelligent nodes for handling each attached device, a straightforward economic and easily expanded system can be created (Fig. 8.16).

Optional multiplexer communications system

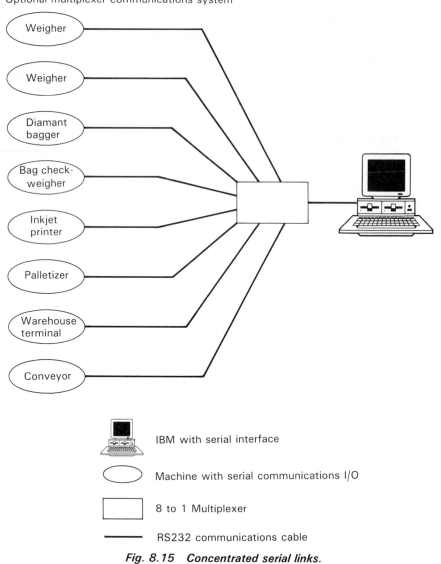

Fig. 8.15 Concentrated serial links.

A modern local area network (LAN) system offers several advantages in an application such as this, with none of the disadvantages inherent in the alternative option, a multiplexed port system. With a LAN system each machine is configured into a 'data highway' analogous to a 'ring main' electricity supply.

Once machines (nodes) have been identified by the system they can be installed anywhere in the 'ring main', which includes the computer. The cable can be extended (up to a few kilometers between machines) to incorporate more nodes or

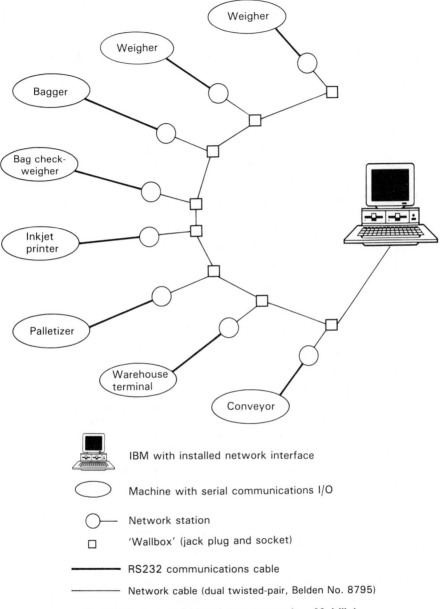

Weigher

Weigher

Bagger

Bag check-
weigher

Inkjet
printer

Palletizer

Warehouse
terminal

Conveyor

IBM with installed network interface

Machine with serial communications I/O

Network station

'Wallbox' (jack plug and socket)

RS232 communications cable

Network cable (dual twisted-pair, Belden No. 8795)

Fig. 8.16   Industrial local area network − Multilink.

resite nodes. The buffering of local machine data is supported by the specific system discussed.

Each plant device (including the supervisory PC) is to be connected to an addressable network node via short RS232/V24 cable runs. Each node is configured to meet the protocol needs of the network and its particular machine. Nodes contain a controlling microprocessor and firmware, together with RAM buffer stores, allowing temporary storage of up to 255 data bytes sent from a station prior to its transmission to the file server PC, alleviating certain timing problems. Up to 125 nodes may be connected to the network.

The LAN is based on a twisted pair, operating at 19.2 kbps for 12 simultaneous channels, over distances up to 3 km. The medium is not prone to noise-induced errors and does not require screening for use in an industrial environment. Use of this type of LAN would also reduce the software overhead on the supervisory PC in terms of channel control and protocol, most of this being carried out by the network nodes. The LAN would, additionally, be ideal for personal computer input to alter process variables or status. Software is still required to configure the network initially, together with the data/file-handling procedures mentioned earlier.

DECISION   Adopt a system as per option (c) since it provides economic, early expandable communications that can be configured for different protocol and data formats from a wide variety of plant devices.

Procedure

Plant devices

Each PLC requires a family communications module, since no serial communications facility on the basic PLCs is supplied. The installation of the cards is straightforward, but each must be programmed to configure the serial ports and data format. Depending on the manufacturer, the programming may be done via a standard terminal or personal computer, or require a specific programming panel which must be purchased or hired from the manufacturer.

All other associated intelligent devices have serial communications hardware on board, but also require programming or configuring via internal selector switches.

A 'standard' data format was chosen that fell within the capability of all devices to be networked. The Multilink network supports baud rates from 300 to 19 200, independently selectable for each station on the LAN.

Communications example   *Pallitizer/Klockner Moeller PLC – resident software.* The pallitizer is controlled by a PS22/12 PLC made by Klockner Moeller. This PLC is to be equipped to provide status and diagnostic information on the ongoing operation of the pallitizer and process. This section includes details of the necessary PLC communications card that must be fitted, and communications software to be resident in the card/PLC.

Type   Klockner Moeller PLC Communications Module – PSW 004 Serial
            converter interface, plus controlling software with above options.

Supplier   (a) PSW 004 or PSW 1 communications card: Klockner Moeller
           (b) Controlling software: software house/consultants (software to provide message-passing facilities between KM PS22/12 and the network system.)

KM module description   There are several modules available for the PS22/12 that provide serial communications facilities, as listed above. The PSW 1 is similar to the PSW card, but has an integral power supply. These modules all allow the controller to be linked into a computer-based data-gathering system using either RS 232 or 20 mA current-loop signaling, once installed and programmed to respond to selected requests for status information. The communications module may be configured to particular transmission specifications via external selection switches. Several of the available PSW responses are included in Fig. 8.17.

Programming of the communications card is carried out using either a standard VDU or a suitably equipped personal computer, which is required only during the installation and programming stages.

Command D: Total status request

Syntax: D	
Reaction: 32 hex. numbers	The PS Q-signal status of the selected address space are output as hexadecimals
Example: D	0 1 2 3 4 00 .... 00F corresponding to: Q4, Q9, Q12, Q18, Q124, Q125, Q126, Q127 high, remainder low

Command Q: Status request

Syntax: Q addr.	addr. = signal number three digit number (000–127)
Reaction: L or H	the signals status is transmitted as H (for High) or L (for Low)
Example: Q003	H (e.g. displayed on screen)

Command I: Status input

Syntax: Q addr. H/L	addr. = signal number three digit number (000–127) H/L = set signal status high/low
Reaction: protocol only (no information)	
Example: I100L	The I-signal 100 is set to low

Fig. 8.17   *Responses from Klockner Moeller PLC communications card.*

Network – Nine Tiles Multilink LAN

The network hardware was specified with nodes for all plant devices, the management personal microcomputer and a 'spare' node. The node for the microcomputer (an IBM XT) was available as a circuit board to allow internal installation. The network-operating software was EPROM based, contained in all system nodes, together with controlling microprocessors. This software allowed each node to be individually addressed for data transmission, or to request to send once its data buffer was nearly full. Since the PLCs and other plant devices operate at speeds from 3 seconds to over a minute, this flexibility was desirable.

Fig. 8.18 shows the different software modules necessary for this application. The reader should be aware of the necessity to adopt a formal software design methodology in applications such as this, where there will be considerable inter-action between software modules. Thus, from the very outset, a top-down approach to structure is used, followed by bottom-up implementation and testing of all procedures within each module. This ensures each section of program will stand alone, but can also be used with integrity by higher-level modules. The aim is to develop software that is both reliable and maintainable, and the above design route helps to achieve this.

The IBM XT was designated as the destination address for all plant data packets, which were to be separated and stored in respective disk files for future operations. To achieve this, custom software had to be designed and written for the IBM XT. In addition, the plant management wanted a data analysis package to run on the microcomputer at any time, in order to produce summary reports and

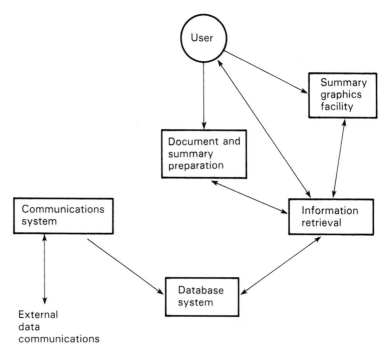

Fig. 8.18   Conceptual model of the system.

information based on the data file contents. dBaseIII by Ashton Tate was chosen, and a system designed to suit the specific requirements. (Other commercial packages such as Lotus 1-2-3 could also have been used.)

Outline descriptions of the file and menu structures are given below to indicate the relationship between the different software modules.

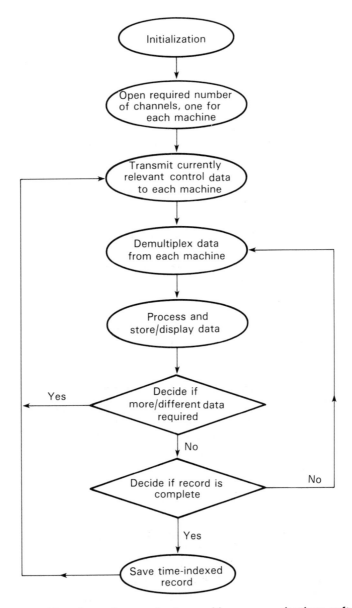

Fig. 8.19 Flowchart of computer-to-machine communications software.

The records stored by the communications system on disk contain fundamental information from the warehouse machines. If useful reports are to be generated, then it is vital that the records from which the reports are generated are sufficiently detailed and well structured.
   The record will contain information relevant to a set interval of time (say one hour or less).
   The record will be made up of a number of fields:

(a) *The time field*
This is the date and time at which the record was logged on to disk.

(b) *The duration field*
This is the time over which the record was compiled; normally this is the usual set time (1 hour), unless new batch information was entered during record compilation.

(c) *The batch field*
This field contains up to four bytes. A null specifies that no batch has ended. A 'b' specifies that the bagging line batch has ended. A 'p' specifies that the PVS line has ended. An 'n' specifies that the 'natural' Octabin batch has ended. A 'y' specifies that the 'yellow' Octabin batch has ended.
   This FLAG allows the database-management software to identify data in the various machine fields as being current, ongoing data or final batch summary data.

The machine fields
These fields are headed by a machine identity byte and are followed by information pertaining to a fixed interval of time (1 hour, say). If any machine is reconfigured for a new batch, then the whole record terminates prematurely and is immediately saved on disk.

Fig. 8.20   Record structure.

(a)   Communications software on the host microcomputer   Fig. 8.19 shows the sequence of operations carried out by a controlling microcomputer when polling data from a number of networked machines.

(b)   Datafile system description   The network communications software logs data from all connected nodes into a disk-based primary data file. This file, PRIM1, is updated by one record at hourly intervals, each record consisting of 26 data fields – a concatenated version of the data from all machines in the warehouse (see Fig. 8.20).

From PRIM1, particular data fields are read into specific line files on a daily basis (Fig. 8.21). These files BAG, OCT1 and OCT2 are built up during the day, then summarized into respective summary files at the end of a day. The summary files SBAG, SOCT1 and SOCT2 form an archive of plant data and performance on a day-to-day index. The full data contained in the specific line files is also appended to archive files, for later analysis. Once daily file data has been appended and/or summarized onto the main files, the daily files are erased to free them for use the next day. (These tasks are housekeeping jobs performed automatically by the software when necessary.)

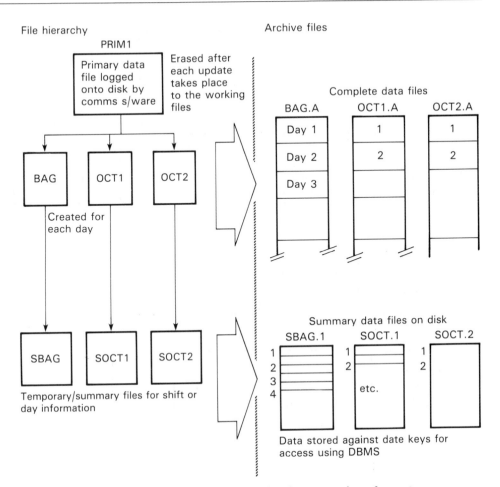

Fig. 8.21   Database operations for the generation of reports.

These files form the source for all user enquiry data on line or machine/man performance over a specified period. Following a user request for data covering a certain period, the database program PENQ.prg accesses the required data files and extracts the process data as necessary. The user is then presented with the requested data, and offered further options through the menu system, including printout, graphic representation, etc. (Fig. 8.22). Other user enquiries will be handled in a similar fashion, indicated on the menu chart by the other options alongside the expanded example for 'total warehouse' production.

In order to achieve compatibility with the factory information system, data files were compiled and created in standard data format (SDF).

It was suggested that a graphics package be integrated, allowing the presentation of 'raw' and summary information as graphs, charts, histograms, etc. as necessary to allow rapid assimilation of trends and values, as shown in Fig. 8.23. A

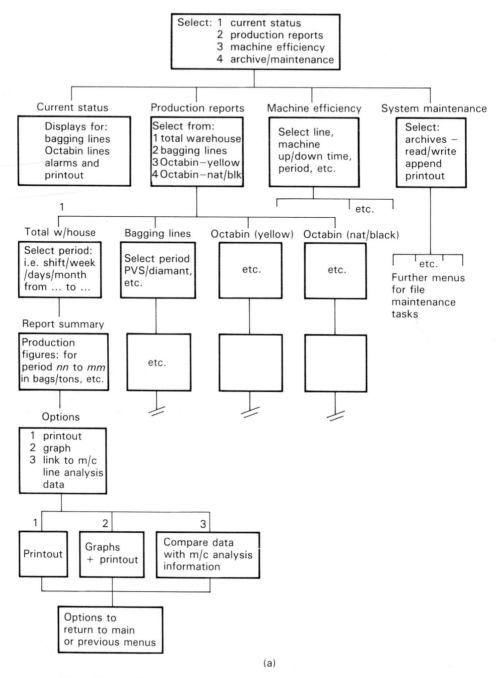

(a)

Fig. 8.22 (a) Menu structure for the overall system; (b) screen layout for main menu.

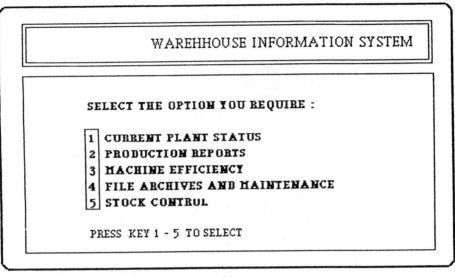

(b)

facility of this nature will lend itself to future expansion into other areas of data analysis, such as production planning and resource distribution. (e.g.: Bar charts for production/maintenance summaries; Gantt charts for machine/man usage).

Network layout

Unscreened twisted pair was specified by the network company as the ideal communications medium, running in protective conduit as a ring configuration to link all network nodes, including the IBM XT located in an adjacent building. Nodes were to be sited close to all plant devices to ensure short interconnecting RS232 cables between node and device. A further consideration when specifying this network was the availability of a 'gateway' to any future MAP systems, providing a standard route to factory-wide integration. Future inclusion of additional nodes can be carried out by simply linking them into the twisted-pair cable. A 'spare' node was available for replacement of any other node in the event of failure, to reduce system downtime.

8.6 SEQUENCE CONTROL USING LOGIC INSTRUCTIONS

EQUIPMENT   A hopper contains a mixture of pellets of two different colors. It is required to separate the hopper contents into two separate bins via a delivery chute

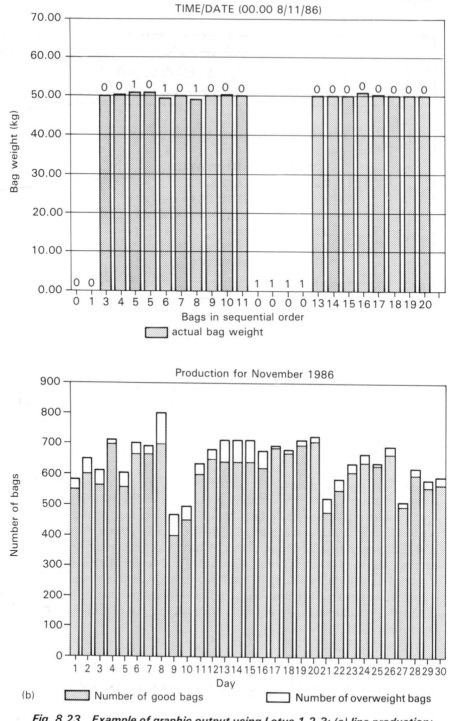

Fig. 8.23 *Example of graphic output using Lotus 1-2-3: (a) line production;*
(b) warehouse.

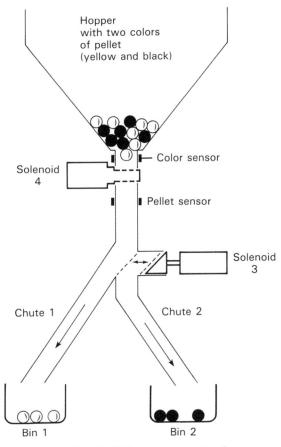

Fig. 8.24   Pallet sorter mechanism.

which is fitted with two solenoids and two sensors (see Fig. 8.24). Solenoid 4 halts pellets immediately below the hopper, allowing the color sensor to detect which type of pellet is stopped there, by measuring the level of light reflected from its surface. Further down the delivery chute is a second sensor to detect when a pellet passes by. Finally, solenoid 3 is positioned at the junction of two chutes leading to separate pellet bins. If solenoid 3 is normal (non-operated), any falling pellet will enter the chute to bin 2. If this solenoid is operated, however, it forms a bridge across chute 2 and causes pellets to be directed into chute 1. The sensors must be reset after each reading, otherwise they remain in the activated state. There is also a buzzer alarm which can be used to signal pellet blockages or an empty hopper.

TASK   Design a ladder program that will control the above devices to sort the pellets into the two bins.

PROCEDURE   Develop a sequence diagram or chart that defines all necessary steps in

Table 8.2   Desired sequence. [*]

Start	Solenoid 4	T454 Initial delay D_5	T450 Reset sensors and delay D_1	Solenoid 3	T451 Delay D_2	T452 Delay D_3	T453 Error routine and delay
INIT	Sol. 4 on	Start D_5 $\downarrow$ 10 s Stop D_5 (T454) $= T_5$	Reset Start delay 0.1 s D_1				
	Sol. 4 off ◄ ─ ─ ─ ─ ─ ─ ─		End delay ─ ─ ─ (T450) $= T_1$	on or off ─ ─ ─ ➤ depending on color input (X400)	Start delay D_2 | 0.1 s $\downarrow$ Stop		
	Sol. 4 on ─ ─ ─ ─ ─ ─ ─ ─ ─ ─ ─ ─ ─ ─ ─ ─ ─				delay ─ ─ ─ ─ ─ ➤ (T451) T_2	Start delay D_3 | 0.3 s $\downarrow$	
			1f X401 = 1 Reset └─ (loop)	Check each sensor X401		Stop delay (T452) T_3	If X401 = 0 Sound buzzer and delay
			Reset └─(loop)				$\downarrow$ 1.0 s End delay (T453) $= T_4$

[*] Easier to define entry exit/conditions than using a flowchart

the control of the machine. Table 8.2 shows one form of sequence chart. Note the entry columns for each output (solenoids and reset), together with several 'delay' elements. These delays are used to control operate/release times of the solenoids (D_2), to allow time for a pellet to fall (D_3), and to reset the sensors (D_1). The flow-chart in Fig. 8.25 also describes the sequence.

Once the desired sequence is established, one has to state the Boolean expression for each output device, including the delays. This is done in terms of inputs and outputs as described in Chapter 4, using information available from the sequence charts. Table 8.3 shows these equations for each output.

We then allocate suitable I/O points, timers, counters, etc. on the chosen programmable controller. Table 8.4 lists I/O points for a Mitsubishi A-series PLC.

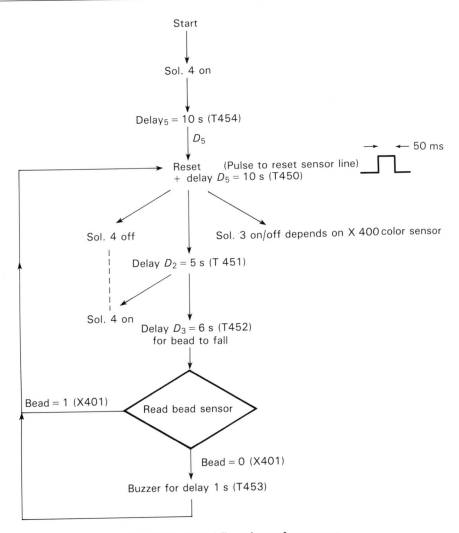

Fig. 8.25   Initial flowchart of sequence.

Following the development of the Boolean terms, it is a straightforward matter to translate these into a ladder program. Fig. 8.26 contains the ladder equivalent, with M relays used to represent the different 'states' of the sequence. It may be of benefit to compare this listing with the equations and the stated operating requirement.

In developing this PLC program it was necessary to experiment with the various delay periods, in order to determine suitable values for reliable operation of sensors and solenoids. The ability to simply alter values and contacts in the program highlights the suitability of PLCs to this type of application.

The above program is incomplete in terms of providing a usable piece of

Table 8.3 Boolean equations for sorter outputs.

	Entry	Hold	Reset

Sol. 4. $(S_4) = (D_2) \cdot T_2 \; + \; (S_4) \cdot \overline{T_1}$ $+ \; \overline{INIT}$

             $+ \; (S_4) \cdot \overline{(D_2)}$

Delay 2 $(D_2) = (R_{ST}) \cdot T_1 \; + \; (D_2) \cdot \overline{T_2}$

             $+ \; (D_2) \cdot \overline{(D_3)}$

Delay 5 $(D_5) = (S_4) \cdot \overline{INIT} \; + \; (D_5) \cdot \overline{T_5}$

             $+ \; (D_5) \cdot \overline{(R_{ST})}$

      ENTRY   +   ENTRY   +   ENTRY   +   HOLD/RESET

RESET $(R_{ST}) = (D_5) \cdot T_5 + (D_3) \cdot T_3 \cdot B + (E) \cdot T_4 \;\; + \; (R_{ST}) \cdot \overline{T_1}$

             $+ \; (R_{ST}) \cdot \overline{(D_2)}$

             (B = beam sensor)

      ENTRY   HOLD/RESET

Delay 3 $(D_3) = (D_2) \cdot T_2 \; + \; (D_3) \cdot \overline{T_3}$

             $+ \; (D_3) \cdot \overline{(R_{ST})} + (D_3) \cdot (E)$ Two exits

             $+ \; (D_3) \; (\overline{(R_{ST})} + \overline{(E)})$ (minimized)

Sol. 3 $(S_3) = (\overline{(R_{ST})} \;\;\; + (S_3)) \cdot C$ (C = color sensor)

       or $(T_1 + (S_3)) \cdot C$

Error $(E) = (D_3) \cdot T_3 \cdot \overline{B} \; + \; (E) \cdot \overline{T_4}$
(buzzer)

             $+ \; (E) \cdot \overline{(R_{ST})}$

      ENTRY   + HOLD − *no* RESET

CANCEL OF $(INIT) = (D_5) \cdot T_5 \; + \; (INIT)$ (Will lock-up after

INITIAL (D_5) and thus allow (D_3) to

ENTRY   follow (S_4) in main loop)

Table 8.4 Input/output assignments for sequencer.

Outputs	Point	Inputs	Point	Delay	Timer
Sol. 3	Y432	Color	X400	T1	T 450
Sol. 4	Y433	Bead sensor	X401	T2	T 451
Sol. reset	Y434			T3	T 452
Buzzer	Y435			T4	T 453
				T5	T 454

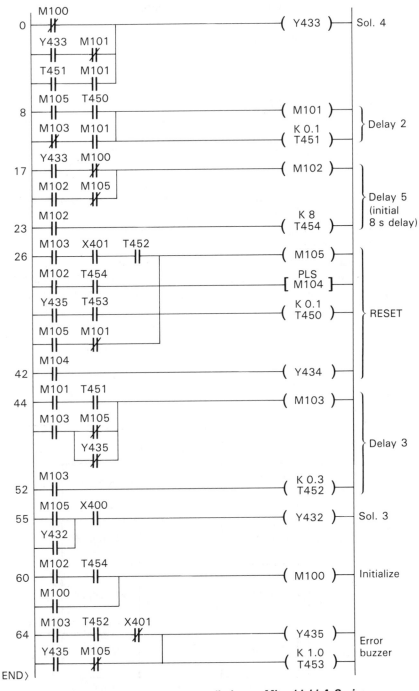

Fig. 8.26 **Working sorter listing – Mitsubishi A Series.**

software. Before installing it in plant, it is necessary to make additions for safety reasons, manual overrides, status indicators, etc. Also, counter circuits can easily be included to provide data on product throughput or maintenance status.

8.7 PLC APPLICATION IN THE BREWING INDUSTRY

(Presented here with the permission of GEC Industrial Controls Ltd)

GEM 80 controllers have recently been installed in the brewing industry for major applications using the latest GEM 80 designs of hardware and software.

One of the applications involves the use of eight GEM 80/140 controllers connected to a GEM 80/250 series supervising controller as shown in Fig. 8.27. The GEM 80/140 controllers are used to provide partial control and full monitoring facilities for the plant, which consists of two lines each consisting of 22 keg washing and filling machines.

The GEM 80/250 controller uses two serial links, which are configured for multidrop operation, each link being multidropped to four GEM 80/140s, to obtain the plant information/status from the plant controllers to enable collation of this information for display and printing.

Display of information is on a color VDU, and a serial 'QWERTY' keyboard

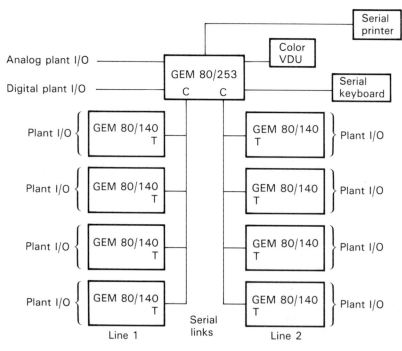

Fig. 8.27 Brewery plant overview.

is used to select the information to be displayed. Printing is by use of the printer serial port facility included in the GEM 80 video processing unit, which allows printing of whatever is displayed on the screen.

The control system ensures that kegs are distributed to washing and filling lanes in a manner which keeps lane starvation at the minimum level. All of the flowmeters which control the filling process are connected to the GEM 80 controllers. The system automatically resets the flowmeter set points and carries out standard deviation calculations to ensure that filling is accurate. Automatic recalibration facilities are provided to take account of the need to wash and fill different sizes of kegs during the operating periods. The gas pressure in every keg is checked against master transducers to ensure that the required pressures are achieved.

Comprehensive on-line monitoring facilities are provided, which include the number of kegs washed, filled, and rejected. Full details of the reasons why kegs are rejected are provided in tabulated form to enable production problems to be quickly identified.

The GEM 80 system makes full use of the latest software available from GEC. The fact that the plant consists of 44 identical lanes has made it possible for array-handling subroutines to be utilized. Standard deviation requirements have been met by the use of new subroutines. By the use of these techniques it has been possible to save many valuable programming instructions.

8.8   BATCH PROCESS – MUSTARD SEED MILLING

(Presented here with the permission of General Electric)

An aging electromechanical system needed replacement. Continuing control failures and growing maintenance costs forced the issue. A PLC was used not only because of its reliability versus relays, but also because it could yield a more consistent product and easily interface with a color graphics system. Another factor was the expandability of the PLC which would allow it to eventually control peripheral systems such as the bottling line.

In this system, ingredients are mixed in two batch tanks. The resulting slurry is then put in a slurry-holding tank. The slurry tank feeds 13 mustard mills. The PLC monitors the mill current to determine the consistency of the grind. After a period of time, the mustard with the correct consistency is passed on to the aerator and the rest is recirculated via the slurry tank for additional milling. The mustard is then cooled in the hot tank and chiller and held for packaging in tanks 2–4 (Fig. 8.28).

A PLC and a color graphics system work together through the use of a high-level communications control module. This gives the operator a picture of the entire batching-system operation. There are real-time displays of the tank levels, the mustard flow, the valve positions, the status of the mills, the status of the pump motors, and the 54 alarm points. The alarm display is set up to look and act like an old-fashioned annunciator with blinking and solid alarm points in green and red.

The PLC is used to control almost all operations of the batching system,

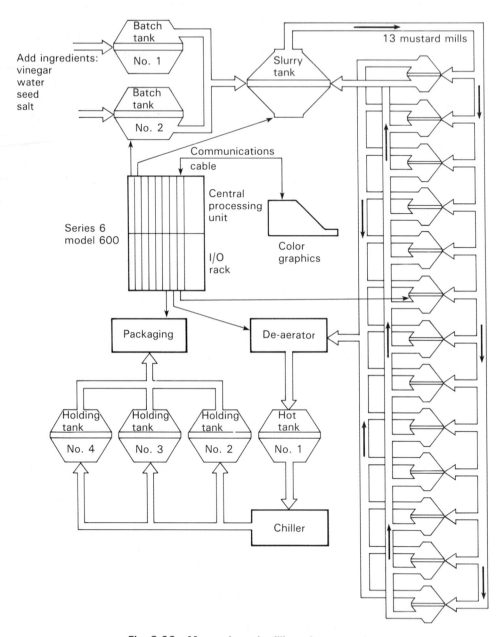

Fig. 8.28   Mustard seed milling plant overview.

including starting, stopping and monitoring temperature and pressure on the 13 mills. It switches tanks, mills and pumps on the basis of pressures, levels, temperatures and individual bottling line requirements.

The PLC does all of the logic control and the color graphics handles all of the operator interfacing. The graphics system is transparent to the PLC and does not affect the general operation speed of the system. Tests after startup found the PLC scan time to be only 8 ms and the update time on the graphic screen to be less than 300 ms.

8.9 PRODUCTION MONITORING – FRUIT FLOW CONTROL
(Presented here with the permission of General Electric)

Fruit processing is a seasonal operation. Depending upon the fruit, a plant may operate 24 hours a day for about 4 months and not operate at all for the rest of the year. Down time is very expensive and any controls must be extremely reliable. This is difficult in light of the poor environmental conditions which result from the summer heat and fruit pulp which seems to find its way into everything. The PLC is built to withstand the temperature and industrialized operator terminals must be able to survive both the environmental factors and washdown.

This application requires the PLC to interface with an image-recognition camera that monitors three conveyors of pre-sized product and transmits the percent-of-loading on each belt to high-density remote I/O. The PLC uses this data to adjust the incoming product flow at the bin dumpers using remote analog outputs to hydraulic drives. It also stages the upstream pitters to prepare them for the amount of product on each size belt. The complexity of the process is increased by the fact that the conveyors feed 12 banks of pitters. The PLC must play 'traffic cop' in order to make sure the correct product by size is fed to the correct pitter bank. The PLC must also coordinate the product with the canning line as 15 different products result from this process.

The operator dynamically monitors and controls the flow process from a remote console with digital displays, vertical bar graph meters, digital thumbwheels, pilot lights, and pushbuttons. From this location he can set alarm points and tune the product distribution for maximum throughput.

The system monitors the actual can output of each can line. It logs this data for dynamic recall through a video display terminal in the production supervisor's office and prints shift reports automatically.

The bin dumps at the receiving end of the process and the empty can feed controls are controlled by two additional PLCs which communicate back to the master central processing unit to coordinate activities. This relieves the system of any bottlenecks due to machine malfunctions or fully loaded machines and guarantees a smooth throughput of product. The result is a continuous flowing line allowing for more product to be processed in each shift.

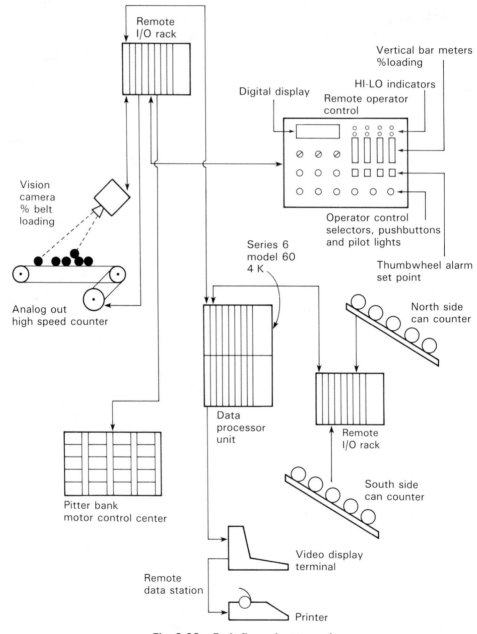

Fig. 8.29   Fruit flow plant overview.

8.10 SUMMARY

In this chapter a range of practical examples and case studies have been presented with differing levels of detail. In the small-scale applications, the aim has been to examine the case at the operational software level. With larger applications, greater emphasis has been given to overall project specification and subsequent design, with only brief details of software and hardware.

In this way it is hoped that the reader will find information and guidance at several levels of specification, design and implementation of PLC systems. Additional program design examples are given in the Appendices.

8.11 EXERCISES

8.1 Explain why programmable controllers are used in a wide variety of industrial applications.

8.2 Discuss areas of application of small, medium and large PLCs. Why are modular systems becoming the norm across the available range of controllers?

8.3 What are the possible areas of application for PLCs in the field of robotic control and system integration with other devices?

8.4 Outline the use of programmable controllers in the Austin–Rover assembly plant case study given earlier in this chapter. Explain the need for, and use of, a distributed, hierarchical control system.

8.5 For a case where is a requirement to link several PLCs and other intelligent machines to permit data collection or similar functions, describe what communications options may be available, and the factors that will decide which medium or product is selected for the task.

8.6 What advantage would there be in considering a MAP networking solution to the given case on factory networked communications? State two disadvantages of this idea.

8.7 Referring to the chemical plating case study, briefly describe the control problems that were presented, and the software design techniques used to implement control.

8.8 In the same example, what was the need for 'job priorities' in the process, and how was this facility provided?

8.9 Comment on the relative importance of system documentation in a programmable controller installation. What features of the chemical plating application documents support your answer?

## NINE

Choosing, Installation and Commissioning of PLC Systems

When embarking on any program of automation, it must be remembered that the controller is a tool or vehicle used to perform the necessary tasks. The tasks are therefore paramount, the controller only a means of achieving them. In Chapters 4 and 5, the importance of defining the control tasks and objectives was stressed, since before any control problem can be solved, it is necessary to develop an accurate definition of the desired system function. Experience has proved that errors existing at this formative stage will be extremely costly to correct at a later stage, when software design and coding will have taken place, despite the relative ease with which PLC programs may be amended or replaced. It remains more economic in terms of time and money to have a correct specification, and therefore a reliable system design.

9.1 FEASIBILITY STUDY

Under certain circumstances an initial feasibility study may be suggested or warranted, prior to any decision on what solution will be adopted for a particular task. The feasibilty study may be carried out either by in-house experts or by external consultants. Often an independent specialist is preferred, having few or no ties to specific vendor equipment.

The scope of such a study can vary enormously, from simply stating the feasibility of the proposal, through to a comprehensive case analysis with complete equipment recommendations. Typically, though, a feasibility/viability study of this nature encompasses several specific areas of investigation:

(a) *economic feasibility*, consisting of the evaluation of possible installation and development costs weighed against the ultimate income or benefits resulting from a developed system;

(b) *technical feasibility*, where the target process and equipment are studied in terms of function, performance and constraints that may relate to achieving an acceptable system;

(c)   *alternatives*, with an investigation and evaluation of alternative approaches to the development of the acceptable system.

Area (a), economic feasibility and worth, can only be addressed fully once the results of areas (b) and (c) are available, with estimated costings, and direct/indirect benefits being considered. Area (b) is detailed in the following sections, with background information for area (a) usually being compiled through liaison with company personnel. The achievement of a complete technical proposal (or proposals) requires us to know what the present and future company needs are in terms of plant automation and desired information systems.

Once the control function has been accurately defined (and agreed by all concerned), a suitable programmable control system has to be chosen from the wide range available. Following the identification of a suitable PLC, work can begin on aspects of electrical hardware design and software design.

9.2   DESIGN PROCEDURE FOR PLC SYSTEMS

Because the programmable controller is based on standard modules, the majority of hardware and software design and implementation can be carried out independently of, but concurrently with, each other. This approach was outlined earlier in Figure 4.1(a). Developing the hardware and software in parallel brings advantages both in terms of saving time and of maintaining the most flexible and adaptable position regarding the eventual system function. This allows changes in the actual control functions through software, until the final version is placed in the system memory and installed in the PLC.

An extremely important (but often neglected) aspect of every design project is the documentation. Accurate and up-to-date documentation of all phases of a project need to be fully documented and updated as the job progresses through to completion. This information will form part of the total system documentation, and can often be invaluable during later stages of commissioning and troubleshooting.

Choosing a programmable controller

There is a massive range of PLC systems available today, with new additions or replacements continually being produced with enhanced features of one type or another. Advances in technology are quickly adopted by manufacturers in order to improve the performance and market status of their products. However, irrespective of make, the majority of PLCs in each size range are very similar in terms of their control facilities. Where significant differences are to be found is in the programming methods and languages, together with differing standards of manufacturer support and backup. This latter point is often overlooked when choosing a suitable make of controller, but the value of good, reliable manufacturer assistance cannot be overstated, both for present and future control needs.

Which manufacturer?

Depending on the task involved, a potential customer may approach manufacturers directly, or he may contact 'system builders', who normally take on the complete project and subcontract various aspects such as rack building and cabling.

System builders tend to work with PLCs from only one or two manufacturers, for reasons of product familiarity and integrity – few suppliers want to deal with a systems house that has little commitment to their particular product! If a system builder is used from the outset, the customer may be offered only one, or at best two, control system options for his plant. However, there are exceptions to this, with some consultancy firms remaining only loosely tied to most of the major manufacturers. (See Fig. 9.1.)

Since most of the decision making in the above route is done by the systems house on behalf of the customer, we will examine instead the case where a customer intends to choose and organize the system on his own behalf, since (I assume) this text will be of most interest to these 'do it yourself' designers!

The information and assistance required from a manufacturer may vary considerably depending on the user's previous experience of programmable controllers and control systems in general. Where a user has experience in this field, it is

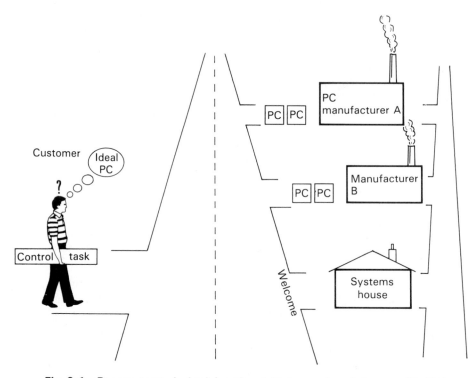

Fig. 9.1 Routes toward obtaining the right form of assistance with PLC systems.

often the areas of technical excellence and performance that are explored, rather than aid with system design and installation work. However, a large number of customers will be experiencing PLCs for the first time, possessing little or no knowledge of the marketplace. In this case, it is worthwhile considering certain questions:

- Can the user obtain assistance with the design work?
- What proportion of the market does the potential supplier/manufacturer hold, and do they have a track record in the necessary area of application?
- Does the manufacturer offer training courses on the type of PLC system likely to meet the user's needs?
- Are all relevant handbooks and manuals available in the required language, to a good standard?
- What compatibility exists between any likely system and other types of PLC from the same or different manufacturers?
- Does the programming method used suit the outline control plan for the application?

For the relatively inexperienced user, it is a great advantage if the supplier or manufacturer can offer assistance with the system design work, together with further help and training in actual program design and coding. Even though PLCs are relatively straightforward to program, it is usually necessary to assist the user through the 'transition' period from one form of control or controller to the new system.

What this has tried to highlight is the need to consider many other factors besides cost and technical performance when attempting to choose a make or type of programmable controller. These are of course, important, but generally the cost of system hardware is only a small part of the complete system, which includes hardware, software, design, training and documentation, plus installation and maintenance.

Size and type of PLC system

This may be decided in conjunction with the choice of manufacturer, on the basis that more than one make of machine can satisfy a particular application, but with the vast choice of equipment now available, the customer can usually obtain similar systems from several original equipment manufacturers (OEMs). Where the specification requires certain types of function or input/output, it can result in one system from a single manufacturer standing out as far superior or cost-effective than the competition, but this is rarely the case. Once the stage of deciding actual size of the PLC system is reached, there are several topics to be considered:

- necessary input/output capacity;
- types of I/O required;
- size of memory required;
- speed and power required of the CPU and instruction set.

All these topics are to a large extent interdependent, with the memory size being directly tied to the amount of I/O as well as program size. As the I/O / memory size rises, this takes longer to process and requires a more powerful, faster central processor if scan times are to remain acceptable.

I/O requirements

The I/O sections of a PLC system must be able to contain sufficient modules to connect all signal and control lines for the process. These modules must conform to the basic system specifications as regards voltage levels, loadings, etc., plus:

- The number and type of I/O points required per module (or unit in the case of self-contained PLCs);
- isolation required between the controller and the target process;
- the need for high-speed I/O, or remote I/O, or any other special facility;
- future needs of the plant in terms of both expansion potential and installed 'spare' I/O points;
- power-supply requirements of I/O points – is an on-board PSU needed to drive any transducers or actuators?

In certain cases there may be a need for signal conditioning modules to be included in the system, with obvious space demands on the main or remote racks. When the system is to be installed over a wide area, the use of a remote or decentralized form of I/O working can give significant economies in cabling the sensors and actuators to the PLC.

Memory and programming requirements

Depending on the type of programmable controller being considered, the system memory may be implemented on the same card as the CPU, or alternatively on dedicated cards. This latter method is the more adaptable, allowing memory size to be increased as necessary up to the system maximum, without a reciprocal change in CPU card.

As stated in the previous section, memory size is normally related to the amount of I/O points required in the system. The other factor that affects the amount of memory required is of course the control program that is to be installed. The exact size of any program cannot be defined until all the software has been designed, encoded, installed and tested. However, it is possible to accurately estimate this size based on average program complexity. A control program with complex, lengthy interlocking or sequencing routines obviously requires more memory than one for a simple process. (Different manufacturers use slightly different methods for obtaining approximate memory size, and should be contacted for this information.) Program size is also related to the number of I/O points, since it must include instructions for reading from or writing to each point. Special functions required for the control task may also require memory space in the main PLC memory map to allow data transfer between cards.

Finally, additional space should be provided to allow for changes in the control program, and for future expansion of the system. (Smaller PLCs normally have fixed memory size, requiring replacement of the main PLC if modifications to the program result in code too large for the resident memory.) (See Fig. 9.2.)

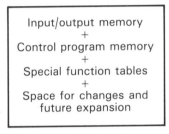

Input/output memory
+
Control program memory
+
Special function tables
+
Space for changes and future expansion

(a)

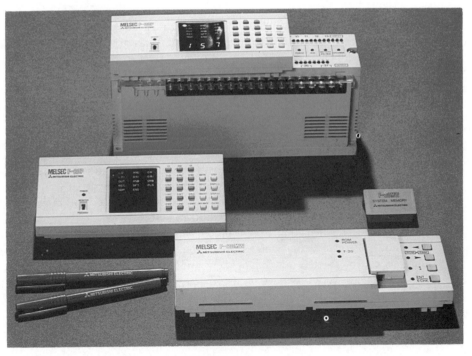

(b)

Fig. 9.2 (a) PLC memory requirements for different tasks; (b) custom EPROM programmer for a Mitsubishi F Series PLC (courtesy Mitsubishi Electric UK Ltd).

There is often a choice of available memory type – RAM or EPROM. The RAM form is the most common, allowing straightforward and rapid program alterations both before and after the system is installed. RAM contents are made semipermanent by the provision of battery-backing on their power supply. RAM must always be used for I/O and data functions, as these involve dynamic data. EPROM memory can be employed for program storage only, and requires the use of a special EPROM eraser/programmer to alter the stored code. The use of EPROMS is ideal where several machines are controlled by identical programmable controllers running the same program. However, until a program has been fully developed and tested, RAM storage should be used.

As mentioned in earlier chapters, microcomputers are commonly used as program development stations. The large amounts of RAM and disk storage space provided in these machines allows the development and storage of many PLC programs, including related text and documentation. Programs can be transferred between the microcomputer and the target PLC for testing and alteration. EPROM programming can also often be carried out via the microcomputer.

Instruction set/CPU

Whatever else is left undefined, any system to be considered must provide an instruction set that is adequate for the task. Regardless of size, all PLCs can handle logic control, sequencing, etc. Where differences start to emerge are in the areas of data handling, special functions and communications. Larger programmable controllers tend to have more powerful instructions than smaller ones in these areas, but careful scrutiny of small/medium machines can often reveal the capability to perform specific functions at surprisingly good levels of performance (e.g. integral PID loops).

In modular programmable controllers there may be a choice of CPU card, offering different levels of performance in terms of speed and functionality. As the number of I/O and function cards increases, the demands on the CPU also increase, since there are greater numbers of signals to process each cycle. This may require the use of a faster CPU card if scan time is not to suffer.

Following the selection of the precise units that will make up the programmable controller for a particular application, the software and hardware design functions can be carried out independently (as described in Fig. 4.1(a)). Software design strategies were discussed in a previous chapter, so the subsequent sections describe the hardware design stages, followed by software testing and installation in the system hardware.

9.3 SYSTEM LAYOUT

The previous sections have described the options that must be considered when attempting to identify a suitable PLC system. Once this process has been completed

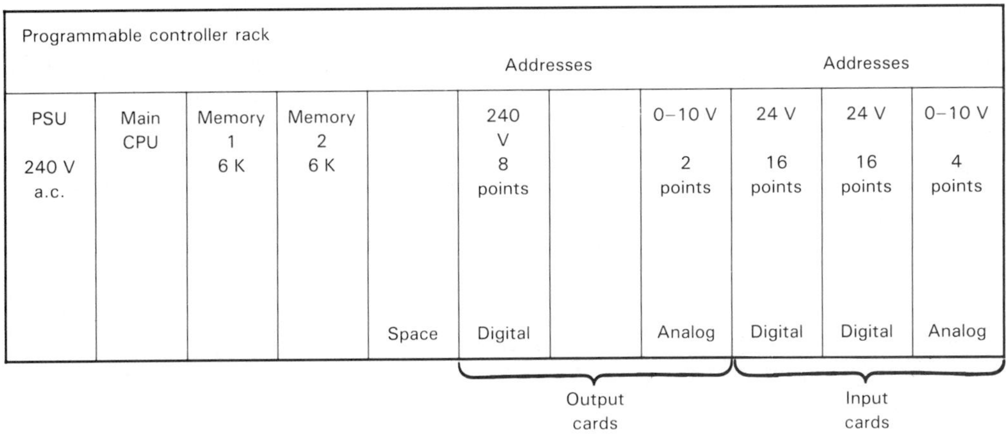

Programmable controller rack											
					Addresses					Addresses	
PSU 240 V a.c.	Main CPU	Memory 1 6 K	Memory 2 6 K		240 V 8 points		0–10 V 2 points	24 V 16 points	24 V 16 points	0–10 V 4 points	
				Space	Digital		Analog	Digital	Digital	Analog	

Output   Input
cards   cards

Fig. 9.3   Typical physical layout for a modular PLC.

and the system chosen, we must consider how it is to be physically laid out. This simply consists of defining where each specific function module is to fit in the rack – information that is required for the next stage when the wiring design and electrical installation are to be considered.

The basic rack or chassis layout depends on the make and model of controller, but will have a number of slots with edge-connector sockets to allow modules to be plugged into the common backplane. Some racks allow virtually any type of card (from the same family) to be placed at any position in the rack, whilst others specify locations for the main cards, with flexibility only for I/O and certain function cards.

The layout is normally described using a simple diagram such as Fig. 9.3, which includes details on position, type of card, signal levels and position in the memory map.

When a larger installation is required, the main rack may become fully populated, requiring the use of subracks to carry additional cards. Fig. 9.4(a) shows a GEM 80 layout using additional subracks. The PLC racks are usually intended to be mounted in a cubicle, and a layout plan should include details of the subrack positions.

Cubicle layout

Many different layouts are possible, depending on the amount and type of equipment to be used, and the physical size of the cubicle involved. It is usual to mount the controller in a reasonable position for access between waist and eye height, to allow:

- insertion and replacement of modules;
- connection of programming panel;
- easy access to rear of panel for wiring, etc.;
- clear sight of status indicators.

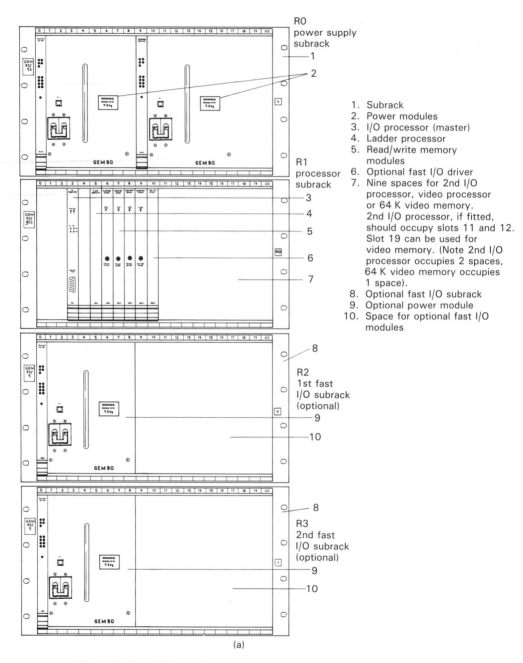

1. Subrack
2. Power modules
3. I/O processor (master)
4. Ladder processor
5. Read/write memory modules
6. Optional fast I/O driver
7. Nine spaces for 2nd I/O processor, video processor or 64 K video memory. 2nd I/O processor, if fitted, should occupy slots 11 and 12. Slot 19 can be used for video memory. (Note 2nd I/O processor occupies 2 spaces, 64 K video memory occupies 1 space).
8. Optional fast I/O subrack
9. Optional power module
10. Space for optional fast I/O modules

(a)

Fig. 9.4   GEM 80: (a) hardware layout; (b) typical cubicle layout (courtesy GEC).

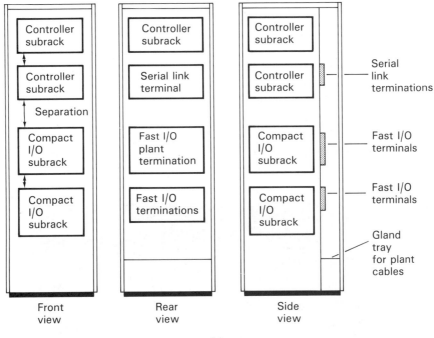

In order to reduce the possibility of noise interference problems, the racks are normally mounted at least as far apart as the minimum specification (typically 80–100 mm). This also gives optimal ventilation for the rack modules.

Electrical wiring design and installation

We are now at the stage where the chosen programmable controller has been defined in terms of the modules to be used and their schematic layout in the rack. The separate routes of software and hardware development can now begin.

The hardware design path uses the layout plan and details of constituent modules to develop a wiring diagram for the rack installation and for associated cabling to plant equipment. It is rare for any additional electrical or electronic design work to be necessary, except where nonstandard interfacing to plant transducers or actuators is required (the use of signal conditioning modules can often satisfy this need). Where inductive loads are to be used on the system outputs (e.g. relays and solenoids, motors), these devices should be specified/fitted with interference suppressors to reduce transient noise spikes during power on/off. The electrical design task includes planning cable routings, connection details for I/O racks, power supplies and communications cards.

Table 9.1 Categories and segregation of wiring.

Very clean	Clean	Dirty
I/O ribbon cables	Serial link cable (ribbon)	Controller PSU cable
I/O Power supply	Analog cables	Plant side I/O and power supplies
PC signal ground (0 V)	Signal ground from subracks	Safety earths
	Other screened plant cables	

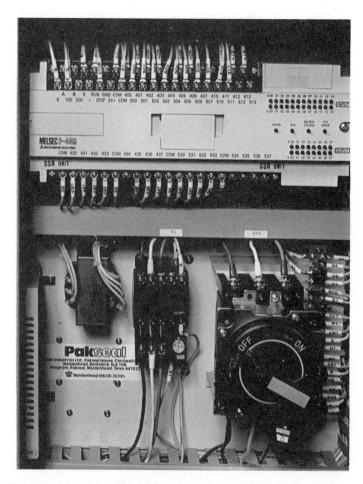

Fig. 9.5 Small PLC installation with associated circuitry. Mitsubishi F40 applied to packaging machine (courtesy Mitsubishi Electric UK Ltd).

Cable routing must be worked out when the cubicle layout is planned, and in order to prevent noise interference, interconnecting cables should be grouped in categories and run with each category physically separated from the others. Table 9.1 gives an indication of wiring categories and their status – clean (of noise) or dirty.

Whenever possible, the design should use screened cables to connect the plant input/output devices to the PLC I/O units. When long runs of cabling are necessary, input and output signal lines should be run on different cables. The cable screens should be earthed at one end only, to avoid earth loop noise problems. (An exception to this can be the serial communications link, which may be specified as unscreened – in this case the fitting of a screened cable may cause signal attenuation and possible earth loop problems. The reader is advised to consult the manufacturer for specific details on wiring recommendations for a particular model.

Connections to the controller racks are made either directly to the various I/O modules, or to terminal blocks mounted in the cubicle. This will decide the type of connectors to be specified for the field wiring and cabling. (See Fig. 9.5.)

Associated with the electrical design is the preparation of information sheets that specify connection details for all I/O points in the system. This will include terminal connections and labels, PLC reference/address, wiring identifier, and I/O device descriptor.

9.4 INSTALLATION

The hardware installation consists of building up the necessary racks and cubicles, then installing and connecting the cabling. Cable runs should be grouped and segregated as discussed in the previous section, and all screening and earthing carried out to the specification. Fig. 9.6 shows an example of a GEM 80 series signal grounding and earthing layout.

The cabinet that contains the programmable controller and associated sub-racks (see Fig. 9.7) must be adequate for the intended environment, as regards security, safety and protection from the elements:

- security in the form of a robust, lockable cabinet;
- safety, by providing automatic cut-off facilities/alarms if the cabinet door is opened (optional);
- protection from humid or corrosive atmospheres by installation of airtight seals on the cubicle. Further electrostatic shielding by earthing the cubicle body.

For maintenance purposes, there must be easy access to the PLC racks for card inspection, changing, etc. Main on/off and status indicators can be built into the cabinet doors, and glass or perspex windows fitted to allow visual checking of card status or relay/contactor operation.

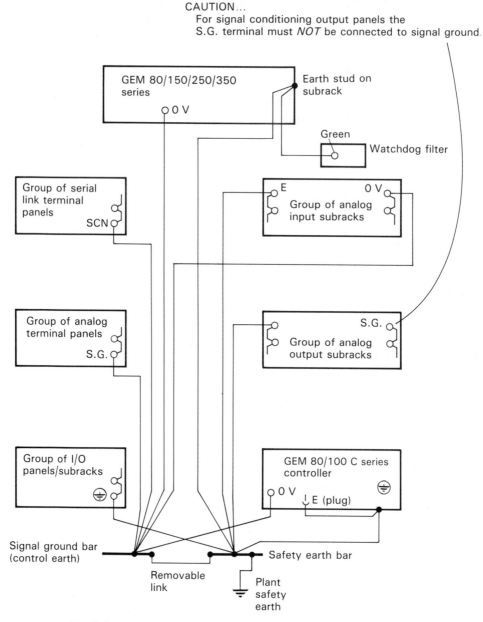

Fig. 9.6 Example of a GEM 80 series signal ground and earthing layout.

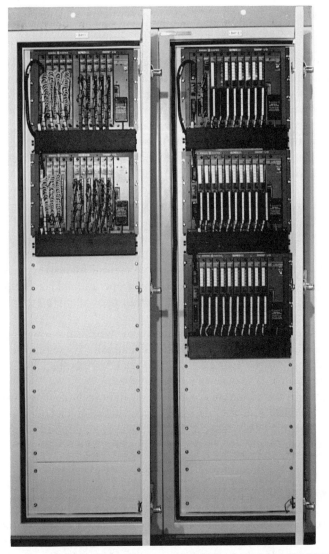

Fig. 9.7 Complete PLC installation and cabinet (courtesy GE FANUC).

9.5 TESTING AND COMMISSIONING

Once the installation work is completed, the next step is to consider the testing and commissioning of the PLC system.

Commissioning comprises two basic stages:

1. Checking the cable connections between the PLC and the plant to be controlled.

2. Installing the completed control software and testing its operation on the target process.

The system interconnections must be thoroughly checked out to ensure all input/output devices are wired to the correct I/O points. In a conventional control system this would be done by 'buzzing out' the connections with suitable continuity test instruments. With a programmable controller, however, the programming panel may be used to monitor the status of input points directly – this is long before the control software is installed, which will only be done after all hardware testing is satisfactorily completed. Before any hardware testing is started, a thorough test of all mains voltages, earthing, etc., must be carried out.

With the programmer attached to the PLC, input points are monitored as the related transducer is operated, checking that the correct signal is received by the PLC. The same technique is used to test the various function cards installed in the system. For example, analog inputs can be checked by altering the analog signal and observing a corresponding change in the data stored in the memory table.

In turn, the output devices can be 'forced' by instructions from the programming panel, checking their connection and operation. The commissioning team must ensure that any operation or misoperation of plant actuators will not result in damage to plant or personnel.

Testing of some PLC functions at this stage is not always practical, such as for PID loops and certain communications channels. These require a significant amount of configuring by software before they can be operated, and are preferably tested once the control software has been installed.

Some programmable controllers contain in-built diagnostic routines that can be used to check out the installed cards, giving error codes on a VDU or integral display screen. These diagnostics are run by commands from the programming panel, or from within a control program once the system is fully operational.

Software testing and simulation

The preceding sections have outlined the various stages in hardware design and implementation. Over the same period of time, the software to control the target process is developed, in parallel, for the chosen PLC system. This development route has been covered in previous chapters, describing the design and development of individual software 'modules' to carry out the various consituent tasks of the total control requirement. These program modules should be tested and proved individually wherever possible, before being linked together to make up the complete applications program. It is highly desirable that any faults or errors be removed before the program is installed in the host controller. The time required to rectify faults can be more than doubled once the software is running in the host PLC.

Virtually all programmable controllers, irrespective of size, contain elementary software-checking facilities. Typically these can scan through an installed program

Display/printout

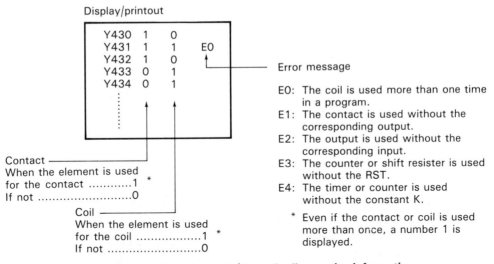

Contact ——————
When the element is used
for the contact ....1 *
If not 0

Coil ——————
When the element is used
for the coil ....1 *
If not 0

Error message

E0: The coil is used more than one time
    in a program.
E1: The contact is used without the
    corresponding output.
E2: The output is used without the
    corresponding input.
E3: The counter or shift resister is used
    without the RST.
E4: The timer or counter is used
    without the constant K.

  * Even if the contact or coil is used
    more than once, a number 1 is
    displayed.

Fig. 9.8 PLC printout of I/O static diagnostics information.

to check for incorrect labels, 'double' output coils (used more than once in a program), etc. Listings of all I/O points used, counter/timer settings and other information is also provided. The resulting information is available on the programmer screen or as a printout (Fig. 9.8). However, this form of testing is only of limited value, since there is no facility to check the operation of the resident program.

In terms of time and cost economies, an ideal method for testing program modules (and later the complete program) is to reproduce the control cycle by simulation, since this activity can be carried out in the design workshop without

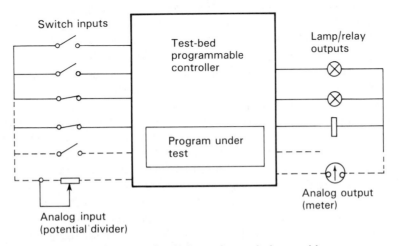

Fig. 9.9 Process simulation using switches and lamps.

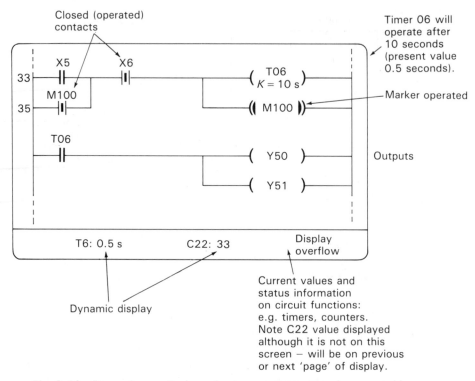

Fig. 9.10   Dynamic monitoring of program contacts using a graphic programming display.

having to actually connect up to the physical process. Simulation of the process is done in a number of ways, depending on the size of process involved.

When the system is relatively small with only a handful of I/O channels, it is often possible to adequately simulate the process by using sets of switches connected up to the PLC as inputs, with outputs represented by connecting arrays of small lamps or relays (Fig. 9.9). This allows inputs to be offered to a test-bed controller containing software under test, checking the action of the control program by noting the operation and sequence of the output lamps or relays. By operating the input switches in specific sequences, it is possible to test sequence routines within a program. Where fast response times are involved, the tester should use the programming panel to 'force' larger time intervals into the timers concerned, allowing that part of the circuit to be tested by the manual switch method.

Most I/O modules have LED indicators that show the status of the channels. These can be used instead of additional test actuators where digital outputs are concerned. Analog inputs can be simulated in part by using potential dividers suitably connected to the input channel, and corresponding analog outputs connected either to variable devices such as small motors or to a moving-coil meter configured to measure voltage or current. Standard sets of input switches and output actuators are normally available from PLC manufacturers.

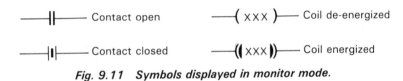

Fig. 9.11   Symbols displayed in monitor mode.

When the system is larger with many input/output channels and longer, more complex programs, the simple form of simulation described above becomes inadequate. Many larger PLC systems are fitted with an integral simulation unit that reads and writes information directly into the I/O memory, removing the need to connect external switches, etc. The simulator is controlled from an associated terminal (often the programming VDU) which can force changes in input status and record all changes in output status as the program runs, for later scrutiny by the test team.

The 'program monitoring' facility provided with most programming terminals should be used in virtually all these proceedings, since it allows the dynamic checking of all elements in the program including preset and remaining values as the program cycles. Figure 9.10 illustrates a monitoring display with status information shown on the bottom of the screen. It is important to realize that the display on the programmer does not update as rapidly as the control program is executing, due to the delays in transmitting the data across to the terminal. Contacts and other elements that are operated for only a few scans are unlikely to affect the display, but since a human observer could not detect this fast a change, this is not a significant disadvantage. To display all changes, the PLC should be run in single-step mode (if provided).

The monitor display shows a selected portion of the ladder program, using standard symbols to depict contacts, outputs and present functions. All elements within the display are dynamically monitored, indicating their status (energized, de-energized, open or closed) as shown in Fig. 9.11.

Installing and running the user control program

Once the control software has been proved as far as possible by the above methods on a test machine (or on the actual PLC when no test machine is available), the next step is to try out the program on the tested PLC hardware installation. Ideally each section of code should be downloaded and tested individually, allowing faults to be quickly localized if the plant misoperates during the program test. If this subdivided testing is not possible, another method is to include JUMP commands in the complete program to miss out all instructions except those in the section to be tested. As each section is proved, the program is amended to place the JUMP instructions so as to select the next section to be tested (see Fig. 9.12).

Where a programmable controller supports single-step operation, this can be used to examine individual program steps for correct sequencing. Again, the programming terminal should be utilized to monitor I/O status or any other area of interest during these tests, with continuous printouts if this is possible.

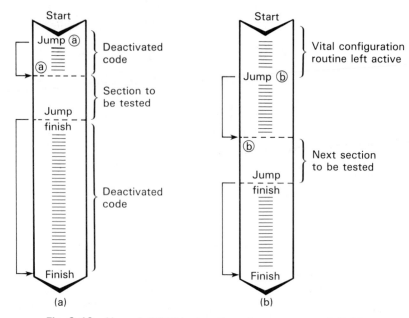

Fig. 9.12 Use of JUMP instructions for program subdivisions.

Software amendments

One of the great advantages of programmable controllers is the ease with which resident programs may be altered. At the design and commissioning stages this allows program variations to be tried and retried until satisfactory control is achieved. This flexibility should not tempt the engineer to effect mass changes at this stage, however, since large program modifications can result in very strange plant-operation sequences. Alterations should be carried out only one at a time, checking the effects on process operation thoroughly before moving on to further alterations.

Software backups

Throughout the testing and modifying stages of software commissioning, adequate copies of each stage must be produced and retained, with attached descriptive labels. As an absolute minimum two copies of the user program should be stored on disk or tape medium, one being the latest version and the second being a backup of the previous stage. A full record of software modifications must be kept and updated as amendments occur, together with reasons and results. If this is not done, it is very likely that if the current program version is lost or corrupted, then all modifications since the last backup will have to be carried out again – and this may represent many days of work!

Once the final applications program is complete, there should be a backup copy available within the plant in case of memory failure in the PLC system. This must be on a medium that will allow simple reloading of the program into the controller RAM. When a personal computer is used as the programming terminal, this backup is on either floppy disk or Winchester hard disk, and can be downloaded to the PLC over an interconnecting cable.

This falls within the area of system documentation, which is discussed in the next section.

9.6 SYSTEM DOCUMENTATION

As soon as a control project is initiated, details and specifications are generated, and must be recorded in such a manner as to allow them to be used in the subsequent project-design stages. New documents are generated at each stage of both the hardware and software design paths, describing a particular aspect of the system from different points of view. These documents and records form a history of the project progress, and are invaluable references for designers and engineers both during and after commissioning the system. All specifications and modifications should appear in the appropriate document, allowing decisions to be taken on the basis of tried and recorded events. This is particularly the case with program documentation for PLC systems, where ease of program alteration can result in undocumented changes, and subsequent confusion.

Documentation is the main 'user guide' to the general day-to-day running of the system, and the main tool used for troubleshooting and servicing. For these reasons, it should be in a format that is easy for maintenance personnel to follow and to understand. Thorough, well-laid-out documentation encourages its use, increasing staff awareness and knowledge of system operation. The system documentation can be of considerable value when any alterations or additions are to be carried out in the future. A complete set of documentation for a programmable controller installation should include the following:

- description of the plant or process;
- description and specification of the control objectives/requirements;
- programmable controller description including equipment manuals;
- description of the software functions, including function/flowcharts, memory maps and allocations;
- program listing with full comments and documentation;
- software backups and/or EPROMS;
- electrical installation diagrams and description;
- lists of input/output connections including all VDUs, printers, etc;
- user/operating manual, including details of system startup/shutdown and alarm references.

9.7 MAINTAINING A PLC SYSTEM

Programmable controllers are designed to operate reliably for long periods of time in adverse industrial environments, this being one of their main advantages over both traditional electromechanical controls and microcomputer-based controllers. Due to robust electronic construction and effective electrostatic screening, the internal hardware can safely be used in locations that are both physically and electrically hostile. The PLC is protected from potentially damaging surges on all input/output lines by opto-isolated circuits. The use of battery-backed RAM or EPROM to hold the applications software ensures vital production time is not lost due to program loss or corruption after a plant power failure or a similar event. In summary, all feasible measures have been taken in the design and construction of most programmable controllers to achieve as high a reliability as possible, at a reasonable cost.

However, high reliability does not mean that a system will never fail. Although high-quality microelectronic components are used throughout each system, and assembled to very high standards, there can still be failures of components or connections. Programmable controllers are normally 'soak tested' before release from the manufacturer, running them continuously with test programs over long periods of time. This testing causes 'infant mortalities' to occur within the factory, rather than once installed in a control system. Infant mortality is a term used to describe component or system failure within the first few hours of use – a feature that occurs in a small percentage of most electronic devices. This effect is shown on the 'bath-tub' diagram in Fig. 9.13.

This illustrates the relatively high failure rates that occur in the first few hours of use, followed by a very low failure rate for a considerable time if the devices does not fail at the infant stage. Failures increase again as the graph reaches the end of the normal lifespan of a device. Thus, systems tested successfully for several hours effectively enter the high-reliability zone, reducing the probability of their failing in the field.

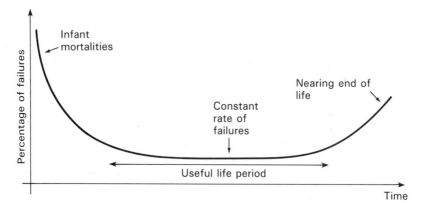

Fig. 9.13   Bath-tub diagram showing percentage failure rate against time.

Internal PLC faults

Programmable controllers provide several built-in fault procedures that are activated by the occurrence of certain internal conditions. The actual facilities vary between PCs, but typically include some or all of those listed in Table 9.2.

Critical faults cause the CPU to halt, whereas others may allow the controller to continue running, displaying a fault code on a terminal or seven-segment display on the PC cabinet. Table 9.3 shows fault code listings for a GEM 80 system.

When self-test failures occur, the controller is halted and will not restart until the fault is cleared and the system re-started. Faults are reported as a code number and (possibly) an accompanying text message:

#702 Object RAM CRC test failure on RAM No 2

This means a RAM test has failed with possible faulty memory cells in the RAM board. The recommended action (see Table 9.3) is to try powering-up the PLC again to re-initialize the RAM, or to change the module if this does not cure the fault. Tests of this nature are normally provided for all resident modules.

Maintenance personnel can use the error message listings to help localize the fault. Fault-finding procedures rely on substituting known good cards for suspected faulty items in the PLC rack. For this reason it is recommended that spare modules be held in the plant, with functional testing every few months.

Table 9.2.

Type of test	*Action on failure*
Self-test routines e.g. program changed	Controller halted, error LED or messages displayed
System status errors e.g. CPU error; failed RAM test	Controller halted, error LED or messages displayed

Faults external to the PLC

The PLC is only part of any control system, however, requiring the addition of sensors and actuators, field wiring, power supplies and software programs to form a complete system. As shown in Fig. 6.23, a large proportion of system failures are due to faults located outside the main programmable controller. For example,

- faulty field I/O device – transducer or actuator;
- faulty field wiring;
- faulty communications wiring;
- power source disturbances – noise or break in supply.

Table 9.3   Examples of typical PLC error messages (from a GEM 80).

CODE	Fault message and possible causes	Recommended recovery
#701	(proc) EPROM CRC TEST FAILURE On-line test failure. Power-up failure will cause system to be dead. Suspect faulty processor module.	Try power off/on. If problem persists then substitute spare module.
#702	OBJECT RAM CRC TEST FAILURE ON RAM NO. (data) On-line corruption, probably faulty object RAM board. Less likely: faulty ladder processor.	Power off/on will re-initialize the RAM, but it may fail again. Substitute spare RAM. If problem still persists then substitute spare processor.
#703	USER PROGRAM CRC TEST FAILURE ON RAM NO. (data) Source code corrupted. Faulty data table RAM board.	Try clear store power off/on and reload. If problem persists then substitute spare RAM board. Note: power off/on without clear will result in #718.
#704	V-TABLE CRC TEST FAILURE On-line RAM failure. Faulty RAM board.	Power off/on (will give #603) then reload. If problem persists then substitute spare RAM board.
#705	(proc) ON-BOARD RAM FAILURE AT (data). WRITTEN (data), READ (data) Failed processor module.	Substitute spare.
#706	SHARED RAM FAILURE AT (data). WRITTEN (data), READ (data) ON RAM NO. (data) Faulty data table RAM board.	Substitute spare.
#707	LDP BIT LOGIC TEST FAILURE BIT (data), TEST (data) Failed ladder processor module.	Substitute spare.
#708	(proc) WATCHDOG TEST FAILURE Suspect failed processor module. Could be a faulty backplane but more likely to get #719 first in this case.	Substitute spare processor. If problem persists, or #719 occurs, then suspect backplane and substitute spare subrack.
#709	PSU TEST FAILURE M-IOP hardware not to latest mod-state. Suspect faulty PSU. Remote possibility of master I/O procedure failed. Very remote possibility of the link between the two being faulty. *Note*: watchdogs do not trip.	Update M-IOP hardware. Substitute spare PSU. If problem persists then substitute spare master I/O processor and check connections.

The internal diagnostics may give an indication where a fault lies, or the control program itself can often give error information. Some manufacturers provide extended fault location facilities through intelligent I/O systems which perform continuous tests on the constituent units in the plant-to-PLC chain, such as the Genius system discussed in Chapter 6.

At the software design stage, consideration is given to possible misoperations and sequence errors in the control cycle. A simple example is the situation where two limit switches are to be closed, and a third remain open before a guillotine operates. If the machinery operation is such that the third switch can only be in the closed position at this time if a foreign object is jammed in the mechanism, then this event should result in an error report or alarm, and the system halted. In the majority of cases, manual emergency stop buttons are included in the program rungs to halt specific sections of a process, and in the watchdog/output enabling relay circuit. (This latter circuit is discussed in a later paragraph.)

When certain process operations are completed in a known time, it is common and straightforward to include watchdog timers in that portion of the program. When an operation starts, the timer is also started, having been set to a specific period. If the operation is not completed within that time it can be assumed that some form of error or fault has occurred and the watchdog timer trips, setting off an alarm and causing a partial or total system halt.

Programmed error/halt sequences

It is a straightforward matter to program additional ladder rungs to detect the type of condition described above, resulting in a jump or branch from the normal program to a special error routine. This routine may be associated with one particular failure, or may be a general error sequence that is 'passed' a value (via a data register) by the section of code that called it. As shown in Fig. 9.14, the error value passed can be used to produce one of several possible responses from the routine – e.g. different text messages to operators. Simpler systems can be programmed to output a failure code alone, which the operator interprets on a look-up chart. These aspects are in addition to normal alarm actuation.

Programmable controllers are often provided with a special function relay which, when operated, disables all outputs. This relay would be included in error-handling program rungs as necessary, to disconnect the output devices. This can be a useful software facility (in addition to independent emergency stop hardwiring), providing the actuators are wired to stop safely for this condition occurring.

Depending on the nature of the failure or error, the system may be stopped and the CPU placed in a halt state. Once the fault is cleared, the PLC can usually be restarted in the normal manner. If the process is likely to be halted part-way through a sequence, the control program must ensure all status data is retained in registers. This allows the process to be resumed from the same point after shut-down. If the process involves moving machinery and the machines are not to start functioning again until an operator has intervened, then it must be arranged that the plant

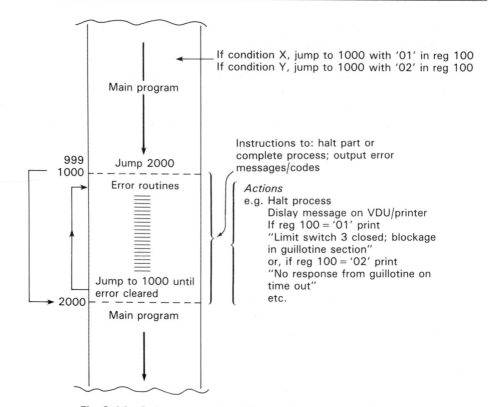

Fig. 9.14 Software error-handling routines – general concepts.

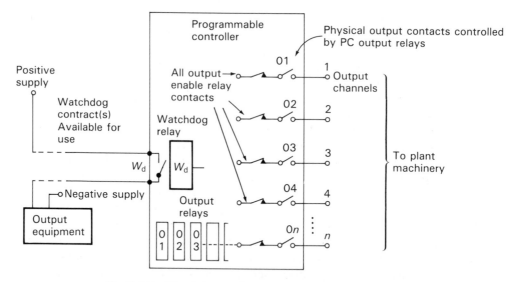

Fig. 9.15 Watchdog and output enabling circuit arrangements.

machine power supplies are not switched in until some operator-controlled relay circuit has been operated.

Watchdogs

A common form of internal PLC error handling involves the use of watchdog circuits. Most programmable controllers include an internal watchdog relay that can be used to control the power supply to one or more external devices. When the PLC has started up correctly, the watchdog relay operates, powering-up any units connected through the watchdog contacts. If an internal fault develops in the PLC, the watchdog will automatically release, and the PLC halts. Any equipment wired through the watchdog contacts also stops (Fig. 9.15).

Safety

No matter how complete and sophisticated the programs are in terms of failure handling, these will only work correctly so long as the controller is running correctly. Since no controller can ever be 100% reliable, the only way to ensure a safe system is by using hardware safety circuits rather than by software.

This text does not pretend to offer instruction or reference in the design or construction of fail-safe circuits, although the general principles have existed for many years; this is the task of an experienced system designer, who must analyze the particular process and installation with regard to its safety requirements, and ultimately take responsibility for any accidents resulting in damage to the plant being controlled or injury to personnel.

Hardware safety circuits

These circuits must be wired independently from the PLC, but should incorporate controller watchdog contacts to ensure the hardware safety circuits trip out if the PC itself fails and trips out.

Normal start and stop push-buttons may be connected via the controller I/O units (i.e. detected by software), whereas emergency stop switches and push-buttons to stop moving or potentially dangerous equipment must be hard-wired. For example, any push-buttons which have to ensure that the machinery does not move whilst maintenance staff are working on it must function correctly whether the controller is operational or not – they must be hard-wired, not fed into PC input channels or over communications links between two or more controllers. A common hardware approach to emergency stop requirements is to use a series of normally closed pushbuttons which lock off (open) when pushed (see Fig. 9.16). These push-buttons should be mounted locally, close to the moving machinery, so that any maintenance personnel can be confident that once that emergency stop button is

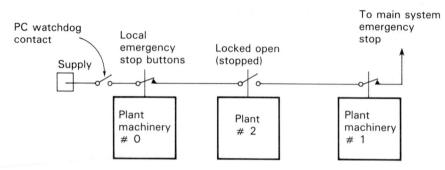

Fig. 9.16   Common hard-wired emergency stop circuit.

locked in, no-one can restart the machine until the local button is reset. Note the watchdog contact shown in series with these manual push-buttons to ensure the system also trips out if the programmable controller halts.

Troubleshooting

In the event of the control system not starting or running correctly, the source and cause of the problem must be identified and corrected:

- note and study the symptoms
- identify the problem area;
- isolate the problem;
- correct the fault.

Note and study the symptoms exhibited by the system. This includes any diagnostic information provided by the programmable controller as described above – i.e. controller status LEDs, error codes or messages, etc. Examination of the I/O status LED arrays will indicate signals coming into and out of the PLC. Where no PLC fault is indicated, this type of I/O comparison with the state of field devices can quickly pick up defective sensors or actuators.

Identify the problem area and isolate the problem. All programmable controller systems consist of a series of components, as in Fig. 9.17. By checking each area in turn, a fault can soon be localized and identified. If the PLC has halted, maintenance personnel can make use of the programmer monitor facility to check program position, possibly giving an indication of events leading up to the halt. The monitor facility can also be used when the PLC is running but the controlled process is malfunctioning. This will ascertain whether particular rungs are being solved as the program runs, and the state of relevant input and output points at that time. The provision of a single-step mode is very useful here. If a rung shows the correct input continuity but the driven output coils are shown to be not responding, checking of any associated master control relays (MCRs) or JUMP conditions may show them to be the cause of de-energized outputs.

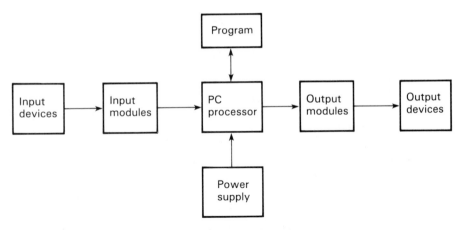

Fig. 9.17   PLC system components.

Correcting the fault. If the fault has been identified to within a PLC card or unit, the unit in question is changed for a healthy replacement. Depending on the card function, it may be necessary to set up selection switches, etc., to configure the card for that system. If a possible software programming fault is indicated, the system designer should be involved before any alteration is attempted, since there may be other aspects of operation that are related to the offending lines of code. When the fault lies outside the PC, the offending device or wiring is repaired or replaced.

Utilizing the PLC as an equipment monitor

There are many advantages in using the facilities of programmable controllers to carry out monitoring and recording operations on items of plant. Most industrial equipment requires periodic routine maintenance, either after set lengths of operating time or after so many operations. Conventional recording methods can be expensive and difficult to operate, and there is a significant market for applying PLCs to these needs.

When a programmable controller is presently installed on or near equipment to be monitored, it is likely that only a few of its total complement of software timers and counters have been used. Most PLC programs use less than 50% of such facilities, leaving several timers and counters for additional monitoring applications, if so desired. Typical monitoring requirements might be:

- to signal 'maintenance due' after 10 000 contact operations
- maintenance due after 1000 hours of machine running.

When the PLC is already used to control the plant in question, it is a trivial matter to add sections of ladder code to carry out the monitoring functions. This is done using the aforementioned counter or timer circuits, driven by appropriate physical contacts in the plant. Cascading of the timers/counters may be necessary to achieve the desired magnitude. In turn, these PLC functions can activate 'prompts' to maintenance personnel when the present limits are reached. In critical cases a 'time-out' or 'count-out' may be programmed to halt the machine until maintenance is effected.

The addition of such trivial circuitry to an existing control program is unlikely to increase overall scan time by a significant amount, since only a few timers or counters are involved. This means that in the majority of cases, it is possible to 'retrofit' this facility at any time during or after the system installation.

In a similar manner, other PLC facilities can be utilized to control alarms, etc. if plant or process inputs move outside preset limits. For example, the arithmetic 'greater than' and 'less than' functions can be used to test for parameters beyond set limits. They can be programmed to effect some counter action together with the triggering of a maintenance or fault alarm.

Both the above aspects of additional applications can be implemented using small program routines very similar to those covered in previous chapters, thus not warranting the inclusion of explicit examples.

Example of working procedures

(Source: SattControl UK Ltd.) See Fig. 9.18, pp. 346–50.

Procedure when constructing a PBS program for sequence control
The following examples show the procedure for gradually transferring an industrial problem to a complete PBS program.

Here we consider some of the factors which are important when solving problems, for example, the writing of a functional description before starting a program.

Note also the addition of interlocks to ensure greater reliability.

Sequence control is used when the controller machine works in a loop comprising a set number of stages or conditions. The PBS program is written so that an order is given for each stage by activating particular outputs.

The order is acknowledged via signals given when the program continues to the next stage in sequence. The advantages of this procedure include clarity and ease of making program changes.

Procedure:
1. functional description;
2. route/time diagram;
3. disposition of inputs/outputs;
4. functional plan;
5. interlocks;
6. relay ladder diagram (Boolean equations);
7. PBS program.

Fig. 9.18 Working procedure (courtesy SattControl, UK Ltd).

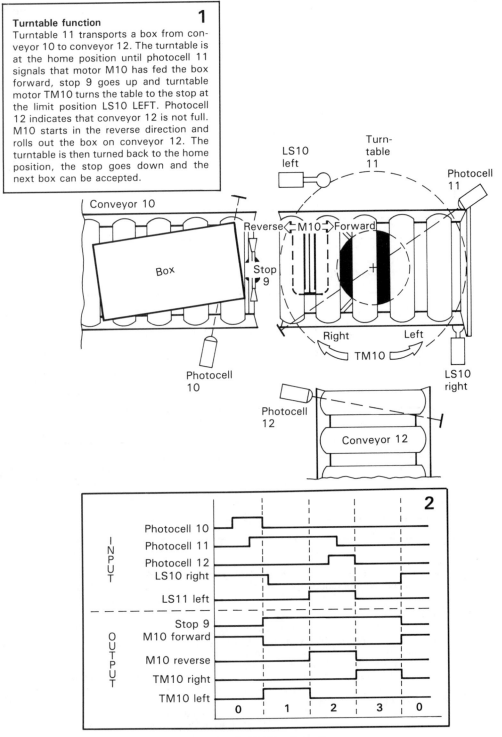

Turntable function

Turntable 11 transports a box from conveyor 10 to conveyor 12. The turntable is at the home position until photocell 11 signals that motor M10 has fed the box forward, stop 9 goes up and turntable motor TM10 turns the table to the stop at the limit position LS10 LEFT. Photocell 12 indicates that conveyor 12 is not full. M10 starts in the reverse direction and rolls out the box on conveyor 12. The turntable is then turned back to the home position, the stop goes down and the next box can be accepted.

Fig. 9.18 (Continued)

3

INPUTS

Photocell 10	000
Photocell 11	001
Photocell 12	002
LS10 right	003
LS10 left	004
Emergency stop	005

SattCon 05

OUTPUTS

040	Stop 9
041	M10 forward
042	M10 reverse
043	TM10 right
044	TM10 left
045	Ind. home position
046	Ind. emergency stop

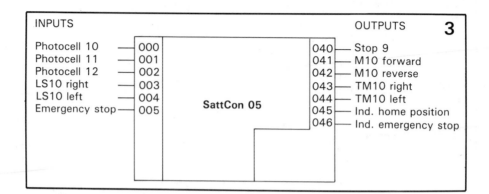

4

Sequence registers

Acknowledgements — Active outputs

LS10 right on 003	→	Step 0 020	→	Stop 9 off $\overline{040}$ / M10 forward on 041 / TM10 right on $\overline{043}$
Photocell 10 on 000 / Photocell 11 on 001	→	Step 1 021	→	Stop 9 on 040 / M10 forward on $\overline{041}$ / TM10 left on 044
LS10 left on 004	→	Step 2 022	→	M10 reverse on 042 / TM10 left off $\overline{044}$
Photocell 11 off $\overline{001}$ / Photocell 12 off $\overline{002}$	→	Step 3 023	→	M10 reverse off $\overline{042}$ / TM10 right on 043

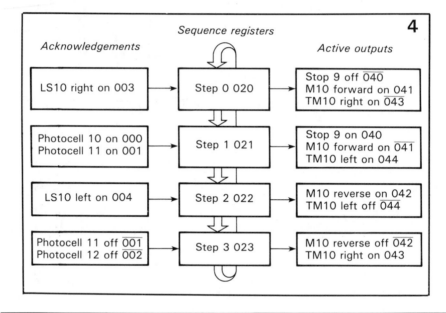

5

Interlocks
The operational sequence must normally be supplemented with interlocks for special purposes.

Emergency stop. When the emergency stop (input 005) is pressed, the sequence register and all outputs are reset. (A). Note that the emergency stop must primarily be implemented outside the SattCon 05, i.e. by breaking the supply to the electro-mechanical devices, and that input 005 is only information for SattCon 05.

Automatic initialization. When the emergency stop is reset and *if external* interference results is an undefined state, the turntable goes back to the home position (B) automatically.

Forward/reverse. To increase operational safety, VM10 RIGHT/LEFT must be interlocked with each other and their corresponding limit position. (C)

Indicators. When the turntable is in its home position, this must be indicated by a flashing light. An activated emergency stop is also indicated via a flashing light. Flashing achieved using I/O cell 007 (Frequency 1 Hz). (D).

Fig. 9.18   (Continued)

6

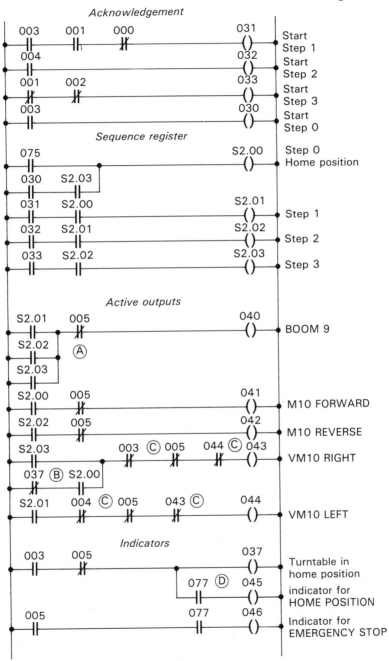

Fig. 9.18 (Continued)

```
--------------------------------------------                    7
SattCon 05 – SattControl
User: .SattControl....................
Date: .20-01-85......................
Project: .Styrning av vridbord 11......
--------------------------------------------

PBS
        *                                              *
000 : 00    – I I–   A       003      010 : 00    – I I–   A      S2.01
      01    – I I–   A       001            01    + I I–   O      S2.02
      02    –I/ I–   AN      000            02    + I I–   O      S2.03
      03    –( )–    =       031            03        )    RP
        *                                   04    –I/ I–   AN     005
001 : 00    – I I–   A       004            05    –( )–    =      040
      01    –( )–    =       032              *
        *                             011 : 00    – I I–   A      S2.00
002 : 00    –I/ I–   AN      001            01    –I/ I–   AN     005
      01    –I/ I–   AN      002            02    –( )–    =      041
      02    –( )–    =       033              *
        *                             012 : 00    – I I–   A      S2.02
003 : 00    – I I–   A       003            01    –I/ I–   AN     005
      01    –( )–    =       030            02    –( )–    =      042
        *                                     *
004 : 00             NPW              013 : 00    – I I–   A      S2.03
        *                                   01    +I/ I–   ON     037
005 : 00    – I I–   A       075            02    – I I–   A      S2.00
      01    + I I–   O       030            03        )    RP
      02    – I I–   A       S2.03          04    –I/ I–   AN     003
      03    –( )–    =       S2.00          05    –I/ I–   AN     005
        *                                   06    –I/ I– ‘ AN     044
006 : 00    – I I–   A       031            07    –( )–    =      043
      01    – I I–   A       S2.00            *
      02    –( )–    =       S2.01    014 : 00    – I I–   A      S2.01
        *                                   01    –I/ I–   AN     004
007 : 00    – I I–   A       032            02    –I/ I–   AN     005
      01    – I I–   A       S2.01          03    –I/ I–   AN     043
      02    –( )–    =       S2.02          04    –( )–    =      044
        *                                     *
008 : 00    – I I–   A       033      015 : 00             NPW
      01    – I I–   A       S2.02            *
      02    –( )–    =       S2.03    016 : 00    – I I–   A      003
        *                                   01    –I/ I–   AN     005
009 : 00             NPW                    02    –( )–    =      037
                                            03    – I I–   A      077
                                            04    –( )–    =      045
                                              *
                                      017 : 00    – I I–   A      005
                                            01    – I I–   A      077
                                            02    –( )–    =      046
```

Fig. 9.18   (Continued)

9.8 SUMMARY

This final chapter has attempted to address the essential factors in choosing, installing and commissioning a programmable controller system, looking at topics from basic rack layout through to system diagnostics and documentation. This chapter should not be read in isolation, but with consideration being given to the preceding chapters that have dealt with the individual aspects of PLC hardware and software.

As mentioned in the text, this book is *not* a handbook to any particular controller – it provides general information and guidance only. Readers should consult manufacturers' guides and data sheets when detailed, specific information is required, especially in the areas of electrical regulations and safety procedures.

9.9 EXERCISES

9.1 List the factors that must be considered when choosing the size and type of programmable controller required for a given application.

9.2 Describe the different ways in which a customer might obtain a PLC or complete installation. Comment on the advantages and disadvantages of each route.

9.3 Explain the use and worth of 'feasibility studies' in potential applications. What type of information may result from such a study?

9.4 Describe the sequence of events that should occur once a suitable PLC system has been chosen, leading to a successful installation. Your answer should include: system layout, wiring design, testing, software, etc.

9.5 Draw a fully annotated diagram to clearly show the hardware and software development paths in the implementation of a complete solution.

9.6 Describe the use of simulation aids during the test and commissioning stages.

9.7 What part do 'software backups' and amendment details play in the overall documentation of the project? Describe how these aspects should be dealt with.

9.8 Describe what is meant by 'internal diagnostics', and discuss the facilities and limitations of these software tools.

9.9 Explain how 'force' and 'monitor' functions can be used to aid troubleshooting sessions.

9.10 Describe the use of a 'watchdog' circuit.

9.11 Explain how a PLC can be used to monitor the operation and maintenance needs of plant machinery.

9.12 List the steps to be taken when attempting to troubleshoot a control problem.

9.13 Describe the facilities provided by a programmable controller that can assist in locating system faults.

Technical Details of Typical Programmable Controllers

The material presented on the following pages is selected and reproduced from manufacturers' manuals. Readers interested in details not given here, or wishing to follow up references should refer to the appropriate manual. The content and order of this Appendix is as follows:

GEC Industrial Controls Limited

Controllers

GEM 80/301, 302, 303, 311, 312 and 313

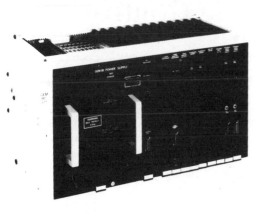

DESCRIPTION

Introduction

Following the highly successful GEM 80/250 series controllers, GEC introduces GEM 80/300 series controllers. This new series is a development of the 250 series and has all the facilities of that series and more. GEM 80/300 series are powerful controllers using 16 bit architecture processors and can handle up to 8192 I/O points.

I/O

A wide choice of GEM 80 I/O equipment is available for connection including Compact and Fast I/O equipment for digital I/O, operator interface units, analog I/O, pulse train handling, remote I/O systems and main/standby changeover systems. This controller uses a "Concentrated Basic I/O system" and interfaces with the standard Basic I/O highway via Basic I/O Expanders.

Communications

One of the most comprehensive ranges of communications facilities is available with this controller. Up to four built-in serial ports are fitted providing RS 422 and 20mA, 2 or 4 wire communication options. These powerful serial ports each offer direct multidrop capability with up to 8 other GEM 80 controllers in supervisory or co-ordinated operation with distributed control schemes.

Other facilities included are remote programming, down the line loading of user programs and serial printer output for report generation. These controllers accept on-line program changes.

In addition to the built-in serial ports. extra 20mA serial ports, programmable serial communications, STARNET or CORONET (LAN) Systems are optionally available for GEM 80/300 series.

Video

A new system has been introduced for graphically displaying the plant information generated from GEC's advanced GEM 80 — called IMAGEM graphic video. GEM80/300 accommodates IMAGEM.

Data Logging or Recipe Storage Facilities

Up to one megabyte of memory may optionally be fitted for off-line data storage and the controller includes full instructions for the control of this data movement.

Memory Options

All controllers are available in four memory options 192K, 256K, 384K or 512K. Over and above this, space is available for expansion to approximately one and a half megabytes. This memory may be optionally RAM or RAM/EPROM. The RAM/EPROM memory module carries both RAM and EPROM.

GEM 80 PRODUCT DATA   *GEC*

GEM 80/301, 302, 303, 311, 312 and 313

This allows program development on RAM and since the modules include their own built-in EPROM blowing circuits, in situ EPROM blowing is facilitated. Simply by plugging in different EPROM sub-modules with different programs, one controller will quickly adapt to operate other control programs.

Programming

These controllers run the well established user friendly GEM 80 ladder diagram language. GEM 80/301, 302 and 303 include a comprehensive range of instructions covering the following functions:-

relay replacement, logic, maths, signal processing, closed loop control (PID), numerical and binary test and manipulation, data control, B.C.D. handling, serial port, output functions, STARNET, off-line data storage.

Program Capacity and Processing Time

The program capacity depends on the memory size and the number of data tables. The 192K memory accommodates about 5,000 user instructions, 256K memory about 12,000 instructions, 384K memory about 24,000 instructions and the 512K memory about 36,000 instructions.

These controllers have a fast processing speed of $2^{1}/_{2}\mu s$ per word of source code and this relates to an average of 5ms per 1000 instructions.

GEM 80/311, 312 and 313

GEM 80/311, 312 and 313 include several additional instructions in their instruction set including PIDCON, TCONST, and RAMPGEN. These are new user friendly special functions whose settings operate independently of changes in the scan time.

Construction

These ruggedly constructed controllers are designed for use in harsh industrial environments. A range of modules plug into a 19″ rack mounting metal subrack. They include a new power module which is cooled by natural air, without fans. The plug-in modular construction allows a module substitution approach for repair and maintenance and easy upgrading of the controller.

Applications

GEM 80/300 series consists of power supply, memory, I/O processor, ladder processor mounted in a subrack with facilities for additional modules and equipment which you can specify for your application. These controllers are well suited to applications in complex industrial plant either as stand alone units or in distributed control schemes.

FEATURES

● Automatic self test and automatic compilation with English, German, Spanish fault diagnostic messages.

● Memory size field upgradable by plug-in modules.

● Built in clock for time and date.

GEM 80

GEC Industrial Controls Limited

GEM 80/141 142 Controllers

DESCRIPTION

Introduction

The GEM 80/141 and 142 controllers are powerful programmable controllers featuring new generation 16-bit architecture. Designed for complex industrial plant the GEM 80/141 or 142 can handle extensive digital and analog control, either standing alone, or in larger distributed systems.

The controllers consist of a 19" subrack with a power supply, two processors, memory module, Basic I/O interface, and space for up to six Compact I/O modules (providing 192 I/O points when using 32-channel modules). Each controller has two serial ports for operation in distributed control systems, remote programming, down the line loading of user programs, or for serial printer connection for report generation. The controllers can accept on-line program changes and have a fast processing speed of 5ms per 1000 instructions with a 20,000 instruction capacity when using a 256K memory module. Both the GEM 80/141 and /142 controllers have a 512 Basic I/O capacity, although the capacity of the GEM 80/142 can be doubled to 1024 by using an 8173 Basic I/O expander module. Both the controllers use ladder diagram language and are fully compatible with the rest of the GEM 80 family.

Hardware Structure

The GEM 80/141 and 142 controllers are multi-processor systems. Two processors are utilised. The ladder processor module executes the ladder diagram user program. A separate I/O processor handles I/O activities simultaneously as parallel tasks. The user program is stored in two forms:

"Compiled" for execution by the ladder processor;

"Source" for program monitoring and editing.

An 8272 RAM/EPROM module is optionally available. On controllers fitted with this option the compiled user program and P-tables can be transferred to EPROM for more permanent storage.

GEM 80 PRODUCT DATA   *GEC*

GEM 80

Controller Operation

The controller operation divides into three phases:

- a start-up procedure;
- an operating cycle;
- a fault shut-down procedure

The start-up procedure includes comprehensive self-tests and compilation of the user program. On satisfactory completion the system is allowed to be run and the first operating cycle begins.

The operating cycle is a continuous cycle where input data is read and the output data is derived by the controller based on the input data and the user program. The operating cycle also includes self-test checks. In the event of a fault, the controller enters a fault shutdown phase and is designed to shut down in an orderly fashion.

The operating cycle time may be either undefined, in which case the program free runs and resets as soon as it can, or it may be preset to a regular defined interval.

Monitoring, De-bugging, Automatic On-line and Off-line, Self-test and Watchdog Systems

The GEM 80/141 and 142 have built-in self-test routines to indicate possible faults within the controller. Capacity is available for the user to program plant fault analysis, and these systems are in addition to monitor points and indicator lamps already provided on many signals within the controller. Single cycle facilities are available for instruction by instruction running. The write instruction allows inputs or outputs to be forced on or off for test purposes, and the programmers include a monitoring mode which displays, in either a ladder diagram form or data table form, the results of running.

Error messages indicate any compilation errors and inter-communication errors, and an automatic self-test system fitted on all controllers in the GEM 80 range trips out a watchdog relay if serious faults occur. The watchdog relay should also be wired into the plant to provide a safe emergency shutdown procedure.

FEATURES

- 64K to 256K battery back-up RAM or EPROM memory module options
- Each controller can be upgraded to a larger memory size (up to 256K byts max.) by replacing the memory module with one of a larger capacity
- GEM 80/141 site upgradable to GEM 80/142 without any change to the I/O equipment already installed.
- Each serial port can be multi-dropped to up to 8 other GEM 80 controllers.
- User control of serial ports.
- Built-in clock for time and date.
- Automatic compilation with text error message.

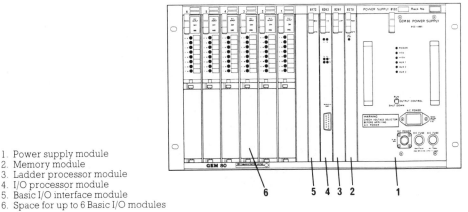

1. Power supply module
2. Memory module
3. Ladder processor module
4. I/O processor module
5. Basic I/O interface module
6. Space for up to 6 Basic I/O modules

Figure 1 - GEM 80/141 and 142 Subrack Layout

GEM 80

1. INTRODUCTION

1.1 General

GEM 80/141 & 142 controllers are designed for industrial applications controlling and supervising plant and processes. This flexible controller is well suited to a wide variety of applications and GEM 80 has been successfully applied in many industries, Automotive, Mining, Electricity Generation, Water Treatment, Food, Oil and Steel to name but a few, in many countries throughout the world.

To make a working system it is necessary to:-

(1) Select a Controller.

(2) Select I/O Equipment.

(3) Mount and wire the equipment and connect plant services and output devices to I/O.

(4) Hard wire watchdog and safety circuits.

(5) Connect a programmer and enter a program.

This manual details how this is to be achieved and includes fault finding, maintenance and spares information.

Before handling the GEM 80 your attention is drawn to Section 18 Safety and Handling Information.

1.2 GEM 80/141 & 142

1.2.1 Description

This controller offers up to 1024 digital I/O points and 2 serial links, on-line program changes and a fast processing speed. There is a choice of memory size, 64K, 128K, 192K or 256K and memory may be RAM or RAM/EPROM.

The large instruction set covers relay replacement, four function maths., timers, counters, sequencers, logic functions, closed loop control, group, array and analogue signal handling, B.C.D. handling and also block instructions.

Though the controller can be operated in the stand-alone mode, it can also operate in distributed controller schemes with data interchange via serial communication links for supervisory or co-ordinated operations.

GEM 80/141 & 142 are user programmable in GEM 80 control language with either System Programmer or Portable Programmer. This controller is compatible with the whole GEM 80 family.

GEM 80/141 is site upgradable to GEM 80/142 without any changes to the I/O equipment already installed.

1.2.2 Specification Summary

I/O	Basic I/O	: 141 up to 512 I/O points
		: 142 up to 1024 I/O points
	Serial Links	: 2 ports
Memory	User memory	: 64K - 6,000 instructions
	size and	: 128K - 12,000 instructions
	approximate	: 192K - 18,000 instructions
	instruction	: 256K - 25,000 instructions
	capacity	
	Memory	: either battery supported
	support	RAM or RAM/EPROM
Program	Typical	: 5ms/1000 instructions
	execution	
	time	
	Format	: GEM 80 rung ladder
		diagram

Instruction set: 64 instructions

Power Supply
110 or 220/240V a.c.
50 or 60 Hz (single phase)
or 24/48V d.c. option available

Environ-ment	Temperature	: 0 to 60°C operating
		: -20 to 70°C storage
	Humidity	: 10 to 90% non-condensing
Mounting	19 inch (483mm) rack mounting	

Additional Facilities
(1) On-line program changes
(2) Remote programming
(3) Built-in text editor for generation of printer port messages
(4) Flexible data table sizes
(5) Built-in clock for time and date
(6) Comprehensive monitoring and diagnostics with English text error messages

1.2.3 Hardware Layout

The basic controller hardware layout is shown in Figure 1.1. The controller consists of a Subrack, with Power Module, I/O, Processor Module, Ladder Processor Module, Memory module and Basic I/O Interface Module.

1. INTRODUCTION

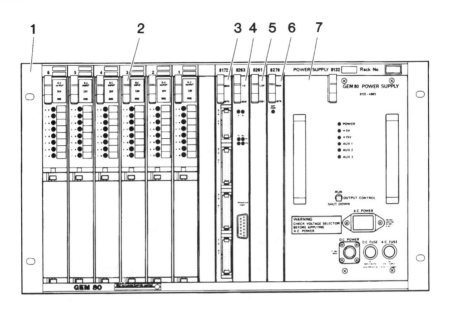

1. Subrack
2. Space for up to 6 Compact I/O Modules
3. Basic I/O Interface Module
4. I/O Processor Module

5. Ladder Processor Module
6. Memory Module
7. Power Supply Module

Figure 1.1 - GEM 80/141 & 142 Hardware Layout

1.3 Typical GEM 80/141 & 142 System

Figure 1.2 shows a typical GEM 80/141 & 142 system. The controller communicates with the outside world through:

(1) Basic I/O highway
(2) Serial Communication Links
(3) Programmer Connection Ports
(4) Watchdog Contacts

(1) Basic I/O Highway

The basic Input/Output highway connects to Compact I/O and Operator Interface Units through basic I/O Interface Modules.

The Compact I/O is designed for connection to such devices as pushbuttons, relays, lamps, solenoids, interlocks and limit switches, etc.

It gives a compatible and protected interface to the controller for electrical plant devices.

The Operator Interface Units include LCD units, thumbwheel switches etc. These allow operators to input control commands to GEM 80 and monitor plant status.

Full details of Compact I/O and Operator Interface Units are in the Product Data Sheets.

Further information is also available in Section 5.

(2) Serial Communication Links

The controller ports output serial link data for RS422 interfaces. Data can be sent serially to a printer for text message printouts (e.g. for alarms) or to other GEM 80 controllers for data interchange in supervisory or co-ordinated controller operations.

The serial links allow data to be interchanged between equipment up to several miles apart. Cable costs are very low as only a few wires are needed. 2 wire, 4 wire or fibre optic serial links can be set up and the two wire and four wire systems operate either 20mA current loop (where 20mA = 1, 0mA = 0) or RS422/V24.

Further information on serial links is in Section 6.

(3) Watchdog Contacts

These are designed to provide an output to shut down the plant safely in the event of a problem on the plant or in GEM 80, see Section 9 for further details.

1. INTRODUCTION

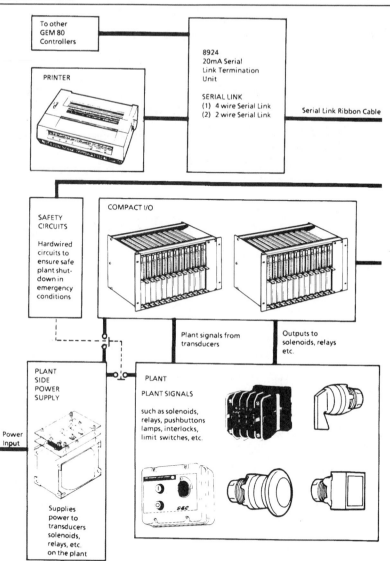

Figure 1.2 - GEM 80/141&142 Series System Diagram

1. INTRODUCTION

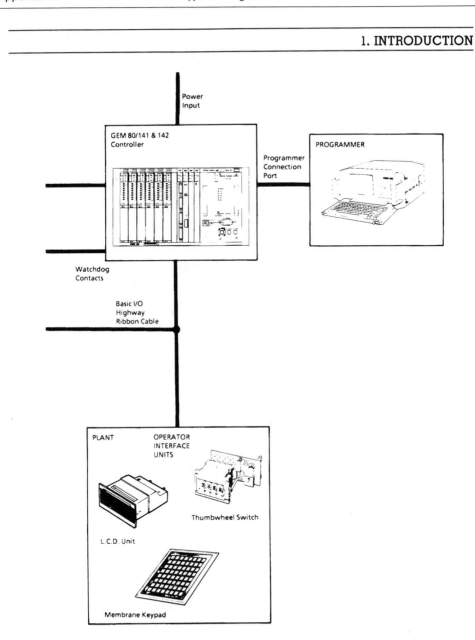

2. HARDWARE IDENTIFICATION

2.1 Subrack Labelling

Figure 1.1 shows identification fitted to the subrack. The model number is mounted on the left hand side flange. The upper rail holds a label marking the module positions and subrack reference number (the subracks are usually numbered R0, R1, R2 and R3).

2.2 Subrack Mechanical Structure

The plug-in modules slide along guide rails from the front of the subrack and push fit into connectors at the back of the subrack. These connectors are mounted on printed circuit boards referred to as "backplane p.c.b.'s". Figure 2.1 shows the guide rails.

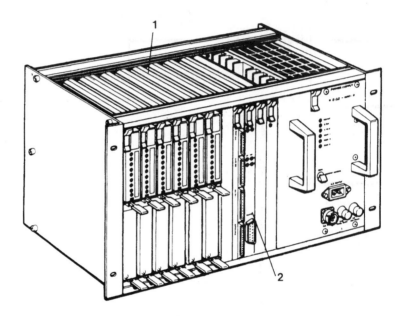

1. Guide rails
2. Plug-in modules

Figure 2.1 - The Controller

2. HARDWARE IDENTIFICATION

2.3 Front Panel Indicators and Controls

(1) Power Module

The plug-in power module front panel is detailed in Figure 2.2.

(2) I/O Processor

The I/O processor is detailed in Figure 2.3.

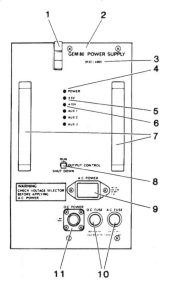

1. Ejector Handle
2. Module Function
3. Ordering Code
4. Power Input Lamp (on for power available)
5. +5V Lamp (on for supply available)
6. +15V Lamp (on for supply available)
7. Lifting Handle
8. Output Control
9. A.C. Power Input Socket
10. Fuses
11. D.C. Power Input Socket

Figure 2.2 - Power Module Front Panel

(3) Ladder Processor Module

This module is shown at Figure 2.4.

(4) Read/Write Memory Module

This module is shown at Figure 2.5.

(5) Basic I/O Interface

This module is shown at Figure 2.6.

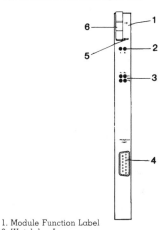

1. Module Function Label
2. Watchdog Lamps
3. Serial Communication LED's
4. Connector for Programmer
5. Module Type Number
6. Ejector Handle

Figure 2.3 - I/O Processor Front Panel

3. CONTROLLER ARCHITECTURE AND OPERATION

3.1 Controller Architecture

Figure 3.1 shows the controller architecture. It has several memories and several micro- processors each performing different duties and all connected to the central highway so that data can be interchanged between processor and memories.

The I/O processor connects to the Basic I/O highway, Serial Communications Ports and the programmer port and it controls data movement from the equipment connected to it (i.e. basic I/O highway, serial ports and programmer ports) to the read/write memories.

The ladder diagram processor executes the compiled user program which is stored in the 1st read/write memory.

The user memory holds three categories of information:

(1) Program instructions in "compiled" form. See section 3.2.3 for further information.

(2) Program instructions in "uncompiled" form referred to as "source" program. See section 3.2.3 for further information.

(3) Data tables, Section 4 details this.

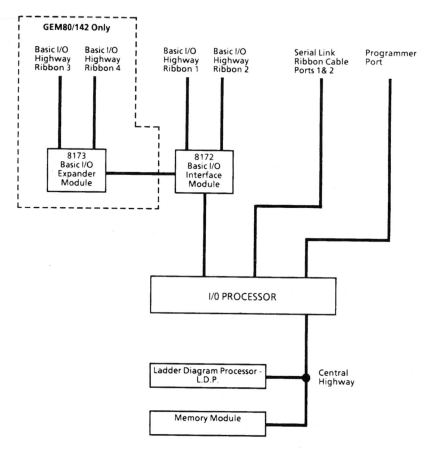

Figure 3.1 - Controller Architecture GEM 80/141 and GEM 80/142

3. CONTROLLER ARCHITECTURE AND OPERATION

3.2 Controller Operation

3.2.1 Controller Operating Cycle

Controller operation divides into three phases, a start-up procedure, an operating cycle and a fault shutdown procedure. The start-up procedure includes checking the controller (self-test) and compilation of the user program and on satisfactory completion allows the system to be run. The operating cycle is a continuous cycle where input data is read and the output data is derived by the controller based on the input data and the user program. The operating cycle also includes self-test checks. In the event of a fault, the controller enters a fault shutdown phase and is designed to shut down in an orderly fashion. Figure 3.2 shows the operating cycle in more detail.

The cycle time may either be undefined, in which case the program free runs and repeats as soon as it can, or else it may be preset to a regular defined interval.

If a repetition interval has been set which is shorter than either the program execution time or the I/O scan time, the repetitive cycle will not be regular, and the controller will free run. If the program is short, such that its execution time is shorter than the I/O scan time, then the user program will wait at the end of its execution until the I/O scan is complete before repeating. If, as is more usual in practical applications, the program takes longer to execute than the I/O scan, the next I/O scan does not begin until after the program execution has been completed.

The output scan is the first item in the cycle. If a regular repetition interval has been preset, a waiting period is inserted after the program execution at the end of the cycle. The program execution time itself will vary due to blocks within the user program and so the waiting period will also vary so as to make the cycle time up to the preset interval. The timing of the output scan is therefore held to a regular interval, varying very little, an advantage for continuous analogue control applications.

Data written by the user into the P Data Table determines whether the controller is free running or preset, see Section 4.6 for details.

Cycle time calculations are discussed in Section 10.9.

3.2.2 Self-test

Information on the self-test system is detailed in Section 15.7 - Fault Finding.

3.2.3 Compilation

The user program is stored in two forms, one for use in editing called the "source" program and the other for execution by the controller, called the "compiled" program. Conversion between the two is called Compilation and is carried out:

(a) at power up (automatically)

(b) on halt/run transitions (automatically)

(c) when the user issues a recompile command with the program running

(d) on single cycle when the program has been changed since the controller was last running.

Compilation not only converts the program into an executable form but also checks the source program to ensure it can be executed.

Any error detected during compilation will prevent the program being executed and if the controller was previously running a version of the program it will continue to run this "old" program. The errors are reported by text messages and a detailed discussion of compilation and associated problems is given in Section 15.5 - Fault Finding.

3.2.4 On-line Program Changes

As stated in Section 3.2.3, there are two versions of the program held in memory, compiled and uncompiled versions. While the compiled version of program is running and controlling the plant, the user can edit the source version without halting the controller. He can also obtain a print-out of his modified version, or dump it onto tape, again without halting the controller.

To incorporate the changes made to the source version of program into the running version, the user can give a "recompile" command from the Programmer. This causes the controller to pause for a short period (6 seconds maximum) while compilation is in progress. During this time, all outputs will freeze and all inputs will be ignored. Assuming that compilation is successful, the controller will then continue to run but using the modified program. If compilation is unsuccessful, the controller will resume running the previous compiled version of program.

There is only one version of the P-table and any changes to the data contained in the P-table affect the running program immediately, without recompilation being necessary.

3.2.5 Serial Links

The operation of the serial link is discussed in Section 6.

---

3. CONTROLLER ARCHITECTURE AND OPERATION

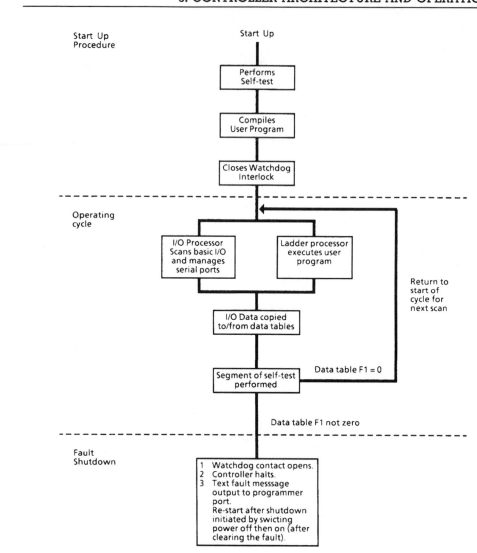

Figure 3.2 - Controller Operating Cycle

---

4. DATA TABLES

4.1 Introduction

The Data Tables consist of 16 bit binary words of data stored in the memory and accessed by their address.

4.2. Data Table Addressing

Each word is stored in the memory at a specific address. These addresses consist of a letter of the alphabet followed by a number, for example, A20, B200, P41, etc.

If, for example, we were to look at the content of data table address "P1" it may be:

```
BIT NUMBER   15 14 13 12 11 10  9  8  7  6  5  4  3  2
             1  0
ACTUAL VALUE  0  0  0  0  0  0  0  0  1  0  1  1  0  0
             1  0
```

Alternatively, we may only be interested in the content of one bit of P1 for example, bit 4. In this case we would look at data table address P1.4 and we would find P1.4 = 1.

4.3 Decimal, Binary and Hexadecimal Numbers

The data table content can be interpreted as:

(a) an equivalent decimal number.

(b) 16 individual bits each of value 1 or 0.

(c) an equivalent hexadecimal number. For example, address P2 may contain:

```
BIT NUMBER   15 14 13 12 11 10  9  8  7  6  5  4  3  2
             1  0
ACTUAL VALUE  1  0  1  0  1  1  0  0  0  0  1  1  0  1
             0  1
```

which can be interpreted as:
BINARY 1 0 1 0 1 1 0 0 0 0 1 1 0 1 0 1
DECIMAL -11,317
HEXADECIMAL @AC35

Decimal

As shown in the above example, when converting from binary to decimal bit 15 is taken as the sign bit and the number displayed will be within the range +32767 to -32768.

When bit 15 = 1 a negative number is represented and when bit 15 = 0 a positive number or zero is represented.

Hexadecimal

The numbers are to base 16 and can be distinguished from decimal by the fact that they are preceded by an @ symbol. The range of numbers covered is from @0000 to @FFFF.

4.4 Data Table Content and Addresses

Table 4.1 shows the addresses for the controller and also what information is to be found in these memory locations.

4.5 Data Table Sizes

Data tables E and F are of a fixed size (as shown in Table 4.1). Other Data table sizes are computed by the controller from the user program such that if for example, the highest G table address in the user program is G345 then G0 to G345 will be reserved.

Where addresses are implied (e.g. when using LOCATE and MOVE) then the user must include a dummy rung or rungs to declare the highest address, see Section 10 for full details.

When constructing the user program it should be borne in mind that memory wastage will occur if large gaps of unused data tables are left with only high numbered and low numbered tables used. Also, even if only one or two bits of a word are used (e.g. G104.1) the whole word G104 is reserved by the controller.

Programmers type 8922 with software version 1.8 or earlier and all 8920 programmers cannot work with bit addresses above 999.15. This restriction means for example, G999.15 can be used, but address G1000.1 cannot be used, as individual bits cannot be accessed. However, word G1000 upwards can be used where access to individual bits is not required.

4.6 P Data Table

Data in the P (Preset) Data Table falls into three categories:-

(1) Data entered by the user to control program repetition interval, serial link operational data and fast I/O self-test.
(2) Preset Data values entered by the user for use in his program.
(3) Storage of Printer Text Messages. The user should access printer text messages through the editor system detailed in Section 10.6.

Table 4.2 details the P table content.

The P Data Table is not cleared at power off or at halt/run transition, although a clear store command will clear address P51, 52 and P100 upwards. The content of the other P tables can be altered by over-writing.

Like other data tables, the size of the P table is derived by the controller as described in Section 4.5.

4. DATA TABLES

Table 4.1 - GEM 80/141 & 142 Data Tables

FUNCTION		Data Table	Content	Minimum Table Size	Maximum Table Size	Cleared to Zero at Halt/Run Transit	Cleared to Zero at Power Off	Cleared to Zero by Clear Store Command	For Further Details see Section
	Basic I/O Highway Data	A	This table of addresses stores the states of INPUT transducers connected to Basic I/O highways	0	64 words (each 16 bits)	YES	YES	YES	5
		B	This table of addresses stores the OUTPUT states (derived from the execution of the user program) for output to the Basic I/O Highways						
I/O	Serial Commun-ication Links Data	J	This table stores data RECEIVED by the controller through the serial communication links	0	512 words (each 16 bits) with one I/O processor	YES	YES	NO	6
		K	This table stores data derived by execution of the user program for TRANSMISSION on the serial communication links						
	Serial Link Control	I	These tables enable the user program to control serial link exchanges	0	18 words	YES	YES	NO	6
PRESET DATA		P0 to P99	Controls Program Repetition Interval, System Identification Codes, Serial Link operational data (signalling rate, printer data, etc.)	100		NO	NO	NO (except P51 & P52)	4.6
		Remai-nder of P table	Available for use in the user program for special functions printer text messages			NO	NO		YES
GENERAL WORK SPACE	Data retained through power down	R	These tables are available for the user program to use for storing counter values, sequencers and general data. Note tables only cleared when power first applied to system.	1		NO	NO	YES (except for R0)	
		W		0		NO	NO	YES	
	Data cleared	G O H Q N U	These tables are available for the user program to use for storing counter values, sequencers and general data.	0		YES (except when R0.0 = 1)	YES	YES	

4. DATA TABLES

Table 4.1 (continued)

FUNCTION	Data Table	Content	Minimum Table Size	Maximum Table Size	Cleared to Zero at Halt/Run Transit	Cleared to Zero at Power Off	Cleared to Zero by Clear Store Command	For Further Details see Section
TIMING AND FLAGS	E	These tables are written to by the controller and indicate overflow, time date and timing flags. They are cleared when power first applied.	7	7	NO	NO (except E0)	NO	4.7
FAULT CODES	F	These tables contain fault codes, which are written by the controller as a result of the built-in self-test routines or as a result of execution of the user program (Some tables are available for the user to use as he wishes).	200	200	NO	YES	NO	15.11
MAINTAINER	M	Available for user programmed tests.	0		YES	YES	YES	

Table 4.2 P Data Table

Address	Content	Remarks
P0	Available for user's system identification code	
P1	Program repetition interval (ms). Range 10 to 9500, zero for free running	Resolution 10ms
P2	Length of preset message area in bytes	Should not be written to by user. Automatically updated by message editor
P3 - P14	Serial link configuration data for port 1	See Sections 6.5, 6.6 and 6.7 for details
P15 - P26	Serial link configuration data for port 2	
P27 - P99	Reserved for future system use	
P100 upwards	Available for use in the user program for special functions	

4.7 E Data Table

The E Data Table contains timing markers and flags and the table is available for the user program to read. The information it contains is shown in Table 4.3.

4.7.1 Time and Date

This is generated from a crystal controlled oscillator mounted on a small backplane behind the power module. The accuracy is better than one minute per month and the clock runs while the controller is halted but stops during controller power interruption. The built-in calendar is correct for the lengths of all months up to the year 2099. Addresses E1 to E6 store the time and date (see Table 4.3). To set the correct time and date, simply write the correction into these tables with the programmer (programmer connection in Section 10.2).

4. DATA TABLES

Table 4.3 E Data Table

Address	Content	Remarks
E0.0	0.1 second timing marker	set on for one program cycle at 0.1 second (E0.0) and 1 second (E0.1) intervals see GEM 80 Programming Manual Ladder Diagram Format for further information.
E0.1	1 second timing marker	
E0.7	Limit flag set ON or OFF after each ADD or SUB instruction.	
E0.8 E0.9	Port 1 Buffer filling flags for printer ports. Port 2 Set ON when insufficient room in printer buffer for one complete line of output.	
E0.10 E0.11	Not used	
E0.12 E0.13	Port 1 Buffer empty flags for printer ports. Port 2 Set ON when all pending printer output has been transmitted by a port.	
E0.14 E0.15	Not used	
E1 E2 E3	Seconds (0 - 59) Minutes (0 - 59) Hours (0 - 23)	Time
E4 E5 E6	Days (as appropriate to month) Month (1 - 12) Years (0 - 99 or 1901 - 2099)	Date

5. BASIC I/O HIGHWAY

5.1 GEM 80/141

GEM 80/141 uses a standard Basic I/O Highway. This allows direct connection of Compact I/O Equipment and Operator Interface Units to the Basic I/O Interface Module BIOI (8172) by standard 26 way Basic I/O ribbon cable. Up to 512 I/O points can be accomodated and Figure 5.1 shows a block diagram of the system with data table addresses for each ribbon cable.

From Figure 5.1 it can be seen that the Basic I/O Highway connections are PL2 and PL3 and these can be accessed by withdrawing the 8172 module from the controller subrack. Figure 5.2 shows the location of connectors PL2 and PL3.

5.2 GEM 80/142

GEM 80/142 uses a concentrated Basic I/O Highway structure. The Basic I/O Interface Module (BIOI) 8172 in this controller is different than that fitted in GEM 80/141 controllers.

Like GEM 80/141, the GEM 80/142 includes circuitry to allow direct connection of Compact I/O equipment and Operator Interface Equipment to this controller for the first 512 I/O points. These 512 I/O points can be connected to PL2 and PL3 by standard 26 way Basic I/O ribbon cables.

Up to 1024 I/O points can be accomodated on GEM 80/142 and the remaining I/O points are available by the addition of an externally mounted Basic I/O Expander Module 8173. This module is designed to mount in a standard Compact I/O Subrack 8869, in the place of the right-hand blanking panel. The 8173 Basic I/O Expander module connects to the 8172 Basic I/O Interface Module by a Concentrated I/O Highway ribbon cable from PL1 in 8172 to PL1 in 8173.

Compact I/O Equipment and Operator Interface Units can then be connected to PL2 and PL3 on the 8173 by a standard 26 way Basic I/O Highway ribbon cable. Figure 5.3 shows a block diagram of this system and data table addresses for each ribbon cable with the select switch set to "0" on the 8172 and "1" on the 8173.

Figure 5.2 shows the location of the connectors and switches for the 8172 while Figure 5.4 shows the location of the connectors and switch for the 8173.

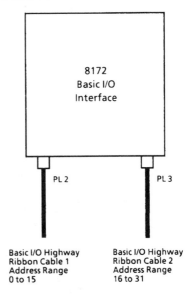

PL 2   PL 3

Basic I/O Highway   Basic I/O Highway
Ribbon Cable 1   Ribbon Cable 2
Address Range   Address Range
0 to 15   16 to 31

Figure 5.1 GEM 80/141 Block Diagram of Basic I/O

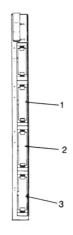

1. PL3
2. PL2
3. PL1 (not fitted on GEM 80/141)

Figure 5.2 8172 Basic I/O Interface Module BIOI

5. BASIC I/O HIGHWAY

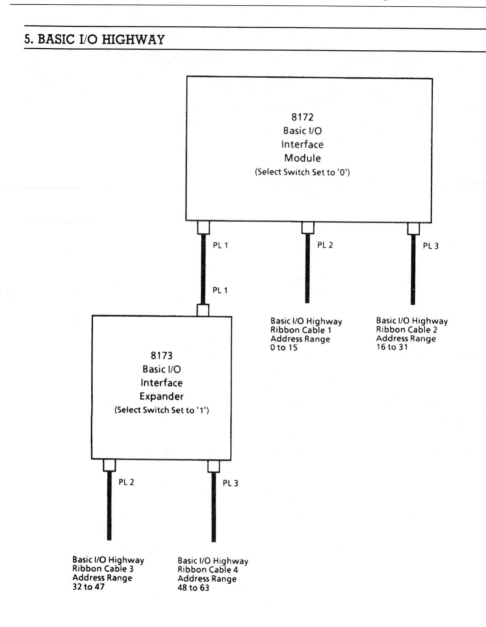

Figure 5.3 GEM 80/142 Block Diagram of Basic I/O

5. BASIC I/O HIGHWAY

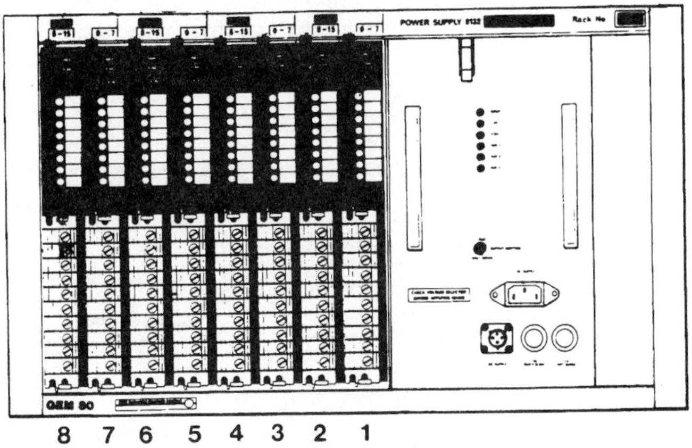

8 7 6 5 4 3 2 1

Annotation References	8	7	6	5	4	3	2	1
8 point module address	3	3	2	2	1	1	0	0
	HB	LB	HB	LB	HB	LB	HB	LB
16 point module address	7	6	5	4	3	2	1	0
32 point module address	15 & 14	13 & 12	11 & 10	9 & 8	7 & 6	5 & 4	3 & 2	1 & 0
Slot Number	8	7	6	5	4	3	2	1

HB = High Byte LB = Low Byte

Figure 5.6 8867 Compact I/O Address

Table 5.1 A/B Data Table Addresses for Decoder Settings 0 to 7 for 8 Bit Modules.

**A/B Data Table Addresses
Decoder Switch Settings**

Ribbon Cable	0	1	2	3	4	5	6	7
1	0 & 1	2 & 3	4 & 5	6 & 7	8 & 9	10 & 11	12 & 13	14 & 15
2	16 & 17	18 & 19	20 & 21	22 & 23	24 & 25	26 & 27	28 & 29	30 & 31
3	32 & 33	34 & 35	36 & 37	38 & 39	40 & 41	42 & 43	44 & 45	46 & 47
4	48 & 49	50 & 51	52 & 53	54 & 55	56 & 57	58 & 59	60 & 61	62 & 63

5.3.3 Data Table Addresses for Each I/O Point

Figure 5.7 shows a GEM 80/141 controller. Module numbers 1 & 2 are 8 channel Compact Input Modules and input addresses are A table, so module numbers 1 and 2 are address A. Each module has 8 channels so the two modules have 16 channels and form one 16-bit word A0 as shown in Figure 5.7. Word A0 will store the input states for these 16 channels. Module numbers 3 and 4 are 8 channel Compact Output Modules and output addresses are B table, so data table B1 stores the 16 output states for these channels.

In this example, address A0 is input module, so address B0 does not exist and also B1 is output module, so address A1 does not exist. Addresses are allocated to words 0 to 63 as A or B according to whether input or output modules are fitted. Also, module pairs form one word, so each pair of modules must be the same, either a pair of input modules or a pair of output modules. Figure 5.8 shows an example of two basic I/O highway ribbon cables on a GEM 80/ 141 controller.

5. BASIC I/O HIGHWAY

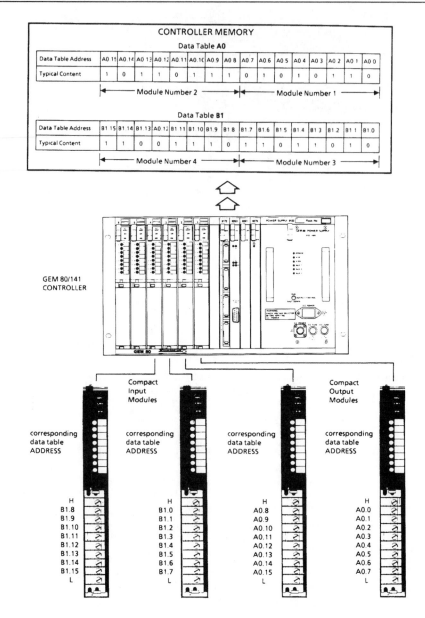

CONTROLLER MEMORY

Data Table A0

Data Table Address	A0.15	A0.14	A0.13	A0.12	A0.11	A0.10	A0.9	A0.8	A0.7	A0.6	A0.5	A0.4	A0.3	A0.2	A0.1	A0.0
Typical Content	1	0	1	1	0	1	1	1	0	1	0	1	0	1	1	0

|← Module Number 2 →|← Module Number 1 →|

Data Table B1

Data Table Address	B1.15	B1.14	B1.13	B1.12	B1.11	B1.10	B1.9	B1.8	B1.7	B1.6	B1.5	B1.4	B1.3	B1.2	B1.1	B1.0
Typical Content	1	1	0	0	1	1	1	0	1	1	0	1	1	0	1	0

|← Module Number 4 →|← Module Number 3 →|

GEM 80/141
CONTROLLER

Compact Input Modules

Compact Output Modules

corresponding data table ADDRESS

corresponding data table ADDRESS

corresponding data table ADDRESS

corresponding data table ADDRESS

H		H		H		H	
B1.8		B1.0		A0.8		A0.0	
B1.9		B1.1		A0.9		A0.1	
B1.10		B1.2		A0.10		A0.2	
B1.11		B1.3		A0.11		A0.3	
B1.12		B1.4		A0.12		A0.4	
B1.13		B1.5		A0.13		A0.5	
B1.14		B1.6		A0.14		A0.6	
B1.15		B1.7		A0.15		A0.7	
L		L		L		L	

Figure 5.7 GEM 80/141

5. BASIC I/O HIGHWAY

Table 5.5 Basic I/O Ribbon Cable Blank Form Example

Item No.	RIBBON CABLE NUMBER Equipment	No. of Bits	Where Used	Rack No./Location	Address Range	Worst Case +15V Consumption mA
1	8110(9-66V dc Input)	8	Conveyor Interlock	R1 slot 1	A0.0/A0.7	8.4
2	8110(9-66V dc Input)	8	Conveyor Interlock	R1 slot 2	A0.8/A0.15	8.4
3	8111(110V ac Input)	8	Conveyor Tracker	R1 slot 3	A1.0/A1.7	6.5
4	8111(110V ac Input)	8	Conveyor Tracker	R1 slot 4	A1.8/A1.15	6.5
5	8114(110V ac Output)	8	Track Sensor 9-12	R1 slot 5	B2.0/B2.7	81.0
6	8114(110V ac Output)	8	Track Sensor 9-12	R1 slot 6	B2.8/B2.15	81.0
7	8111(110V ac Input)	8	Safety Cubicle	R2 slot 1	A3.0/A3.7	6.5
8	8111(110V ac Input)	8	Safety Cubicle	R2 slot 2	A3.8/A3.15	6.5
9	8114(110V ac Output)	8	Operator Desk	R2 slot 3	B4.0/B4.7	81.0
10	8114(110V ac Output)	8	Operator Desk	R2 slot 4	B4.8/B4.15	81.0
11						
12						
13						
14						
15						
16						
17						
18						
19						
20						
21						
22						
23						
24						
25						
26						
27						
28						
29						
30						
31						
32						

Total +15V power used by this ribbon cable — 358.8

Length in metres of this ribbon cable — less than 10 m.

Check that +15V consumption is less than maximum ribbon cable currents given in Table 5.3. — YES / NO

6. SERIAL COMMUNICATION LINKS

6.1 General

The 140 series controllers have two serial ports. The signalling rate is user adjustable between 110 and 9600 bits per second.

Serial link connections are provided for 26-way ribbon cable on connector PL2. Access to the connector is obtained by withdrawing the two basic I/O expander modules and the I/O processor module.

Connector PL2 is located on the upper backplane to the right of the basic I/O expander connections, as shown in Figure 6.1. This one connector carries data for both ports 1 and 2.

6.2 Applications

The serial ports are used for four types of application:
(a) Data table exchange with other GEM 80 controllers, for supervisory operations or co-ordinated operation of GEM 80 controllers. Details in Section 6.3
(b) Output and input of alpha-numeric characters for printers, VDU's or keyboards. Details in Section 6.8.
(c) Programming the controller (remote programming). Details in Section 6.8.
(d) Program uploading or downloading (Block Messages). Details in Section 6.9.

6.3 Data Table Exchange with other GEM 80 Controllers.

By data set into the P (preset) Data table, any of the four ports can be configured as one of the following:
(a) Point-to-Point Control Port.
(b) Multidrop Control Port.
(c) Tributary Port.

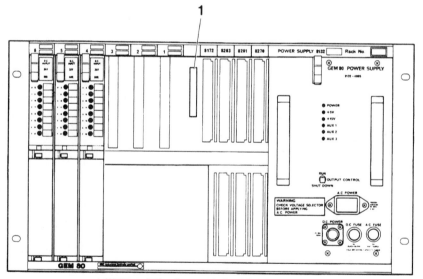

1. PL2 26 way Ribbon Connector for Serial Link Ports 1 & 2.

Figure 6.1 GEM 80/141 and 142 Serial Port Connections.

6. SERIAL COMMUNICATION LINKS

6.3.1 Point-to-Point Control Port

Figure 6.2 shows a point-to-point serial link, here two controllers interchange data. One controller's serial communication port is set up as a point-to-point control port and the other as a tributary port.

The control port is responsible for all activity on the serial link and the tributary responds only when a message is addressed to it inviting a reply.

Figure 6.4 shows an example of a point to point serial communication link between two 140 series controllers (4 wire) 20mA. Up to 32 input and output words can be exchanged and the serial link can be run in either the free running mode or the user control mode depending on data set in the control port P Data Table. Serial links can be 2 wire or 4 wire connection to 8924 and full connection details are in 8924 Product Data Sheet.

Sections 6.5 and 6.6 detail the data which needs to be written into the P table and where data arrives and is transmitted from, for a tributary port.

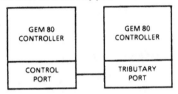

Figure 6.2 - Point-to-Point Serial Link

6.3.2 Multidrop Link

Figure 6.3 shows a multidrop serial link. Here up to 3 controllers can interchange data when the GEM 80/141 or 142 powers the 20mA current sources within the 8924. When GEM 80/200 or /250 or /350 or /150 controllers power the 20mA current sources within the 8924 up to 8 controllers can interchange data (see Section 6.14 for further details). One controller's serial communication port is set up as a multidrop control port and the others are set up as tributary ports. The control port communicates with each tributary on a cyclic basis.

Communication directly between tributaries is not possible though data can be exchanged between tributaries via the control port.

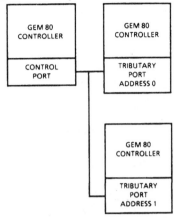

Figure 6.3 - Multidrop Serial Link

Figure 6.5 shows an example of a typical multi- drop link.

Up to 32 input and output 16 bit words of data can be exchanged and the serial link can be run in either the free-running mode or the user control mode depending on data set in the P- tables. Full connection details for 4 wire and 2 wire links are shown in 8924 Product Data Sheet.

Sections 6.5 and 6.6 detail the data which must be written into the P table and where data arrives and is transmitted from.

6. SERIAL COMMUNICATION LINKS

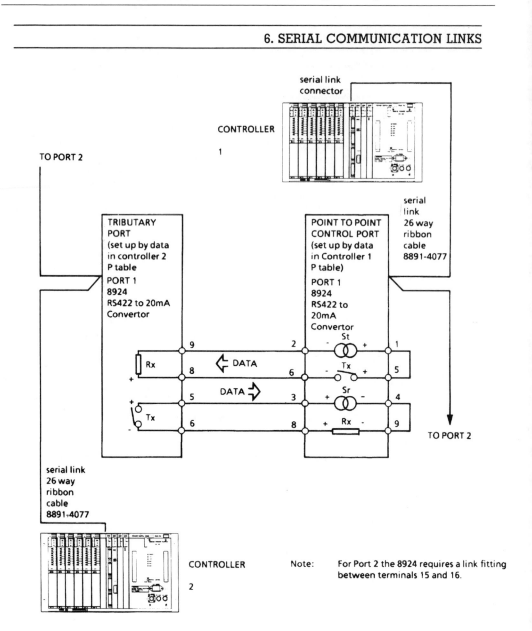

Figure 6.4 - Point to Point Serial Communications Link Between Two 141 or 142 Controllers 4-wire 20mA

6. SERIAL COMMUNICATION LINKS

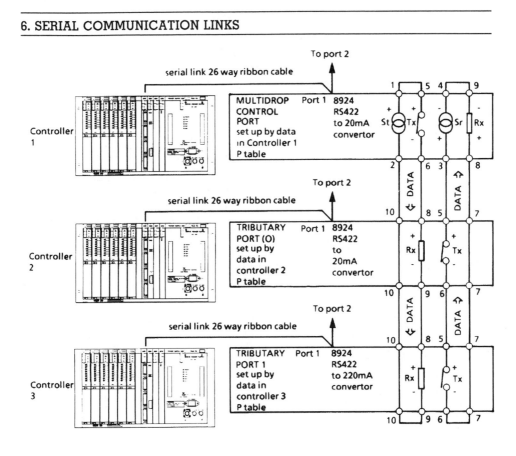

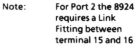

Note:　　For Port 2 the 8924
　　　　requires a Link
　　　　Fitting between
　　　　terminal 15 and 16

Figure 6.5 - Multi-drop Serial Communication Link Between Three Controllers 4 wire 20mA

6. SERIAL COMMUNICATION LINKS

6.4 Data Tables Used by Serial Links

The Serial Communication Links read and write data in data tables, J, K, P, I and F. Table 6.1 details these.

Data written into the P table by the user determines whether the port is a Tributary Port, Control Port or a Printer Port and Table 6.2 shows how to tell which mode the port is set to operate in. For example, a 0 on P4 indicates port 1 is set as a printer port.

6.5 Tributary Port

6.5.1 General

Table 6.3 shows the address and what data the user must write into these addresses to enable the port to operate as a tributary port.

When the port is configured as a tributary port, data is received at the data table addresses shown in Table 6.4. Data to be transmitted by the controller is at data table addresses shown in Table 6.5 and I Table information to operate the serial link in the user control mode is shown in Table 6.6.

Recompile to implement changes to P table (no need to power down). Changes to transmission length are effective immediately.

Table 6.1 - Data Table Used by Serial Communications Links

		Address	
Data Table	Content	Port 1	Port 2
J	holds data sent to the controller	J0 to J255	J256 to J511
K	holds data to be sent out by the controller	K0 to K255	K256 to K511
P	data entered by the user into this table forms operation parameters for the serial port (signalling rate, control port or printer port, etc.)	P3 to P14	P15 to P26
I	holds information about control of serial link for user control mode	I0 to I8	I9 to I17
F	fault diagnosis table	See Section 13 for further information.	

Table 6.2 - Determining Operating Mode of a Serial Port

Port	Data Table Address	Content	Mode	Further details in paragraph
1	P4	0	PRINTER	6.7
		2	CONTROL PORT	6.6
		any other value	TRIBUTARY PORT	6.5
2	P16	0	PRINTER	6.7
		2	CONTROL PORT	6.6
		any other value	TRIBUTARY PORT	6.5

6. SERIAL COMMUNICATION LINKS

Table 6.3 - P Table for Tributary Port

Function	Content	Port 1	Port 2
signalling rate	enter 110 for 110 bits/sec. enter 300 for 300 bits/sec. enter 600 for 600 bits/sec. enter 1200 for 1200 bits/sec. enter 2400 for 2400 bits/sec. enter 4800 for 4800 bits/sec. enter 9600 for 9600 bits/sec. enter 0 for port not in use enter -ve numbers for extended time out see paragraph 6.6.2(c)	P3	P15
operation mode	enter any value but 0 or 2 for tributary port	P4	P16
free-running/user control	enter 0 for free running enter non-zero for user control. See Section 6.5.2 for explanation	P5	P17
tributary address	enter 0 to 14 (GEM 80 control ports 0 to 7). Where other computers generate the control port, addresses up to 14 can be used.	P6	P18
number of data table locations to be transmitted (Free Running Mode Only)	enter 0 to 32	P7	P19
	not used	P8-P14	P20-26

Table 6.4 - J Table for Tributary Port

	Port 1	Port 2
J Table (Data Received by the Controller)	J0-31	J256-287

Table 6.5 - K Table for Tributary Port

	Port 1	Port 2
K Table (Data to be Transmitted by the Controller)	K0-31	K256-287

Table 6.6 - I Table for Tributary Port

Function	Content	Port 1	Port 2
Flags	SEND FLAG is bit 0 of the words shown on the right (I0, I9 etc.) RECEIVE FLAG is bit 8 of the words shown on the right (I0, I9 etc.)	I0	I9
number of words received or transmitted	bits 8-15 received length bits 0-7 length to transmit	I1	I10

6.5.2 Operation as a Tributary Port

Data Table transfers only occur between user program scans and the I table is also only examined between scans in the user control mode.

For J/K table exchanges between controllers the user can select either user control mode or free running mode.

In either mode an indication is provided in F41 bits 0 to 3 when a port has not received a valid message for a period of approximately 30 seconds.

(a) Free Running Mode (Tributary Port)

When Free-running mode is selected the reply to a received message is generated immediately, at the end of the program scan the quantity of locations for the reply being taken from the P- table.

(b) User Control Mode (Tributary Port)

User-control mode uses data in the I-table to control the generation of reply. I-table is set to indicate reception and length of received message. The transmission of the reply is initiated at the end of the program scan in which the I-table bit is set. The length of the message transmitted is taken from the I-table. The following description is for port 1 but applies equally to all ports if the relevant data tables are substituted.

6. SERIAL COMMUNICATION LINKS

Table 6.9 - I-table for Control Port

				Port 1	Port 2
Flags	SEND FLAGS are bits 0 to 7 of the 16 bit words shown to the right (I0, I9 etc.). RECEIVE FLAGS are bits 8 to 15 of the 16 bit words shown to the right (I0, I9 etc.). Each bit applies to tributary addresses as shown below:-			I0	I9

SEND FLAGS bit	tributary address	RECEIVE FLAGS bit	tributary address
0	0	8	0
1	1	9	1
2	2	10	2
3	3	11	3
4	4	12	4
5	5	13	5
6	6	14	6
7	7	15	7

			Port 1	Port 2
Number of words received or transmitted	bits 0-7	tributary address 0	I1	I10
	transmit	tributary address 1	I2	I11
	length	tributary address 2	I3	I12
	bits 8-15	tributary address 3	I4	I13 receive
		tributary address 4	I5	I14 length
		tributary address 5	I6	I15
		tributary address 6	I7	I16
		tributary address 7	I8	I17

6.6.2 Operation as a Control Port

Data table transfers only occur between user program scans and the I table is also only examined between scans in the user control mode.

The control port can be operated in the user control mode or the free running mode and the user can select either of these modes.

(a) Free-running Mode

Transfers occur automatically on a cyclic basis to tributaries in use. Transmission occurs between user program scans and message length is in the P table. When free-running mode is selected the controller cycles round the routes indicated in the preset data to be in use and attempts a J/K exchange, transmitting the quantity of locations requested in the P-table. If a valid reply is not received after the system of retries has been exhausted the route is classified as failed and the corresponding bit in F42 or F43 is set ON. In any one cycle a message exchange will be attempted to only one route which is known to be failed. If a failed route recommences communication the F- table bit is set OFF and the route is no longer classified as failed.

(b) User-control Mode

A transfer only occurs when an I table bit is set. The destination of the data is determined by which bit of the I table is set.

User-control mode allows the user to vary the transmitted message lengths and also to select dynamically which routes will be used. To initiate a message transfer on a particular route the following procedure should be followed: (example refers to port 3 route 3)

(i) Set bits 0 to 7 of I22 to number of locations to transmit.

(ii) Set desired data from K608 onwards.

(iii) Set I18.11 OFF and set I18.3 ON. When the message has been generated I18.3 will be set OFF. The user could now begin assembling the next message for this route in K608 onwards.

(iv) If the transfer fails F43.3 will be set ON. If the transfer succeeds the data received will be placed in J608 upwards and bits 8 to 15 of I22 will contain the number of locations written. I18.11 will be set ON to indicate that the reply has arrived.

It is possible to request transfers on several routes simultaneously. In this situation the requests are handled in order of route number: a transfer on a particular route being completed before the next is commenced.

6. SERIAL COMMUNICATION LINKS

When a message is received the data is placed in J0 onwards as when Free-running. The number of locations written is placed in bits 8 to 15 of I1 (this may be extracted using SWAP (S4) and masking the result) and bit I0.8 is set ON. The user is thus notified of the arrival of a message. Appropriate action may then be taken and the data for the reply placed in K0 onwards. The length of this data is placed in bits 0 to 7 of I1. When the uer has set bit I0.8 OFF and I0.0 ON the reply will be generated. I0.0 will be set OFF to indicate this.

6.6 Control Port

6.6.1 General

Table 6.7 shows the P table addresses and data the user must write into these addresses to enable the port to operate as a control port.

When the port is configured as a control port data is received and transmitted at data table addresses shown in Table 6.8.

I table information to operate the serial link in the user control mode is shown in Table 6.9.

Recompile to implement changes to P table (no need to power down). Changes to transmission lengths and tributary selections are effective immediately.

Table 6.7 - P-table for Control Port

Function	Content	Port 1	Port 2
signalling rate	enter 110 for 110 bits/sec. enter 300 for 300 bits/sec enter 600 for 600 bits/sec. enter 1200 for 1200 bits/sec. enter 2400 for 2400 bits/sec. enter 4800 for 4800 bits/sec. enter 9600 for 9600 bits/sec. enter 0 for port not in use enter negative values for extended time out see paragraph 6.6.2 (c)	P3	P15
operation mode	enter 2 for control port (note 0 selects printer port any other value selects tributary port)	P4	P16
free-running/user control	enter 0 for free running enter non-zero for user control.	P5	P17
indication of number of tributaries (Free-running mode only)	bit 0 to 7 set enables data to be sent to tributaries 0 to 7 respectively (bit 8to 15 unused)	P6	P18
number of data table locations to be transmitted (Free Running Mode Only)	Trib address 0 (enter 0 to 32)	P7	P19
	Trib address 1 (enter 0 to 32)	P8	P20
	Trib address 2 (enter 0 to 32)	P9	P21
	Trib address 3 (enter 0 to 32)	P10	P22
	Trib address 4 (enter 0 to 32)	P11	P23
	Trib address 5 (enter 0 to 32)	P12	P24
	Trib address 6 (enter 0 to 32)	P13	P25
	Trib address 7 (enter 0 to 32)	P14	P26

Table 6.8 - J and K Table Addresses for Control Port

	Port 1	Port 2
Tributary address 0	J/K0-31	J/K256-287
Tributary address 1	J/K31-63	J/K288-319
Tributary address 2	J/K64-95	J/K320-351
Tributary address 3	J/K96-127	J/K352-383
Tributary address 4	J/K128-159	J/K384-415
Tributary address 5	J/K160-191	J/K416-447
Tributary address 6	J/K192-223	J/K448-479
Tributary address 7	J/K224-255	J/K480-511

9. WATCHDOG AND SAFETY CIRCUITS

9.1 Introduction

The controller includes automatic self-test routines. These are performed when the power is switched on and also while the controller is running. The diagnostic information obtained from these tests is stored in an area of memory known as the F Table.

The Controller also includes a built-in relay called a watchdog relay and the state of this is determined by the content of a particular location of the F Table, location F1. When the system is running normally and in a healthy condition, F1 is zero and the watchdog relay is energised. When a non-zero value is put in location F1, the watchdog relay is de- energised, the normally open contact opens, and the controller halts. The watchdog relay is also de-energised whenever the controller is halted.

9.2 Safety and the User Program

The program can be arranged to take particular courses of action when certain error conditions arise, but these only work while the controller is up and running correctly. For those exceptional conditions where the controller malfunctions regardless of what the program says, it will not necessarily respond, thus safety circuits must be hardwired.

9.3 Power Up Conditions

When the Controller is first powered up controller output equipment can give spurious outputs until the watchdog relay has pulled in (during which time the Controller is doing its start-up checks). It is therefore essential that the plant side power supplies are not switched onto the output modules and units until after the watchdog relay has pulled in.

9.4 Power Loss, Emergency Stops, Pushbuttons and Moving Machinery

If there is a mains power cut the controller will shut down in a controlled fashion, and restart automatically when the power is re-instated. If, because moving machinery must not start moving again without intervention by the operator, then the plant side power supplies must not be applied until an external relay circuit has been operated.

Where moving machinery must be stopped if the controller malfunctions brakes and braking circuits must be hard wired.

While normal start/stop pushbuttons can be wired via the I/O Emergency stops moving machinery must be hard wired. Use a chain of normally closed pushbuttons of the stop lock- off type for emergency stop circuits, mounted locally near moving equipment so that any maintenance man can have confidence that, if he has pushed his local button down, no-one can start the equipment again until he has reset his local button.

9.5 Typical Watchdog Circuits

Figure 9.1 shows an example of a typical watchdog safety circuit. The input equipment will be energised when the plant side power supply is on. However, the output equipment will only be energised when the plant side power supplies and the watchdog follower relay is energised. Other interlocks such as emergency stop pushbuttons can be connected in this circuit if required (in series with the watchdog contacts).

9.6 Watchdog Contact Ratings

circuit	: volt free contact provided
rating	: a.c. 200VA 1A 250V make/break :d.c. 100W 1A 250V make/break
operation	: Closed when host controller healthy and its status is running with normal inputs. Held closed during user program on-line re-compilation.

9.7 Watchdog Connections

Figure 9.2 shows the location of the watchdog terminals . Access is gained by removing the power supply module and the blanking panel.

Two 6.3mm (1/4") push-on tabs are revealed at the top right hand side on the upper backplane.

9. WATCHDOG AND SAFETY CIRCUITS

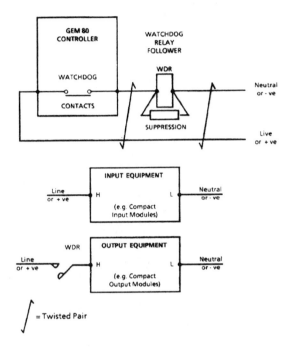

Figure 9.1 - Typical Watchdog Circuit

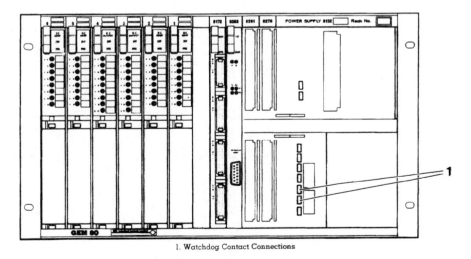

1. Watchdog Contact Connections

Figure 9.2 - GEM 80/141 and 142 Watchdog Connections

10. USER PROGRAM

Function		Instruction	Special Function	Remarks	Number of Instructions	Number of Table Locations used by VALUE	Execution Time 5 μs Average unless otherwise stated below
4. Signal Processing	Debounce Input Signal	DBOUNCE	S0	Provides signal conditioning to logic signals prone to spurious or transient problems	2	5 (See also note 12)	92
	Function generator	FGEN	S30	Four quadrant operation	2	4 + N (See also note 13)	380
	Dead Band	DEDBAND	S31	Includes offset	2	3 (See also note 12)	55
	Limiter	LIMIT	S32	High and low limits	2	3 (See also note 12)	85
	Ramp Generator	RAMP	S33	Comprehensive Ramp generator (includes rounding)	2	15 (See also note 12)	3600
	First Order Time Constant	ANALAG	S37	Smooths fluctuating numerical values	2	5 (See also note 12)	214
5. Closed Loop Control	Absolute 3 Term Controller	PID ABS	S34	Provides absolute proportional, integral and derivative control	2	15 (See note 12)	520
	Incremental 3 Term Controller	PID INC	S35	Provides incremental proportional, integral and derivative control	2	9 (See note 12)	330
	Incremental Output	INC OUT	S36		2	5 (See note 12)	225
6. Binary Tests	High State	HISTATE	S3	Reports bit number of highest bit of 16 bit word set on (1)	1	-	42
	Bits ON	BITSON	S26	Reports number of bits sent on (1) in a group of 16 bit words	2	2	See note 3
	Compare Group	CMPGRP	T21	Compares two groups of 16 bits and reports details of the first difference	2	3	See note 8
	Get Bit	GETBIT	T22	Performs a test of on state individual bits in group of words	2	2	150
7. Binary Manipulation	Swap 1/2 locations	SWAP	S4	Swap 1/2 locations	1	-	1
	Invert	INVERT	-	Inverts each bit of 16 bit word	1	-	-
	Invert Group	INVGRP	S25	Inverts each bit of a group of 16 bit words	2	3	See note 2
	Shift	SHIFT	S27	Left and right shift on a group of 16 bit words	2	3	See note 4
	Rotate	ROTATE	S28	Left and right rotations on a group of 16 bit words	2	3	See note 4
	Put Bit	PUTBIT	T23	Allows setting and resetting bits of a word and generating specified words	2	3	See note 9

10. USER PROGRAM

Function		Instruction	Special Function	Remarks	Number of Instructions	Number of Table Locations used by VALUE	Execution Time 5μs Average unless otherwise stated below
8.Numerical Tests	Compare	COMPARE	T0	These special functions report if:	2	1	52
	Comparison	COMPARISON					
		EQ	T1	(a) X = Y	1	0	7
		NE	T2	(b) X = Y	1	0	7
		GT	T3	(c) X Y	1	0	7
		LT	T4	(d) X Y	1	0	7
		GE	T5	(e) X Y	1	0	7
		LE	T6	(f) X Y	1	0	7
9. Numerical Manipulation	Negate	NEGATE	S5	Changes the sign of a number. Makes a number positive. Outputs all 16 bits if input is non-zero	1	0	4
	Absolute	ABS	S6		1	0	5
	Non-zero	NONZERO	S7		1	0	8
10. Block Instructions	Enclose conditionally executed rungs	START OF BLOCK	-	Conditionally executed block of ladder diagram instructions can be nested to a depth of 16	1		
		END OF BLOCK	-		1		
11. Data Moving	Locate Data Table Address	LOCATE	S20	These two special functions are usually used together for moving groups of data from one area of memory to another	2	1	132
	Move Data Table Address	MOVE	T20		2	0 or 1	See note 7
	Cyclic Store	STORE	T30	first in first out store	2	4 + N See note 15	270
12. B.C.D. Handling	B.C.D. input conversion	BCDIN	S1	Converts up to 4 decades of binary.	1	0	89
	B.C.D. output conversion	BCDOUT	S2	Converts binary to up to 4 decades of binary codeddecimal	1	0	66
13. Serial Port Alphanumeric Input and Output	Print Text	PRITEXT	S38	Enables messages stored in controller memory to be output to serial port	2 Port	(C + 1)/2 See note 14	360
	Print	PRINT	T38	Enables user definable messages to be output on a serial port	2	N See note 16	See note 11
	Character Input	CHARIN	T39	Enables characters input on the serial ports to be stored in controller memory	2	(C + 1)/2 See note 14	315
14. Output Functions	Output	OUT	-	Outputs 16 bit word tospecified data table location	1		
	Output	OUTPUT	S8	Intermediate output from rung	2	1	6

10. USER PROGRAM

Notes on Table 10.1

1. 247 + 166*Qty of locations
2. 214 + 9*Qty of locations
3. Simple operation : 113 Group operation : 128 + 74*Qty of locations
4. 186 + 30*Qty of locations
5. 355 + 32*Qty of locations
6. 390 + 42*Qty of locations
7. 220 + 3*Qty of locations
8. 250 + 24*Qty of locations
9. Simple operation : 60 Group operation : 160
10. 312 + 19*Qty of locations
11. 8*Message number + 1400 for pure text OR 1600 to 2400 for mixture of text and numbers
12. Typically uses stated number of locations in both preset and working tables
13. N = number of tabulated points or array length
14. (message length)/2
15. 4 + store length
16. N = quantity of numbers to be output

10.1.3 Instruction Capacity

The instruction capacity depends on two factors:

(a)   Memory size

(b)   The number of data table words
          used

Table 10.2 shows the approximate instruction capacity for each memory size with 5,000 and 10,000 data table locations.

In fact the controller calculates the size of the data table memory based on the addresses specified in the user program. The space remaining after allocation of data tables is then available for user program instructions.

Each program instruction occupies four bytes of memory while each data table accupies two bytes of memory. If the number of data tables are known then the approximate program capacity can be calculated.

If for example, the controller has a 64k memory and the user program requires 5,000 data tables then:-

Approximate maximum   = (Number of bytes
number of program   available for Data tables
instructions   and instructions)
                                         - (2 x Number of Data
                                         Table Locations)

$$\overline{}$$
                                  4

From table 10.2 the number of bytes available for data table and instructions is 30k for 64k memory. 30k is approximately 30,000.

Approximate maximum   = (30,000) - (2 x 5,000) = 5,000
number of program $\overline{}$
instructions
                                         4

For GEM 80/141 and 142 no more than 62,000 bytes may be used for data table locations.

10.1.4 Rung Diagrams Rules and Restrictions

1) Up to 10 elements are permitted across, including the output element, i.e. coil.

```
|-][ -][ -][ -][ -][ -][ -][ -][ -][ -( ) -|
```

2) Up to 5 elements are permitted down (i.e. up to 5 branches).

```
| -] [- --------------------------- ( ) ----
| -] [- |
| -] [- |
| -] [- |
| -] [- |
```

Figure 10.3 Up to 5 parallel branches

Combinations of series and parallel branches are allowed.

3) Only one output is permitted on any one page of ladder diagram. (Anything that joins to the right hand side is termed an output).

4) Every element must have an address (or numerical value).

5) Any elements requiring two inputs (such as counters, sequencers) must have inputs allocated, i.e. reset lines.

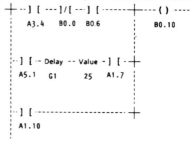

Figure 10.4 Typical Rung Diagram

Compact Analogue Input Multiplexer 8108

DESCRIPTION

Traditionally, the integration of analogue facilities with a programmable control system was an expensive step.

The introduction of the GEM 80 8108 Analogue Input Multiplexer brings analogue capability within reach of even the most limited budget. It offers sixteen highly accurate channels, makes only minimal demands on the host controller's data tables and provides a remarkable degree of flexibility.

The 8108 is a double-width Compact I/O module, designed for inclusion in any GEM 80 Basic I/O subrack. Since the 8108 Analogue Input Multiplexer does not require a Fast I/O capability the benefits of this module can be effectively utilised to extend the I/O range of 100 Series Controllers.

Although each of the channels is independently converted only one input data table word is required in the host controller. Such efficiency is achieved by time-division-multiplexing all converted channel data into this data table word. The module attaches a channel number to the converted data so that the processor knows which channel it is reading. Simple programming techniques, fully described in the 8108 User Information sheet, are used to separate the multiplexed data.

The multiplexing can be under the control of the module (self-scanning) or the host controller (forced-scanning).

Self-scanning

If channel priority is not important, the task of selecting channels for conversion may be left to the module, in which case the channels are converted in a repeated sequence from one to sixteen. If the full complement of sixteen channels is not required, the overall conversion time can be reduced by presetting the number of channels from the front panel.

Forced-scanning

The host controller dictates the channel for conversion using a second data table word, which gives rapid access to data on particular channels as required. If channel requests cease the module returns to the self-scanning mode, constantly updating the data tables until new requests are received from the host controller.

GEM 80 PRODUCT DATA

GEC Industrial Controls Limited

FEATURES

- Offers full compatability with a wide variety of signal conditioning equipment. (See Data Sheet T492 for guidance.)
- Selectable signal range ±5V or ±10V.
- 20mA range available, using the 20mA Termination Panel or Termination Rail.
- The module can be supplied with a Termination Panel or a Termination Rail as requested. All necessary interconnecting cables are included.
- Monitor points are available for the connection of test equipment during commissioning and fault diagnosis. (Termination Panel option only.)
- All inputs are filtered to preserve data integrity.
- The selector switches and the plant wiring connector are conveniently located on the front panel.
- No user power connections are required. Power is supplied via the Basic I/O ribbon.

SPECIFICATION

Channels	: 16 analogue inputs
Ranges	: ±5V or ±10V (nominal) ±5.117V or ±10.235V (full scale)
	: 20mA (using 20mA Termination Panel)
Input impedance	: >10MΩ (5V range) 500kΩ (10V range)
Conversion time	: 70ms ±20ms per channel
Accuracy 25°C	: ±0.2%
0 to 60°C	: ±0.5%
Resolution	: 11 bits plus sign
Filtering	: 4.5ms
Electrical isolation	: 100V between GEM 80 and plant
	: 50V between channels

Power from host controller	: 150mA @ 15V
Installation	: Any Basic I/O subrack, occupying two slots
Temperature	: 0 to 60°C operating
	: −20 to +70°C storage
Humidity	: 10 to 90% non-condensing
Termination Panel dimensions	: 156 x 481 x 32mm
Termination Rail dimensions	: 40 x 483 x 46mm

ORIENTATION

Please note that the module must be mounted vertically, making it unsuitable for applications where rolling may occur, tumblers etc.

ORDERING CODES

8108 Module	Termination Panel	Rail	5/10V range	20mA range	Ordering Code
●	●			●	8108-4011
●	●			●	8108-4012
●	(includes 3m cable harness)				8108-4015
●	(module only)				8108-4003
(rail only)		●	●		8891-4010
(rail only)		●		●	8891-4014

Note: Modules supplied with Termination Panel include 2m of ribbon cable.

ADDITIONAL INFORMATION

If you require any further information or wish to discuss an application in detail, do not hesitate to contact our GEM 80 representative.

GEC Industrial Controls Limited.
Programmable Controllers Division,
Kidsgrove, Stoke-on-Trent,
Staffordshire ST7 1TW, England.
Tel: Kidsgrove (07816) 3511
FAX (Group 3) Kidsgrove (07816) 76329
Telex: 36293 GECICK

Also at:
Mill Road, Rugby,
Warwickshire CV21 1BD.
Tel: Rugby (0788) 2121

Registered at London No. 540682
Holding Company: The General Electric Company p.l.c.,
1 Stanhope Gate, London, W1A 1EH.

Publication No. T.513   Issue 1   7.86

GEC Industrial Controls Limited

Compact I/O Modules (Digital)

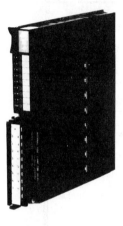  

DESCRIPTION

Introduction

Progressive development of this excellent range of modules has resulted in the addition of 16 and 32 point modules and also an output enable circuit. These modules interface on/off signals to GEM 80/100/200/300 and 700 series controllers. Signals from equipment such as pushbuttons, lamps, solenoids, limit switches, relays, contactors, TTL and CMOS may all be interfaced through these modules.

The Range

Both a.c. and d.c. plant side supplies are catered for. A.C. supplies from 24 to 220V at 50 or 60 hertz are accepted. D.C. supplies from 5 to 110V are accepted. Relay Output Modules function up to 380V.

Electrical Design

The modules provide a high noise immunity interface using optical isolators for each plant signal. An indicator lamp is provided for the state of each channel, and the design includes built in fuse protection on output modules.

The output modules can be connected to inductive loads. Interface to the controller is through the Basic I/O highway.

Mechanical Design

Practicality, toughness and quality have been the watchwords of the design of these modules, and indeed throughout the GEM 80 range. The modules include moulded side covers, and connectors are mechanically coded to prevent incorrect replacement.

Output Enable Modules

Some output modules now include the output enable circuit. This provides two additional features. Firstly, at power up this circuit ensures that all outputs are held off until the host controller initiates output. Secondly, two terminals are available for connection to an output enable signal. This will hold all outputs off until needed. The output enable circuit operates in the same manner for all the modules except the 8167. This module has its own output enable terminals, unlike the remaining modules where the output enable circuit is only accessed via the backplane of subracks which provide this facility (e.g. 8869-4002). For more information on the output enable, see the Compact I/O User Information Document (T438).

GEM 80 PRODUCT DATA

GEM 80

Mounting

The plug-in modules fit in compact subracks such as:

   8869 12-slot Compact I/O subrack
   GEM 80/100C/130C/141 and 142 Controllers

Application

The high electrical isolation, wide operating temperature and excellent noise immunity make these modules ideal for today's industrial environment.

FEATURES

● Up to 32 I/O points per module.

● The plug-in mechanically coded plant connector with locking mechanism allows quick and foolproof module service exchange.

● The module is powered through the Basic I/O ribbon cable usually from the host controller. This minimises power supply requirements and power supply wiring.

SPECIFICATION

Temperature	: 0 to 60°C operating
	: −20 to 70°C storage
Environment and Tests	: GEM 80 standard
Weight	
Input Modules	: 0.5kg (approximately)
Output Modules	: 0.75kg (approximately)
Dimensions	: 233mm high x 28mm wide x 250mm deep
Common Mode Isolation Voltage	: 2kV peak transient except 8116 and 8143 which is 4kV and 8163 and 8164 which is 1.5kV peak transient 325V r.m.s. continuous

Load Current (Output Modules)

The load currents quoted are only valid up to operating temperatures of 40°C. Thereafter the derating of Figure 1 applies.

Further, these load currents, and also surge currents, are only valid when the TOTAL CURRENT ON ALL CHANNELS DOES NOT EXCEED 6A RMS.

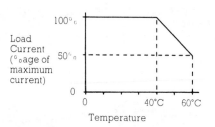

Figure 1 - Output Module Deratings at Operating Temperature above 40°C

Power Requirements

Optical isolation separates the plant signals from the GEM 80 controller so two power sources are required. One on the plant side of the optical isolation (i.e. the plant side power supply). The other on the GEM controller side of the optical isolation. This is provided by a +15V supply drawn from the host controller (usually through the Basic I/O highway ribbon cable).

Plant Connections 8 and 16 Channel Modules

Plant wiring is terminated onto screw terminals of a removable edge connector socket on the front of the module. Conductors up to 1.5mm² can be accommodated.

Plant Connections 32 Channel Modules and 8130 Relay Module

Plant wiring is terminated onto a terminal rail which is connected by a 3 to 5 metre cable harness and socket to the module.

8140 Strobe Output Module

The 8140 has a strobed output facility on each channel. Once an output channel has been set "ON", GEM 80 I/O scans will produce a delayed strobed output.

The strobed output for each I/O scan when a channel is set "ON" is shown below:

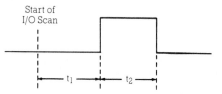

Where: t_1 = Delay from setting bit "ON" to output changing

      t_2 = Period for which output pulse is present

         t_1 = 11.7 to 18.4mS
         t_2 = 10.0 to 15.8mS

GEM 80

COMPACT I/O MODULES INDEX SHEET

Compact I/O modules currently available and contained within this combined Product Data are listed below. The T numbers identify the specification sheet where more information on the modules can be found.

8110	8 Channel 9-66V DC Input Module	(T533)
8113	8 Channel 9-66V DC Output Module	(T530)
8111	8 Channel 110V AC Input Module	(T534)
8114	8 Channel 110V AC Output Module	(T531)
8112	8 Channel 220V AC Input Module	(T534)
8115	8 Channel 220V AC Output Module	(T531)
8116	8 Latched Normally Open Contacts 380V Relay Output Module	(T532)
8143	8 Unlatched Normally Open Contacts 380V Relay Output Module	(T532)
8117	4 Channel 110V AC Isolated Output Module	(T531)
8167	4 Channel 110V AC Isolated and with Output Enable Output Module	(T531)
8118	8 Channel 110V DC or AC Input Module	(T534)
8168	8 Channel 110V DC or AC Isolated Input Module	(T534)
8119	8 Channel 110V DC Output Module	(T530)
8120	8 Channel 9-66V DC Inverse Input Module	(T533)
8121	8 Channel 9-66V DC Inverse Output Module	(T530)
8123	8 Channel 24V AC Input Module	(T534)
8124	8 Channel 24V AC Output Module	(T531)
8125	8 Channel 48V AC Input Module	(T534)
8126	8 Channel 48V AC Output Module	(T531)
8130	8 Change over Contacts 50V Relay Output Module	(T532)
8138	8 Channel 4-15V DC TTL CMOS Input Module	(T533)
8139	8 Channel 4-15V DC TTL CMOS Output Module	(T530)
8140	8 Channel 9-66V DC with Output Strobe Output Module	(T530)
8147	8 Channel 9-66V DC Isolated Input Module	(T533)
8148	8 Channel 110V DC or AC Inverse Input Module	(T534)
8149	8 Channel 110V DC Inverse Output Module	(T530)
8152	8 Channel 24V DC Input Module	(T533)
8153	8 Channel 24V DC 0.5A Output Module	(T530)
8162	8 Channel 24V DC 0.05A Output Module	(T530)
8163	32 Channel 24V DC Input Module	(T533)
8164	32 Channel 24V DC with Output Enable Output Module	(T530)

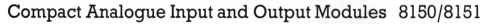

GEC Industrial Controls Limited

Compact Analogue Input and Output Modules   8150/8151

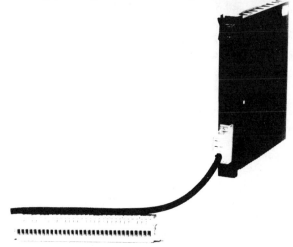

DESCRIPTION

The 8150 and 8151 Compact Analogue Modules have been designed specifically for applications where control over a limited number of analogue functions is required and where the 16 channel capacity or the Fast I/O requirement of the 8840 and 8841 modules is uneconomic.

Since the 8150 and 8151 Compact Analogue Modules do not require a Fast I/O capability the benefits of these modules can be effectively utilised to extend the I/O range of 100 Series Controllers.

The 8150 and 8151 modules locate securely into the 8869-4002 Compact I/O Subrack. Each module appears to the user as two consecutive Basic I/O data table words.

Because the 8150 and 8151 modules can be accommodated in the same subrack as numeric modules and digital modules the number of subracks required can be significantly reduced thereby affording generous economies of space and capital.

Physically identical, both modules have tough, moulded covers and front panels, supporting rugged connectors, individually coded to prevent incorrect replacement. Together with their superior electrical isolation, wide operating temperature tolerance and excellent immunity to noise, these features make the modules ideal for today's industrial environment.

FEATURES

● Both channels are electrically isolated from each other, the plant side power supply and the GEM 80 controller, with common mode isolation up to 200V peak.

● Voltage or current operation (either unipolar or bipolar) on both the output and input modules, allowing a wide variety of transducers, actuators, meters etc. to be connected.

● Overall accuracy is better than 0.2% FSD (at 25°C).

● An ejection handle allows easy withdrawal of the module without stressing the connectors and also locks the module in place when fully seated in the subrack.

● On each module, plant wiring is terminated at the logically arranged and clearly labelled termination rail (supplied). Three metres of multicore cable connect the rail to the easily accessible 'D-type' plant connector on the front panel. The plug is securely held for maximum stability and personalised with coding pins to prevent its connection to the wrong module. The module itself is similarly coded allowing its insertion in the correct subrack slot only. These features mean that no special skills are required to replace a module, freeing trained personnel for more important tasks.

GEM 80 PRODUCT DATA

GEC Industrial Controls Limited

SPECIFICATION 8150

Type	Compact Analogue Input
Channels per module	Two
Nominal signal ranges	$\pm 10V$ 0-20mA User selectable ranges see Order Codes for details
Input impedance (Voltage ranges)	$500k\Omega$ Overloading - no damage up to 250V d.c. or a.c. rms
Input impedance (Current ranges)	50Ω Overloading - no damage to $\pm 30mA$ d.c. or a.c. rms
Conversion time	24ms
Accuracy 25°C 0-60°C	$\pm 0.2\%$ full scale $\pm 0.5\%$ voltage $\pm 1.0\%$ current
Resolution	12 bits (unipolar) 11 bits plus sign (bipolar)
Plant side power requirements	24V d.c. $+20\%$ - 10% @ 400mA max
Power supply available for plant devices	60mA per channel @ $\pm 15V$
Power from host controller	8mA @ 15V
Temperature	0-60°C (operating) -20 - +70°C (storage)
Humidity	10-90% non-condensing
Module mounting	single width module.

SPECIFICATION 8151

Type	Compact Analogue Ouput
Channels per module	Two
Nominal signal ranges	$\pm 10V$ 0-20mA User selectable ranges see Order Codes for details
Output load (Voltage ranges)	$2k\Omega$ minimum Short circuit to signal 0 volts continuous no damage
Output load (Current ranges)	500Ω maximum using internal $+15$ volt supply Compliance up to 24V using external power supply
Conversion time	2ms
Accuracy 25°C 0-60°C	$\pm 0.2\%$ full scale $\pm 0.5\%$ voltage $\pm 1.0\%$ current
Resolution	12 bits (unipolar) 11 bits plus sign (bipolar)
Plant side power requirements	24V d.c. $+20\%$ - 10% @ 400mA max
Power supply available for plant devices	60mA per channel @ 15V
Power from host controller	6mA @ 15V
Temperature	0-60°C (operating) -20 - +70°C (storage)
Humidity	10-90% non-condensing
Module mounting	single width module.

ORDERING CODES

Equipment	Ordering Codes	
	Input	Output
Module for 0-20mA signal range only Terminal rail 3m Cable Connector & Coding Pin	8150-4052 8891-4037 8891-4094	8151-4052 8891-4036 8891-4094
Module for $+/-10v$ signal range only Terminal rail 3m Cable Connector & Coding Pin	8150-4012 8891-4037 8891-4094	8151-4012 8891-4036 8891-4094
Module with user selectable ranges $\pm 10v$, $\pm 5v$, 0-10v, 0-5v, 0-20mA, 0-10mA. Terminal rail 3m Cable Connector & Coding pin	8150-4002 8891-4028 8891-4094	8151-4002 8891-4026 8891-4094

ADDITIONAL INFORMATION

If you require any further information or wish to discuss an application in detail, do not hesitate to contact our GEM 80 Representative.

GEC Industrial Controls Limited.
Programmable Controllers Division.
Kidsgrove. Stoke-on-Trent.
Staffordshire ST7 1TW. England.
Tel: Kidsgrove (07816) 3511
FAX (Group 3) Kidsgrove (07816) 76329
Telex: 36293 GECICK

Also at:
Mill Road. Rugby.
Warwickshire CV21 1BD.
Tel: Rugby (0788) 2121

Registered at London No. 540682
Holding Company: The General Electric Company p.l.c.,
1 Stanhope Gate. London. W1A 1EH.

S.P. E8594

Publication No. T.510   Issue 1.   6.86
© 1986 The General Electric Company p.l.c. of England.

GEC Industrial Controls Limited

STARNET   8816

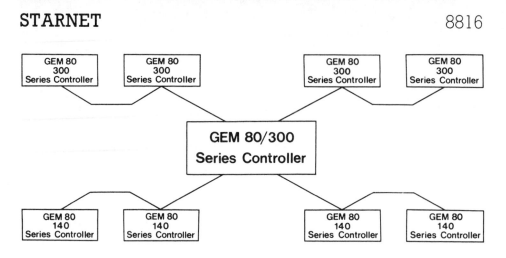

DESCRIPTION

STARNET modules may be fitted into central highway spaces in GEM 80/300 or 700 controllers. They provide two modes of communication, either:-

a. High performance serial communications between controllers, by 2 wire RS485 links using HDLC - NET protocol at a data rate of up to 180k bit/s and distances up to 2km.

b. 2 or 4 wire 20mA current loop serial ports for communication with standard GEM 80 controllers with ESP protocol.

Each Starnet module provides 4 serial ports and occupies 2 slots in the central highway.

FEATURES

● Fast turn round times

● High data security - automatic error detection and recovery

● Controller is isolated from serial link

● Choice of configuration of point to point STARNET links, multidrop STARNET links or cross coupled STARNET links

● Compatible with early GEM 80 controllers

SPECIFICATION
Performance Figures

Maximum Turnround Time (Seconds)	ESP or Starnet HDLC-NET	Data Rate	Conditions
0.084	HDLC-NET	180K bit/s	Transfer of 32 data tables between 8 controllers primary controller scan time 6ms, secondary controller scan time 5ms.
0.168	HDLC-NET	48K bit/s	
0.630	ESP	9.6K bit/s	
0.7	HDLC-NET	180K bit/s	Transfer of 32 data tables between 8 controllers primary controller scan time 100ms, secondary controller scan time 80ms.
0.7	HDLC-NET	48K bit/s	
1.4	ESP	9.6K bit/s	

8816

GEC Industrial Controls Limited

HDLC-NET Protocol

Bit rates	: 48 or 180k bit/s
Encoding Method	: NRZI method normally used on STARNET. FMO and FM1 also available.
Cable Length	: up to 2km at 48k bit/s NRZI coding : up to 1.5km at 180k bit/s NRZI coding
Isolation	: Transmission cable is isolated from GEM 80 electronics and can withstand 1.5kV transient
STARNET Link cable	: Screened twisted pair normal impedance 150 ohms capacitance 28.9pf/m resistance less than 46 ohms/km (typical cable Belden 9182 screened twisted pair 19 x 34 stranded conductors foil shield)
Electrical characterstics of signals	: Balanced 2 wire to RS 485 standard
Status of Ports	: Each port can be configured as primary or secondary : Primary organises all link data flow. Only one primary per link.
Maximum Number of secondaries	: Up to 31
Quantity of data transfered	: Up to 777 consecutive data table
Control of transfer (two methods)	: Special functions CONFIG, STATS, SEND and RECEIVE allow user program to control the operation of STARNET : Available table swop using N, O, P and U tables

ESP Protocol

Bit rates	: 9.6k, 4.8k, 2.4k or 1.2k bit/s
Electrical characteristics of link	: 2 or 4 wire 20mA current loop

In other respects the ESP protocol operates in the same way as the serial ports built in to the controller.

ORDERING CODES

STARNET module and backplane	8816-4001
HDLC serial link termination unit (provides terminals for one port)	8587-4001
20mA Serial Link termination panel (provides terminals for one port)	8924-0000
Ribbon cable for connecting the above (includes facilities for two ports)	8891-4077

ADDITIONAL INFORMATION

If you require any further information or wish to discuss an application in detail, do not hesitate to contact our GEM 80 Representative.

GEC Industrial Controls Limited.
Programmable Controllers Division,
Kidsgrove. Stoke-on-Trent.
Staffordshire ST7 1TW, England.
Tel: Kidsgrove (07816) 3511
FAX (Group 3) Kidsgrove (07816) 76329
Telex: 36293 GECICK

Also at:
Mill Road. Rugby.
Warwickshire CV21 1BD.
Tel: Rugby (0788) 2121

Registered at London No. 540682
Holding Company: The General Electric Company p.l.c.,
    1 Stanhope Gate. London W1A 1EH.

ID0226G   S.P.I1662

Publication No. T521   Issue 1   9/86

GEC Industrial Controls Limited

Analogue Input   8840

DESCRIPTION

The 8840 analogue input module has been designed to provide a simple solution for applications requiring high speed analogue conversion of plant signals.

Connecting directly into the Fast I/O highway of a GEM 80 subrack, the analogue input module can interface sixteen analogue signals from transducers, transmitters, potentiometers, etc.

All plant connections are easily made into a terminal rail or panel and taken via a cable harness to the module.

No additional power supplies are necessary as the module draws power directly from the Fast I/O highway.

FEATURES

● Suitable for industrial applications.
● Fast conversion, high resolution.
● Overvoltage protection.
● Choice of voltage and current ranges.
● Differential amplifier input.
● Analogue signal filtering.
● Mechanical coding of module type.

SPECIFICATION

Number of inputs: 16

Scaling/Filtering : same for all 16 inputs

Nominal signal range : ±10V; ±5V; ±1V; ±0.5V

(options) : ±20mA

Input filtering (options) : 5ms; 0.5ms

Conversion time : 1.5ms per 16 signal scan table

Scaling in data table : ±32000 = ±nominal range

Data table total range : –32768 to +32752

Input circuit : differential amplifier

Input resistance : greater than 10 Megohm

Full scale input options : − 10.235 to + 10.235V
: − 5.1175 to + 5.1175V
: − 1.0235 to + 1.0235V*
: − 0.51175 to + 0.51175V

*applies to 20mA current option

Working input voltage : 14V total (common mode plus signal)

8840

GEC Industrial Controls Limited

No damage input : ± 15V peak maximum
voltage

Localised : 240V a.c. continuous
damage input
voltage

Voltage range : ±0.13% from 0 to 40°C
accuracy
(zero common : ±0.23% from 0 to 60°C
mode)

Current range : ±0.18% from to 0 to 40°C
accuracy
(zero common : ±0.28% from 0 to 60°C
mode volts)

Common mode : 0.015% maximum for
error   ±10V range
: 0.15% maximum for ± 1V range
and current range

Temperature : 0 to 60°C operating
: –20 to 70°C storage

Humidity : 10 to 90% non-condensing

Module : single slot width in GEM 80
dimensions   subrack

Termination panel:
  dimensions : 156 x 483 x 30mm
(height x width x depth)
  mounting : 19in (483mm) rack
: between mounting channels at
5U (222mm) centres.

Terminal Rail :
  dimensions : 40 x 483 x 46mm
(height x width x depth)
  mounting : 19in (483mm) rack or panel
: mounting space required is
133mm (3U) between adjacent
terminal rail centres

Plant terminals : 2 signal terminals plus 1 screen
terminal per input
: up to 2.5mm² conductors
: slotted head screw clamp

Power : 150mA @ + 5V
consumption   180mA @ + 15V
  80mA @ –15V

Monitor sockets : Single pole
: 2mm diameter plug

ORDERING CODES
Equipment for Direct Connection to Plant:
Each of the following sets of equipment includes:
analogue input module.
backplane connector assembly including wire links.
analogue input termination panel/rail.
2m ribbon cable with connectors (Termination panel).
3m wiring harness with connectors (Termination rail).

Input Range	Filter Time	Ordering Codes	
		Termination Panel	Termination Rail
± 10V	0.5ms	8840-4001	8840-4006
±5V	0.5ms	8840-4002	8840-4007
±1V	0.5ms	8840-4003	8840-4008
±0.5V	0.5ms	8840-4004	8840-4009
±20mA	0.5ms	8840-4005	8840-4010
± 10V	5ms	8840-4011	8840-4016
±5V	5ms	8840-4012	8840-4017
±1V	5ms	8840-4013	8840-4018
±0.5V	5ms	8840-4014	8840-4019
±20mA	5ms	8840-4015	8840-4020

Equipment for use with signal conditioning
(excludes termination panel)
  Range ± 5V   Filter 5ms   8840-4022

Optional Extras
Mounting kit comprising:
  1 length of steel channel
  5 cable retention clips (plastic)   8959-0000
  1 set of fixing screws

Spares:
Analogue input module (0.5ms filter)   8311-4001
Analogue input module (5ms filter)   8311-4002
Backplane connector assembly   8062-4001
(excludes wire links)
Analogue input termination panel   8033-4001
for voltage inputs
Analogue input termination panel   8033-4002
for 20mA inputs
Plugs for user connection to   Red   80801/180
monitoring sockets   Black 80801/181.

GEC Industrial Controls Limited.
Programmable Controllers Division,
Kidsgrove, Stoke-on-Trent,
Staffordshire ST7 1TW, England.
Tel: Kidsgrove (07816) 3511
FAX (Group 3) Kidsgrove (07816) 76329
Telex: 36293 GECICK
Also at:
Mill Road, Rugby,
Warwickshire CV21 1BD.
Tel: Rugby (0788) 2121
Registered at London No. 540682
Holding Company: The General Electric Company p.l.c.,
      1 Stanhope Gate, London W1A 1EH.

SP C3224   A0135M/D1796M

Publication No. T. 211   Issue 5   4/87
©1987 The General Electric Company p.l.c.

GEC Industrial Controls Limited

Analogue Output   8841

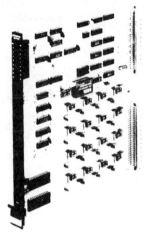

DESCRIPTION

The 8841 analogue output equipment interfaces from a GEM 80 controller to sixteen analogue plant signals for actuators, receivers, meters etc. The high speed analogue convertor resolves signal magnitude better than one part in four thousand.

The equipment comprises a module connected by cable to either a termination panel or alternatively a terminal rail. The analogue output module plugs into the subrack of a GEM 80 controller and connects to its fast I/O highway. A mounting kit for the termination panel is an optional extra.

FEATURES

● Suitable for industrial applications.

● Fast conversion, high resolution.

● Short circuit protection.

● Mechanical coding of module type.

SPECIFICATION

Number of outputs: 16

Nominal Signal
Output Levels   : There is a choice of
                       3 combinations.

1. ±10v

2. 0 to 20mA

3. ±5v

Output current for : ±10mA maximum
±5v and ±10v.

Compliance   : 0 to 5v (maximum output load
voltage for   of 250 ohm)
0 to 20mA

Protection   : continuous short circuit

Output response   : 10ms maximum
time

Accuracy   : ±0.15% from 0 to 40°C
                    : ±0.25% from 0 to 60°C

Resolution   : 1 in 4096 ±5v and ±10v
                    : 1 in 2048 20mA

	Data table range	Corresponding output
±5v analogue output	−32768 +32752	−5.117v +5.117v
±10v analogue output	−32768 +32752	−10.235v +10.235v
0 to 20mA analogue output	−32768 0 +32752	0mA 0mA 20.47mA

Monitor sockets   : single pole
                      2mm diameter plug

No damage output
voltage   : ±15v peak max.

8841

GEC Industrial Controls Limited

Localised
damage
output voltage

: ±5v analogue
output: ±45v max.
: ±10v analogue
output: ±45v max.
: 0 to 20mA analogue
output: +100v (with respect to
GEM 80 0v). For negative voltage
with respect to GEM 80 0v up to
1.5 amps (from GEM 80 source).

Temperature

: 0 to 60°C operating
: −20 to 70°C storage

Humidity
non-condensing

: 10 to 90°₀

Power
consumption
(excluding
output loads)

: 70mA @ +5v
: 150mA @ +15v
: 140mA @ −15v

Plant terminals

: 2 signal terminals plus 1 screen
terminal per output.
: up to 2.5mm² conductors
: slotted head screw clamp

Module
dimensions

: single slot width in GEM 80
: subrack

Termination panel:

Dimensions

: 156 x 483 x 30mm
(height x width x depth)

Mounting

: 19in (483mm) rack

Terminal Rail:

Dimensions

: 40 x 483 x 46mm
(height x width x depth)

Mounting

: 19in (483mm) rack or panel.

ORDERING CODES
Equipment for Direct Connection to Plant

Analogue Output Module	Terminal Rail	Termination Panel
±10v Analogue Output equipment	8841–4004	8841–4001
0 to 20mA Analogue Output equipment	8841–4011	8841–4012

The above order code each comprise the following equipment:-
An analogue output module, backplane and connectors and either:-
a) termination panel with 2M ribbon cable.
b) terminal rail with 3M wire harness.

Equipment for use with signal conditioning comprises:-

analogue output module for ±5v
backplane connector assembly
ribbon cable with connectors
(excludes termination panel)   8841-4002

Optional extras

Mounting kit comprising:-
1 length of steel channel
5 cable retention clips (plastic)
1 set of fixing screws   8959-0000

Spares

Analogue output module 0 to 20mA	8313-4001
Analogue output module ±5v	8312-4003
Analogue output module ±10v	8312-4002
Backplane connector assembly	8062-4001
Backplane connector assembly	8064-4001
Analogue output termination panel	8033-4001

Plugs for user connection to
monitoring sockets

Red   80801/180
Black 80801/181

GEC Industrial Controls Limited.
Programmable Controllers Division,
Kidsgrove. Stoke-on-Trent,
Staffordshire ST7 1TW, England.
Tel: Kidsgrove (07816) 3511
FAX (Group 3) Kidsgrove (07816) 76329
Telex: 36293 GECICK

Also at:
Mill Road, Rugby,
Warwickshire CV21 1BD.
Tel: Rugby (0788) 2121

Registered at London No. 540682
Holding Company: The General Electric Company p.l.c.,
1 Stanhope Gate, London W1A 1EH.

SP C3224   A0135M/D 1796M

Publication No. T212 Issue 6 4/87
©1987 The General Electric Company p.l.c.

SERIES ONE JUNIOR
Programmable Controller
General Description

GENERAL ⊛ ELECTRIC®

The Series One Junior programmable controller is a low-cost, compact offspring of the Series One PC, designed for control applications with 4-60 relays. Its small mounting area allows the Series One Junior to be installed in the same space as four 4-pole relays. The basic Series One Junior provides 24 I/O (15 inputs/9 outputs) with 700 words of CMOS RAM, or EPROM memory. It can be expanded up to 64 I/O using a Series One rack and I/O modules. The expansion I/O can be mixed and matched using AC or DC modules. The expansion rack may be installed remotely up to 100 ft. (30 m) from the Series One Junior PC. The Series One Junior shares other devices with the Series One PC, including the snap-on, hand-held programmer, PROM writer, printer and data communications unit. Also, a program developed in the Series One Junior can be transferred to a Series One PC. One of the input circuits can be used to count pulses up to 2000 counts/second (2KHz), providing a built-in high speed counter. Models are available in various input and output voltages. The Workmaster™ CRT can be used to program and document the Series One Junior, as well as all other GE programmable controllers.

Actual Size 210 mm W x 162 mm H x 56 mm D

Features

Low price, compact size (53 sq. inch 330 mm² mounting area)

Expandable to 64 I/O using Series One rack and I/O modules

Expansion rack may be up to 100 ft (30 m) from the CPU

Built-in 2KHz high speed counter with 20 preset points

Snap-on, hand-held programmer, or CRT Programmer

Time or event-driven sequencers

Common Series One language and support devices such as programmer, printer, PROM writer and data communications module

Benefits

Replace as few as 4 relays cost effectively with a 24 I/O PC system. And, use smaller, less expensive panels.

Use over a broader range of applications. Mix or match AC or DC I/O modules in the expansion rack.

Reduce wiring cost by placing I/O interfaces closer to the machine or process.

Control up to 20 high speed counting or sequencing events using a compact, inexpensive PC.

Plug hand-held programmer into CPU for operator convenience. Use Workmaster™ IBM compatible programmer to create annotated ladder diagram user programs.

Replace a variety of drum sequencers or stepping switches with one product.

Reduce cost by using the same support devices for the Series One and Series One Junior. Able to transfer Series One Junior programs to Series One.

Application Categories

Air Compressors
Annunciators
Assembly Line
Batch Process (wet and dry)
Boxing
Capping
Continuous Process
(waste treatment, etc.)

Crushing
Elevators
Energy Management
Finishing (plate, print, etc.)
Forming (bend, mold, etc.)
Machining (grind, tap, drill, etc.)
Material Handling (wet and dry)
Mechanical Assembly Monitoring

Palletizers
Pressing
Pumping
Relay Retrofit
Sorting
Testing/Gauging
Winding/Weaving (motors, fabric, etc.)

Series Six
Programmable Controller
ProLoop™ System

INSTRUMENTATION SYSTEMS LIMITED

MAUCHLINE · AYRSHIRE KA5 5BR·(0290) 51994

The ProLoop System is a unique, new approach for implementing process control solutions with a programmable controller. ProLoop is based on the concept of distributed, intelligent processing. Existing Series Six programmable control hardware operates as a supervisory controller for the new line of General Electric microprocessor-based process loop controllers. A Loop Management Module in the Series Six I/O structure monitors up to sixteen (16) process control loops and reports process status to the Series Six central processor. Multiple Loop Management Modules may be incorporated into each ProLoop System.

ProLoop provides a custom control system tailored to exact process requirements using off-the-shelf components. Since ProLoop can be implemented a loop at a time, customers are provided the flexibility to prove the system effectiveness before committing to a complete system. Once a ProLoop System is installed, it can be easily and economically expanded to meet additional or changing process requirements.

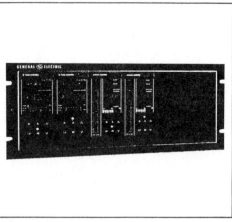

Features:

Dedicated PID Controllers

Menu Driven Configuration

Visual Process Indication

Loop Management Module

Hand-Held Configurator

Redundant Power Supplies

Analog Input Conversion And Scaling

Benefits:

Single Loop Integrity
Stand-Alone Operation
Industry Accepted Technology

Easy to Implement
Configurable with Workmaster™
Programmable Control Information Center
or Operator Interface Terminal

Built-In Process Variable Display
Local Operator Station
No Special Operator Training

Easily Integrated into PC via Direct
Multidrop RS-422 Interface
No Programming Required

Easy Modification of Loop Parameters
Plugs Directly into the Front of Each Controller

Dual AC and/or DC Power Sources Accepted
Protects Critical Process During Power
Interruption or Power Supply Failure

Eliminates Need for Expensive Transmitters
Reduces Maintenance Requirements for Calibration

NEW HIERACHY NETWORK SYSTEM
OF MITSUBISHI

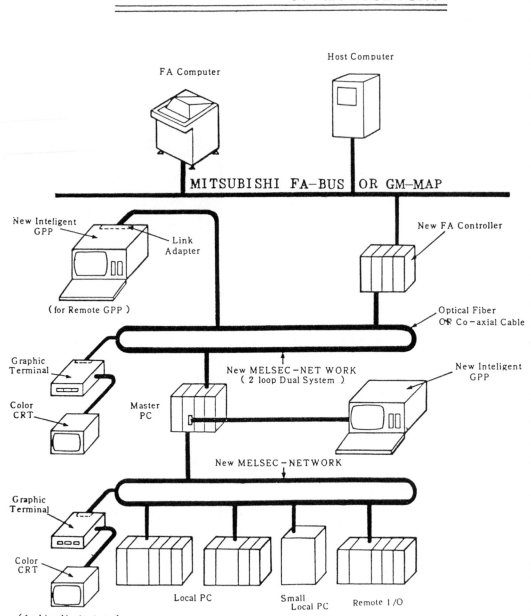

Host Computer

FA Computer

MITSUBISHI FA–BUS OR GM–MAP

New Inteligent GPP

Link Adapter

New FA Controller

(for Remote GPP)

Optical Fiber OR Co–axial Cable

Graphic Terminal

Color CRT

Master PC

New MELSEC –NET WORK
(2 loop Dual System)

New Inteligent GPP

New MELSEC –NETWORK

Graphic Terminal

Color CRT

Local PC

Small Local PC

Remote I /O

(for Line Monitoring)

F2 Technical Specifications

GENERAL SPECIFICATIONS

Power supply	AC 110-120V/220-240V single phase 50/60 Hz
Power failure compensation	≤10m sec.
Ambient temperature	0.55°C
Ambient humidity	45-95%, No condensation
Vibration resistance	10-55 Hz, 0.5mm (maximum 2G)
Impact resistance	10G, 3 timer to XYZ
Soundproofing	1000VP-P, 1μsec. 30-100 Hz (noise simulator)
Insulation withstand voltage	AC 1500V, 1 minu. (between earth terminal and all other terminals)
Insulation resistance	5mΩ, 500V DC (between earth terminal and all other terminals)
Grounding	Less than 100Ω
Environment	No corrosive gas, no conductive debris

FUNCTIONAL SPECIFICATIONS (BASE UNIT)

Execution method		Stored programme cycle execution, collective input/output
Execution speed		Average 7μ second/step
Programme language		Relay and logic symbols
Programme capacity		1000 steps
Instructions	Basic sequence	20 instructions (including MC/MCR, CJP/EJP, S/R)
	Stepladder	2 instructions (STL, RET)
	Function block	47 instructions (including >, =, <, etc.)
Programme memory		CMOS-RAM built in, EPROM cassette (option)
Auxiliary relays	Non-retentive	128 points
	Retentive	64 points
	State (retentive)	40 points (can be used as normal keep relays)
	Special relays	32 points
Timers	0.1 sec timers	24 points, on-delay timers (0.1 – 999 seconds)
	0.01 sec timers	8 points, on-delay timers (0.01 – 99.9 seconds)
Counters (retentive)		32 points, down counter (0 – 999)
Battery back-up		Lithium battery approximately five years life
Diagnosis		Programme check (sum, syntax, circuit check), watch-dog timer, battery voltage, power supply voltage

INPUT SPECIFICATIONS

		F2 – *** – E*	F2***R – UA*
Input type		Non-voltage contact or PNP open collector transistor Photo-coupler isolated	AC input device Photo-coupler isolated
Input voltage		Built-in power DC 24V±4V external power DC 24V±8V	External power AC 100/110V+10% (F2 – ***R – UA1) AC 200/220V – 15% (F2 – ***R – UA2)
Operation current	OFF→ON	DC 4mA minimum	>8mA (* – UA1), >7mA (* – UA2)
	ON→OFF	DC 1.5mA maximum	<3mA (* – UA1), <3mA (* – UA2)
Response time	OFF→ON	Approximately 10ms	~15m sec
	ON→OFF	Approximately 10ms	~8m sec

OUTPUT SPECIFICATIONS

		Relay output	Transistor output	Triac output
Output type		Relay output	Transistor output	Triac output
Isolation		Relay isolated	Optocoupler	Optocoupler
Output load	Resistance load	2A/point	1A per point	1A per point
	Inductive load	35VA/up to 300000 operations	24W	50UA 110/120V AC 100 UA 220/220V AC
	Lamp load	100W	100W	100W
Leakage current		0-55mA – 110V AC 1-1mA – 220V AC	—	—
Response time	OFF→ON	Approximately 10m sec	<1m sec	<1m sec
	ON→OFF	Approximately 10m sec	<1m sec	<10m sec

INDIVIDUAL SPECIFICATIONS

Models	F2-20MR-ES F2-20ER-ES	F2-40MR-ES F2-40ER-ES	F2-60MR-ES F2-60ER-ES
Numbers of inputs	12 points	24 points	36 points
Number of ouputs	8 points	16 points	24 points
Terminal block		Removable terminals	
Power consumption	20VA	25VA	40VA
Input sensor power	0.1A	0.1A	0.2A

F2
Technical Specifications

System Data	GP-80 F2-E
Voltage Requirements	220-240 V 50/60 Hz +10% −15%
Power Consumption	< 15 VA
Ambient Temperature	0°C — 40°C
Storage Temperature	−20°C — 60°C
Ambient Humidity	< 90% RH without condensation
Vibration Resistance	10-55Hz 0.5 mm (max. 2G)
Insulation Resistance	10MΩ (500 V DC)
Insulation Withstand Voltage	1500 V AC. 1 min.
Noise Immunity	1000 V. 1 μsec.
Noise Spike	NEMA-ICS2-230
Dimensions	W = 300 mm. L = 215 mm. H = 70 mm.
Weight	2 kg
Battery	Lithium, capable of 5 year back up service

System Data	F2-20H-DE
Voltage Requirements	220-240 V 50/60 Hz +10% −15%
Power Consumption	< 20 VA
Power Failure Compensation	≦ 20 ms: operation continues
Ambient Temperature	0°C — 40°C
Storage Temperature	−20°C — 60°C
Ambient Humidity	< 90% RH without condensation
Insulation Resistance	10MΩ (500 V DC)
Insulation Withstand Voltage	1500 V AC. 1 min.
Noise Spike	NEMA-ICS2-230
Battery	Lithium, capable of 5 year RAM back up service
Dimensions	W = 129 mm. L = 249 mm. H = 72 mm.

System Data	F20CM-5ESS	System Data	
Voltage Requirements	110/120 V AC (85%, 110%)	Outputs	
	220/240 V AC (85%, 110%)	Relays Voltage Handling	24 V DC 240 V AC
	Dual Supply	Current Handling	2 A
Power Consumption	<12 VA	Transistor Voltage Handling	24 V DC +35% −130%
Power Failure Compensation	≦20 ms operation cont.	Current Handling	1 A
Ambient Temperature	0-55°C		
Storage Temperature	−20°C − 60°C		
Ambient Humidity	90% RH (no condensation)		
Vibration Resistance	10-55Hz 0.5 mm (max. 2G)		
Insulation Resistance	500 V DC, 5MΩ		
Insulation Withstand Voltage	1500 V AC 1 min.		
Noise Immunity	1000 V 1 μsec.		
Noise Spike	NEMA-ICS2-230		
Battery	Lithium, capable of 5 year back up service		
Memory	C-M05 RAM		

Programming the
F2 Panel

F2 Series Special Functions

The special function is executed by element 670 and defined by a constant "K" value.

Any parameters appropriate to the function are defined by the "K" values of elements 671-677.

The following example illustrates the use of one of the more involved special functions – loading a timer from a thumbwheel switch.

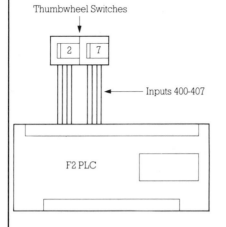

Thumbwheel Switches

Inputs 400-407

F2 PLC

671	K400	Starting address of BCD input
672	K100	(Optional) addition
673	K7	Number format
674	K450	Destination Timer
670	K51	Transfer instruction

Setting timer 450 with values between 100 and 199 seconds using a two digit BCD thumbwheel switch.

K Value assignment for element 670

K

0 Refreshes the image memory for all inputs
1 Refreshes the memory for designated inputs
2 Refreshes the status of all ouputs
3 Refreshes the status for designated outputs
6 Sub routine start
7 Sub routine call
8 Conditional sub routine
9 Sub routine end
10 Resetting of element M473
11 Resetting of element C660
12 Switches external counter input
13 Allows run terminal to be used as a limited extra input
14 Set   M571
15 Reset M571
16 Set   M572
17 Reset M572
18 Set   M573
19 Reset M573
20 2 – 4 bit decode
21 3 – 8 bit decode
22 4 – 16 bit decode
23 4 – 2 bit encode
24 8 – 3 bit encode

K

25 16 – 4 bit encode
26 Resets designated elements
27 Transfers a decimal constant
28 Transfers an octal constant
30 Writes a decimal constant to a timer or counter (data register)
31 Writes a BCD constant from the inputs to a timer or counter (data register)
32 Reads a timer or counter (data register) to designated outputs in BCD
33 Writes a decimal constant to a timer or counter (current value)
34 Writes a BCD constant from the inputs to a timer or counter (current value)
35 Reads a timer or counter (current value) to designated outputs in BCD
40 Comparison of decimal constant (K) with timer or counter Current value (A) $A<K$, $A>K$, $A=K$
41 Comparison of BCD input (K) with timer or counter Current value (A) $A<K$, $A>K$, $A=K$
43 Comparison of zone defined by two constants (K1 AND K2) with timer or counter Current value (A) $A<K1$, $A>K2$, $K1 \leq A \leq K2$
44 As above but with 6 digit counter

Analogue Input/Output Unit F2-6A-E

The F2-6A-E adds to the features of an F1 system and expands its applications to include a variety of analogue signal control functions such as temperature and fluid flow.

The F2-6A-E markedly broadens the range of applications in which the F1 series may be used.

Typical applications where analogue control signals are used include temperature control and fluid flow control.

It accepts four analogue inputs and converts them to digital signals, from which data can then be arithmetically treated and utilized in programs in the F1 base unit.

Once the data has been processed it may then

be converted back into an analogue signal which can be directed to the output terminals of the unit.

The unit may be connected to the extension port of any F1 base unit except the F1-12.

Number of inputs/outputs		4 inputs/2 outputs signals-ended
I/O occupation		20 points/unit (one extension connector)
Input range	VOLTAGE	0 – 10V DC (input impedance 100kΩ)
	CURRENT	0 – 20mA DC (input impedance 250Ω)
Digital output to CPU		8 bit binary
Output range	VOLTAGE	0 – +10V DC (load 500 – 1MΩ)
	CURRENT	0 – +20mA DC (load 0 – 550Ω)
Digital input form CPU		8 bit binary
Absolute voltage and current		±12V/±22mA
Conversion speed	A/D	Approximately 350 μsec.
	D/A	Approximately 200 μsec.
Insulation between I/O		Photo-coupler isolated

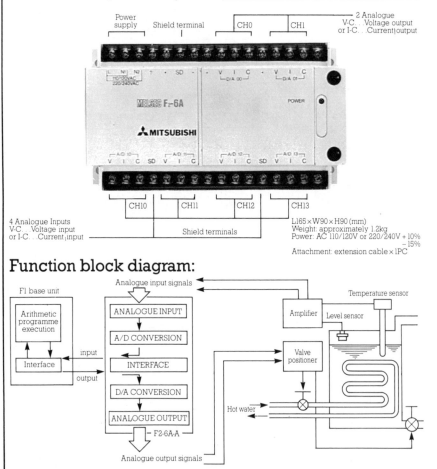

Power supply — Shield terminal — CH0 — CH1 — 2 Analogue V-C. .Voltage output or I-C. .Current\output

MELSEC F2-6A — POWER — MITSUBISHI

CH10 — CH11 — CH12 — CH13

4 Analogue Inputs
V-C. . Voltage input
or I-C. . .Current\input — Shield terminals

L165 × W90 × H90 (mm)
Weight: approximately 1.2kg
Power: AC 110/120V or 220/240V + 10% − 15%
Attachment: extension cable × 1PC

Function block diagram:

F1 base unit — Analogue input signals — Temperature sensor

Arithmetic programme execution — ANALOGUE INPUT — Amplifier — Level sensor

A/D CONVERSION

input — Interface — INTERFACE — Valve positioner

output

D/A CONVERSION — Hot water

ANALOGUE OUTPUT — F2-6A-A

Analogue output signals

Position Controller Module

Introduction

The Positioning Position Control Module (PCM) is an addition to the F-Series of Programmable Logic Controllers and is basically a high speed counter which is able to accept pulse chains from any suitable device, eg, a rotary encoder.

The PCM when linked to an F-Series Programmable Logic Controller, can be used to control additional processes (eg, the action of a drilling machine).

Communication between the Programmable Logic Controller and the Positioning Counter Module is by means of a ribbon cable through the PLC's extension port.

One position controller can be connected to an F – 12R or F – 20M PLC.

Where two axes control is required, two PCM's may be used with an F – 40M or F2 – 40M. Three axes may be controlled using an F_2-60M.

General

1. A maximum of 400 positions can be programmed into the module in 10 blocks of 40 positions.

2. The PLC and the position controller module communicate via "handshake" signals. This enable processes under PLC control and movement under PCM control to be coordinated.

3. Programming is simple, either by the direct keying in of the position data or by taking the machine through the processes manually and 'teaching' the module the relevant positions.

Positioning Procedure

1. The block number to be operated is selected by the PLC.

2. The position controller recognises the actual position by the counter value and switches on the forward or reverse drive according to the difference between the current counter value and the target value.

3. At a designated distance from the target position the drive is switched to low speed.

4. The position controller has the facility to compensate for the inertia in a mechanical system by switching the drive off at a designated distance from the target position.

5. An 'accuracy allowance and judgement' facility is available within a block. This will only allow the next process to occur if the position attained is within the specified tolerances. Should the position be outside tolerances the machine stops.

4. Positioning tolerances can be set within a block. When the actual stop position falls within these tolerances a verify signal to the PLC allows the next process to continue.

5. The RAM in the position controller is backed by a Lithium battery with a five year life. The programming panel includes a cassette interface.

6. The unit is supplied with both transistor and relay output.

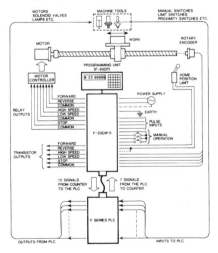

Typical system block diagram.

CPU Unit Performance Specifications

This section describes performance specifications, such as memory capacity, and specifications such as device list and instruction list, of the A0J2CPU unit A1, A2 and A3 CPU's

Table 1.1

A0J2CPU Performance Specification

Item		Specifications
Control method		Stored program, repeated operation
I/O control method		Direct system
Program language		Language dedicated to sequence control (Combined use of relay symbol system, logic symbolic word, SAP language)
Instruction	Sequence instruction	21 types
	Basic instruction	38 types
	Application instruction	21 types
Processing speed		Sequence instruction: 4.4μS to 5.6μS/step
Memory capacity and memory type	Memory capacity	Maximum 7K steps
	4KEROM	3K steps
	4KRAM	3K steps
	4KROM	3K steps
	16KRAM	7K steps
	8KROM	7K steps
I/O points		336 points (Maximum 480 points when extension base unit is used)
Internal relay (M/L)		M/L0000 to 2047 (2048 points. Set M/L with A6GPP or A7PU.)
Link relay (B)		B000 to B3FF (1024 points. Usable as internal relay)
Timer	Points	128 points
	100ms	T0 to T79 (80 points, 0.1 to 3276.7 sec)
	10ms	T80 to T119 (40 points, 0.01 to 327.67 sec)
	Retentive (100ms)	T120 to T127 (8 points, 0.1 to 3276.7 sec)
Counter		C0 to C127 (128 points, 1 to 32767)
Data register (D)		D0 to D511 (512 points, 16 bits)
Annunciator (F)		F0 to F255 (256 points, 1 bit)
Index register (V, Z)		V, Z (2 points, 16 bits)
Pointer (P)		P0 to P63 (64 points)
Special relay (M)		M9000 to M9255 (256 points)
Special register (D)		D9000 to D9127 (128 points)
Comment		Can be created with A6GPP (In this case, comments which may be entered in CPU are only 95 points, F0 to F94.)
Latch (power failure compensation) function		L, B, T, C, D possible (Set range with A6GPP or A7PU.)
Remote run/stop function		Operable with A6GPP or A7PU.
Operation at error		Operation continued when software instruction is abnormal.
Output state at the time of STOP to RUN		Operation result at STOP time is re-output.
Print title entry		Entry to CPU is disabled. May be created with A6GPP.
Self-diagnostic function		Watch dog timer error, battery error, AC DOWN detection, fuse blow detection, etc.
Allowable instantaneous power failure period		Within 10ms
At power on, when power is restored after power failure		Automatic restart when 'RUN' switch is turned on (Initial start)
Backup of IC-RAM latch device		Battery backup system, lithium battery (battery guaranteed for 5 years)
Parameter		Setting of latch range with A6GPP or A7PU
Microcomputer mode		Other than sequence program. Utility FD contents are written to microcomputer area.
Watch dog timer period (WDT)		200ms fixed

CPU Unit Performance Specifications

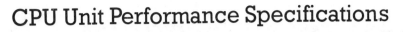

Table 1.2
A1, A2, A3CPU Performance specifications

CPU Item		A1CPU		A2CPU		A3CPU	
		Default value	Parameter setting range	Default value	Parameter setting range	Default value	Parameter setting range
Main program	Sequence programe	6K steps	1 to 6K steps (in units of 1K step)	6K steps	1 to 14K steps (in units of 1K step)	6K steps	1 to 30K steps (in units of 1K step)
	Micro-computer program	OK byte	0 to 10K bytes (in units of 2K bytes)	OK byte	0 to 26K bytes (in units of 2K bytes)	OK byte	0 to 58K bytes (in units of 2K bytes)
Subprogram	Sequence program	OK step	No setting	OK step	No setting	OK step	1 to 30K steps (in units of 1 K step)
	Micro-comnputer program	OK byte	No setting	OK byte	No setting	OK byte	0 to 58K bytes (in units of 2 K bytes)
File register		OK byte	No setting	OK byte	0 to 4K points (in units of 1K point)	OK point	0 to 8K points (in units of 1K point)
Comment capacity		OK byte	0 to 4032 points (in units of 64 points)	OK byte	0 to 4032 points (in units of 64 points)	OK byte	0 to 4032 points (in units of 64 points)

		A1	A2	A3
Offline switch memory		256 points for Y	512 points for Y	2048 points for Y
		2048 points for M, L, S 1024 points for B 256 points for F	2048 points for M, L, S 1024 points for B 256 points for F	
Internal relay (M)		Up to 2048 points (2048 bits)		Internal relay, latch relay, and step relay are commonly used and a total of 2048 points.
Latch relay (L)		Up to 2048 points (2048 bits)		
Step relay (S)		Up to 2048 points (2048 bits)		
Link relay (B)		1024 points (1024 bits)		
Timer (T)	Number of points	256 points		
Counter (C)	Number of points	256 points		
Data register (D)	Number of points	1024 points		
	Specifications	Handled value	Decimal (−32768 to 32767) Hexadecimal (0 to FFFF)	
Link register (W)		1024 points 16 bits/data		
Annunciator (F)		256 points (256 bits)		
File register (R)		—	Max. 4096 words	Max. 8192 words
Accumulator (A)		2 points for storage of intermediate results of data processing. 1 word (16 bits) per point.		
Index register (Z, V)		2 points for qualification of K, X, Y, M, etc. 1 word (16 bits) per points.		
Pointer (P)		256 points for jump destination specification of CJ, CALL, etc.		
Special relay (M)		256 points (256 bits), M9000: fuse blow, M9001: output bus error, etc.		
Special register (D)		256 points (256 bits), D9000: fuse blow, D9001: output bus error, etc.		

SattControl ⑨   SattCon 05-35

SattCon 05 is a family of small PC systems - a family which forms part of the new generation of control systems, developed by SattControl to meet the needs of industry, today and in the future.

SattCon 05-35 is a complete control system capable of controlling machines and small processes both indenpendently and in communication with other systems.

- Analogue and digital I/O in the basic unit.
- Can be expanded with analogue, digital and positioning units.
- PID loops with integrated hand station for presentation and manual operation of loops.
- Fast loop with execution times from 1 ms.
- Communicates with other systems via COMLI, SattControl's standard protocol for serial transmission.
- Time clock control for starting and stopping plant on specific days and at specific times.
- Text printouts with date and time and process variables, possibility to create dialogue programs.
- Easy to program using built-in keyboard, standard VDU or an IBM personal computer with the DOX5 software package.

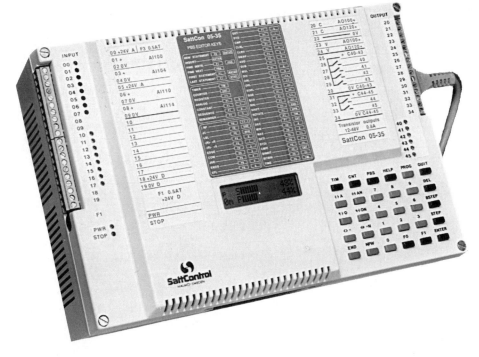

What is SattCon 05?

SattCon 05-20 belongs to a new family of programmable control systems (PC system) from SattControl. When we use the term SattCon 05 in this description we refer to the SattCon 05-20 system.

SattCon 05-20 is a compact unit which contains all the functions necessary in a complete control system: memory, power supply, programming unit, test unit etc.

SattCon 05-20 can be used with the following accessories:

— Digital expansion unit (XD24R).
— Test panel (PTC05).
— Analogue expansion units (XACV-B) — can be used with ATC05 analogue test panels.
— Backup module (BUP05).
— Printer.
— External programming unit (SattProg 100 with DOC05).

These accessories are presented in more detail in their respective data sheets.

SattCon 05-20's serial channel can be used for communicating with computers and other PC systems, for example with SattCon 31 via SattControl's COMLI protocol.

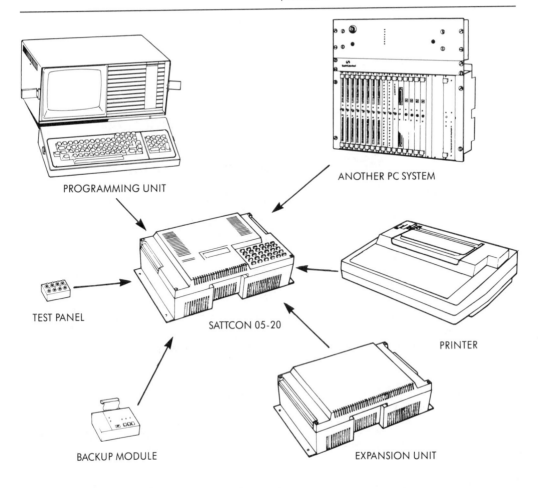

PROGRAMMING UNIT

ANOTHER PC SYSTEM

TEST PANEL

SATTCON 05-20

PRINTER

BACKUP MODULE

EXPANSION UNIT

A

INPUTS
16 inputs for 24 V DC input signals. Can be supplied from internal 24 V. Connection via unpluggable screw terminal blocks. Terminals 09 and 19 are normally linked.

B

STATUS INDICATION, INPUTS
The yellow LED lights when the corresponding input is on.

C

OUTPUTS
12 outputs comprising normally open relays in groups of four with a common pole. Unpluggable screw terminal blocks.

D

STATUS INDICATION, OUTPUTS
Yellow LED lights when corresponding output is ON, i.e. when the relay is closed.

E

SERIAL CHANNEL
For connecting printer, Sattprog 110 and communication with other systems.

F

OTHER INDICATORS
PWR — Green; lit continuously to indicate correct supply voltage. Flashes if battery needs to be replaced.
STOP — Red; lights when PBS program is stopped.

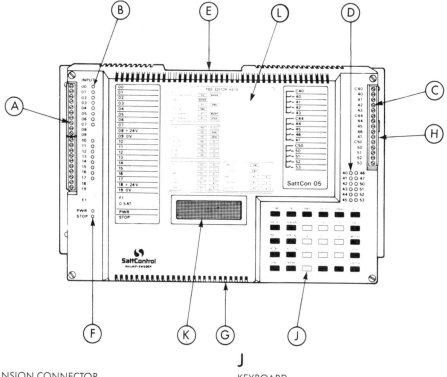

G

EXPANSION CONNECTOR
For connection of expansion units, or modules for external program storage and copying.

H

POWER SUPPLY
3-pole connector and mains fuse for 220 V AC.

J

KEYBOARD
30 function keys for programming etc.

K

ALPHANUMERIC DISPLAY
2 lines each of 16 characters.

L

QUICK GUIDE
Summary of the keys used for various commands and instructions.

Instructions and operands

3.6 Word operands

Word in the I/O RAM

The memory cells in the I/O RAM can be addressed as words, 16 consecutive I/O addresses forming one word. You address the word by specifying the **most significant bit** (MSB). For example, GET 110 means that the value in memory cells 110 to 127 is copied to the word accumulator. 110 is MSB and 127 is LSB.

Analogue inputs/outputs

Up to three analogue expansion units, each with 12 inputs and 4 outputs, can be connected to SattCon 05. You address the inputs as AI (Analogue Inputs) and the outputs AO (Analogue Outputs).

AO n

AI n
n=input or output
number

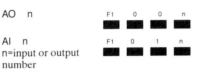

Analogue inputs:

Analogue input signals are stored as 16-bit words in an analogue input memory, in the same way as digital signals are temporarily stored in the I/O RAM. The last digit of an input's address will always be 0 or 4.

Analogue outputs:

In order to handle the analogue output signals quickly, they are temporarily stored in the I/O RAM. The last digits of output addresses will always be 00, 20, 40 or 60.

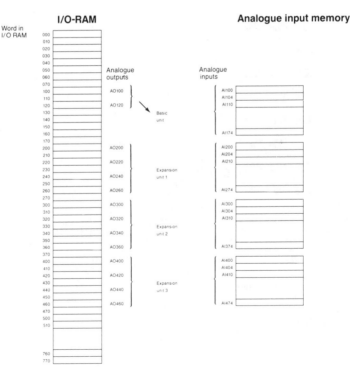

Instructions and operands

Bit operands

Bits in I/O RAM

The I/O RAM consists of 512 memory cells (bits) having I/O addresses 000–777 (octal numbering). The first memory cell therefore has the address 000, and memory cell 512 has the address 777.

Memory cells **010–017** and **040–053** correspond to the digital inputs **010–017** and the outputs **040–053**.

Memory cells **020–037** and **054–067** are battery-backed, and retain their status (0 or 1) if the power fails. You may therefore use them as working memories for intermediate storage of results etc.

If you connect expansion units to SattCon 05, the I/O addresses **200–477** will then correspond to the in-puts/outputs of the expansion units. The addresses used are shown in the data sheet for the relevant expansion unit. I/O addresses which are not used by the expansion units may be used as working memories.

I/O addresses **500–537** are battery-backed working memories.

I/O addresses **540–777** are working memories without battery backup.

Note:
I/O addresses 700–777 are used as start and error bits for COMLI Master. Start and error bits which do not belong to programmed communication areas can be used as working memories.

I/O RAM allocation

Memory cells with special functions

060 Set to one with ⊣/⊢ AN 0 ENTER or CMD0 ENTER
061 Set to one with ⊣/⊢ AN 1 ENTER or CMD1 ENTER
062 Set to one with ⊣/⊢ AN 2 ENTER or CMD2 ENTER
063 Set to one with ⊣/⊢ AN 3 ENTER or CMD3 ENTER
064 Set to one with ⊣/⊢ AN 4 ENTER or CMD4 ENTER
065 Set to one with ⊣/⊢ AN 5 ENTER or CMD5 ENTER
066 Set to one with ⊣/⊢ AN 6 ENTER or CMD6 ENTER
067 Set to one with ⊣/⊢ AN 7 ENTER or CMD7 ENTER

070 One when text is being output.
071 < after CMP
072 =after CMP   See comparison
073 > after CMP used   instructions
074 Set to one if the battery voltage is too low
     (BATTERY LOW).
075 Set to one during the first program cycle after
     a power failure or reset (FIRST SCAN).
076 Clock pulse with frequency 10 Hz.
077 Clock pulse with frequency 1 Hz.

Address	Allocation	
0 – 7	Working memory	
10 – 17	Digital inputs	
20 – 37	Working memory	
40 – 45	Digital outputs	
46 – 57	Working memory	
60 – 77	Memory cells with special functions	
100 – 137	Working memory if analogue outputs are not used	
140 – 177	Working memory	
200 – 277	I/O or working memory	exp. 1
300 – 377	I/O or working memory	exp. 2
400 – 477	I/O or working memory	exp. 3
500 – 537	Battery-backed working memory	
540 – 677	Working memory	
700 – 777	Start and error bits for COMLI master or working memory	

Dynamic status display and I/O forcing

7.4 Analogue inputs/outputs

SattCon 05-35 can display the status of two analogue inputs or outputs simultaneously.

You obtain the display when you give the command

7   ENTER

at basic level.

┌─────────────────────────────┐
│ ┬100IIII   25%   │
│ ┆104   0%   │
└─────────────────────────────┘

Each signal is shown as a percentage of full scale (65535), and as a bargraph. Each increment of the bar represents 6% of full scale.

A small I indicates that it is an analogue input being displayed, and a small o shows that it is an analogue output.

You may select which analogue inputs or outputs are to be displayed.

Move the arrow with

ENTER
█

to the line where you want to change the display.
Using the key

NPW
█

you may select whether an analogue input or analogue output will be displayed on the line marked with the arrow.

The input or output to be displayed is selected with:

STEP   Display next input or output
█

BSTEP   Display previous input or
█   output

n   ENTER   Display the input or output n
█   █

F0   STEP   Automatic stepping
█   █

You return to the basic level with QUIT.

For example, if you want analogue input 104 to be displayed on the upper line, and analogue output 300 on the lower line, this is what you should do:

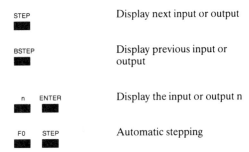

┌─────────────────────────────┐
│ SattCon05-35 1.0   │
│ P   pr   │
└─────────────────────────────┘

7   ENTER
█   █

┌─────────────────────────────┐
│ ┬100   0%   │
│ ┆104IIII   25%   │
└─────────────────────────────┘

STEP
█

┌─────────────────────────────┐
│ ┬104IIII   25%   │
│ ┆104IIII   25%   │
└─────────────────────────────┘

ENTER
█

┌─────────────────────────────┐
│ ┆104IIII   25%   │
│ ┬104IIII   25%   │
└─────────────────────────────┘

NPW
█

┌─────────────────────────────┐
│ ┆104IIII   25%   │
│ ┬120   0%   │
└─────────────────────────────┘

300 ENTER
█   █

┌─────────────────────────────┐
│ ┆104IIII   25%   │
│ ┬300IIIIIIIIIIII   75%   │
└─────────────────────────────┘

Communication

Master/Slave

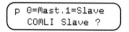
```
p 0=Mast.1=Slave
  COMLI Slave ?
```

0 = Master function
1 = Slave function

Baudrate

```
p  baudrate 9600
  COMLI new   300
```

Step forward to the required baudrate with

STEP

Do not forget to press PROG when you have selected a
new baudrate.

Parity

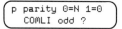
```
p parity 0=N 1=0
  COMLI odd ?
```

N = No No parity
O = Odd Odd parity

Timeout

The following is displayed only if you have selected the
Master function.

```
p timeout (s)  3
  COMLI new ?
```

You may now enter a new time if required.

Identity

The following is displayed if you have selected the
Slave function:

```
p identity   1
  COMLI new ?
```

You may now enter SattCon 05-30's identity (1–247).

Modem delay

```
p modem delay 0
      (×5ms)
```

You may enter a number between 0 and 6. "0" corresponds to a delay of 0ms and "6" corresponds to a delay of 30ms.

If you have selected the Master function, you may now
proceed to the Communications areas with
ENTER. If you have selected the Slave function, you
return to basic level.

Sample Programs and System Documentation

1 EXAMPLES OF DESIGN SCHEMATICS BY SATTCONTROL LTD

Figures B.1 to B.4 show schematic diagrams of SattControl PCS linked to the input/output devices of three different control applications.

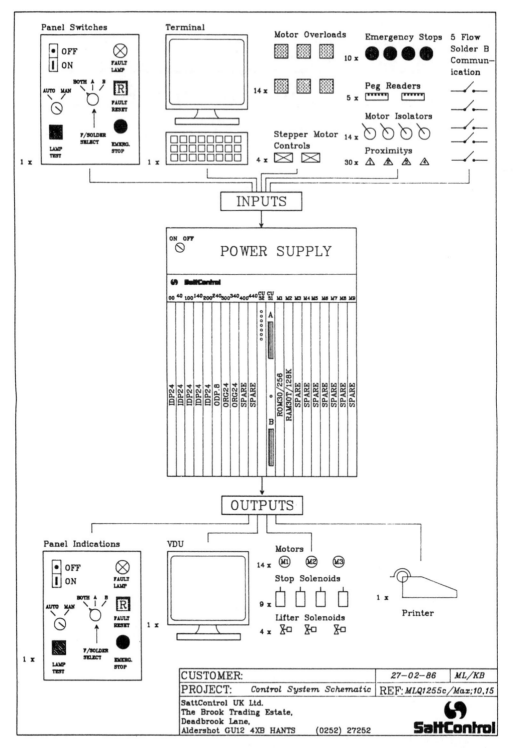

Fig. B.1 *Discontinuous control situation with switched I/O and operator terminals.*

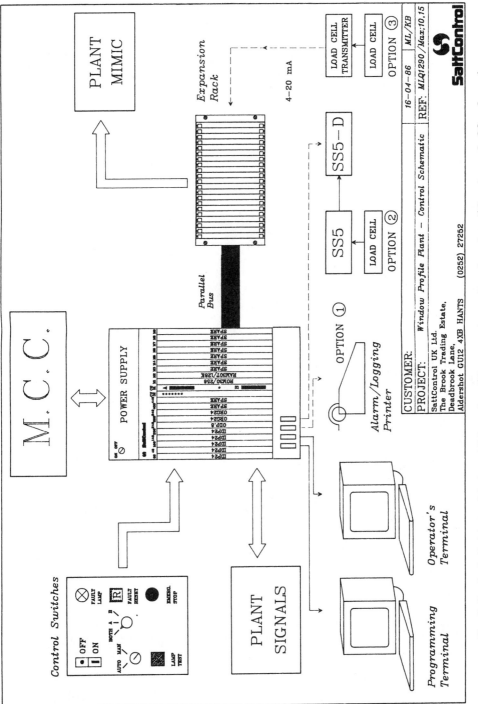

Fig. B.2   As in B.1 but with the addition of an I/O expansion unit and analog I/O using a 4–20 mA card.

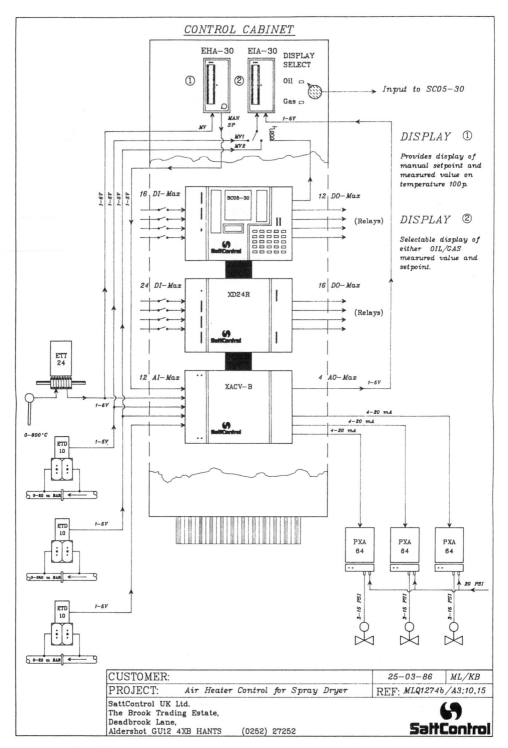

Fig. B.3 *A larger control system than B.1 employing two PLC expansion units: XD24R for digital I/O; XACV-B for analog I/O.*

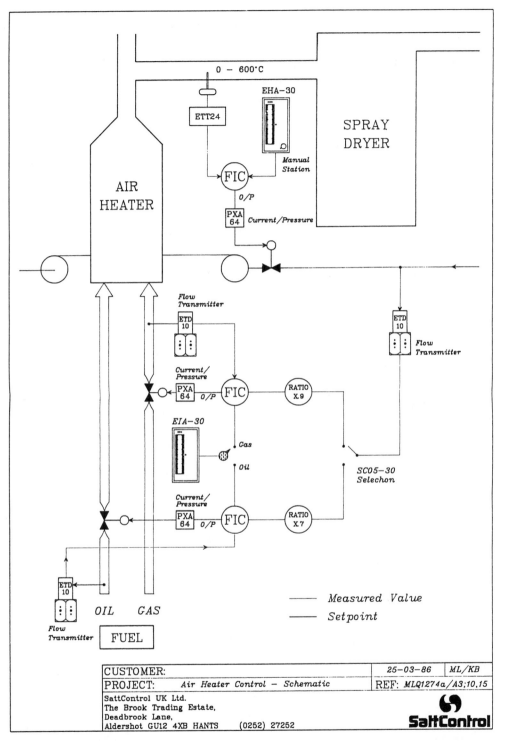

CUSTOMER:		25-03-86	ML/KB
PROJECT:	*Air Heater Control — Schematic*	REF: *MLQ1274a/A3;10,15*	
SattControl UK Ltd. The Brook Trading Estate, Deadbrook Lane, Aldershot GU12 4XB HANTS (0252) 27252		**SattControl**	

Fig. B.4 *An air heater control system with continuous control requirements. This process will require a PLC with analog and digital I/O capability, together with maths functions and PID-type control plans.*

2   COMMENTED PROGRAM LISTING OF (PART OF) THE CHEMICAL PLATING CONTROLLER DETAILED IN SECTION 8.4

(courtesy Instrumentation Systems Ltd)

GE SERIES SIX DOCUMENTATION
Plating Line Automatic Control Ladder Listing
Instrumentation Systems Limited

```
********************
*                  *
* INSTRUCTION SET: ADVANCED
*            CPU ID:  177
*  CPU MEMORY SIZE:  8192
* PROGRAM MEMORY SIZE: 5196
* REGISTER MEMORY SIZE: 8192
*                  *
********************
```

GE SERIES SIX DOCUMENTATION
Plating Line Automatic Control Ladder Listing
Instrumentation Systems Limited

<< RUNG   74 >>

##
This timer is used to inhibit the hoist untill the transporter has been
stopped at the station for the timeset which is 2 seconds.
##

```
On                                                              On
Station                                                      Station
Prox.                                                        Timer.
I0038                                             Const     O0800
—] [—                                            [PRESC]    —(TS)—
                                                   002

On
Station
Prox.
I0038                                             R00400
—] [—                                            [ACCRG]—  —( R)—

        Trans.
        Dest.
        Loaded
00005   00425
—] [————] [—
```

<< RUNG   75 >>

##
The next twenety seven rungs are used to decode the station proximity
switches and set a flag to indicate the station number that the
transporter is presently at.
##

```
                                                                At
                                                               Load
Prox.   Prox.   Prox.   Prox.   Prox.   Prox.                Station
2^4.    2^3.    2^2.    2^1.    2^0.    2^0.                  O0257
I0037   I0036   I0035   I0034   I0033   I0033                —( )—
—]/[————]/[————]/[————]/[————]/[————] [—
```

<< RUNG   76 >>

```
                                                                At
                                                             Station
Prox.   Prox.   Prox.   Prox.   Prox.   Prox.                   F95
2^4.    2^3.    2^2.    2^1.    2^0.    2^0.                  O0258
I0037   I0036   I0035   I0034   I0033   I0033                —( )—
—]/[————]/[————]/[————]/[————]/[————]/[—
```

Transporter is stopped on station and has
been so for at least 2 seconds. This time
will be modified on commisioning the line
to allow the flightbar to become stable
before a raise or lower command is issued.

Plating Line Automatic Control Ladder Listing
GE SERIES SIX DOCUMENTATION
Instrumentation Systems Limited

<< RUNG 129 >>

The next   rungs   of logic are used to control the transporter
movement. The first rungs read in the station number as the
transporter passes the flags, if the destination station is the next one
then control is passed to the transporter PC to decelerate the next one
transporter and bring it to a halt on station.

On   Prox.   Raw
Station 2^0.   station
Prox.   flags.
I0038   I0033   R0302
┤ ├─────┤ I/O   TO REG ┐├─   (   )

<< RUNG 130 >>

Raw   Station   Present
station flag   Station
flags. mask.
R00302 R00303 R00300 Const ┐
┤A AND B   = C   LEN ┐├─   (   )
                       001

<< RUNG 131 >>

###
###
NB:
The next four rungs are only applicable when in computer manual mode.
###
###
This rung is used to cause the transporter to decelerate and stop at the
load station even if the operator does not remove his finger from the
button.
###
###

Present Test   Moving   At 2
Station Reg?   Right   Let go
Station   ?   Button
R0300   R0803 01017   00380
┤A :   B ┐───┐├──────┤ ├──────────────────────────────────( )─

GE SERIES SIX DOCUMENTATION
Plating Line Automatic Control Ladder Listing
Instrumentation Systems Limited

<< RUNG 132 >>

$$
$ This rung is used to cause the transporter to decelerate and stop at $
$ station twenty seven even if the operator does not remove his finger $
$ from the traverse button. $
$ $
$ It has been modified to suit the constraints of the line at present, see $
$ the explanation at rung 134 for the changes required to restore the $
$ availability of the full line. $
$$

Present   Test   Moveing
Station   Reg.= Left.
          23.?

 ┌─ R0300   R0824   0101.6   A00015
 └─ A :   B]────┘[───( )──

GE SERIES SIX DOCUMENTATION
Plating Line Automatic Control Ladder Listing
Instrumentation Systems Limited

<< RUNG 135 >>

$$###$$
$$###$$ This rung is used to disable the traverse left button when the $$###$$
$$###$$ transporter is sitting at station twenty seven. $$###$$
$$###$$ $$###$$
$$###$$ NB. Since this tank is at present out of action the transporter will be $$###$$
$$###$$ disabled at station 25, ie the plc will not allow the transporter to go $$###$$
$$###$$ any further than tank 25 and will force the transporter to decelerate $$###$$
$$###$$ from station 24. The only action required to restore the full line to $$###$$
$$###$$ tank 27 will be to change the referance register in this rung to 828 and $$###$$
$$###$$ in rung 132 to register 827. $$###$$
$$###$$

Present   Test   At 27
Station   Reg=   No Trv
           25 ?   Left
                                                             00383
|—R0300   R0826——( )
   A :   B]

<< RUNG 136 >>

1111111111111111111111111111111111111111111111111111111111111
1$$$1
1$ Disable Automatic Transporter. (DAT) $1
1$$$1
1 This section of the ladder is used to control the transporter 1
1 automaticaly, it will be disabled by the MCR (DAT) when the system is not 1
1 running in automatic mode.   1
1111111111111111111111111111111111111111111111111111111111111

Auto   Reset:
Relay   Flags.
Verify
         I0078   A0126   Const
|———]/[———————]/[—————[MCR]—   ( )
                        016

<< RUNG 137 >>

Trans. Present   Dest =
Dest.   Station   Present
Set Pnt
                                                             00482
|—R0301   R0300——( )
   A :   B]

1111111111111111111111111111111111111111111111111111111111111
1 First rung in control span of MCR DAT.   1
1$$$1
1 This flag will be set if the station that the   1
1 transporter is sitting stopped at is the same   1
1 as the station which has requested a flightbar   1
1 movement, if this is the case then the   1
1 transporter will simply raise the hoist.   1
1111111111111111111111111111111111111111111111111111111111111

```
<< RUNG  138 >>
#####################################################################
#                                                                   #
#  The next three rungs are used to range check the transporter destination  #
#  just loaded in to the setpoint register if this flag is set     #
#  will be reported and the transporter will be inhibited from starting its  #
#  move.                                                            #
#####################################################################

Test      Trans.      Dummy
Reg?=     sDest:      Reg.
27?       Set Pnt               Dest.
                                >= 27.
                                       AQ0560
[ R0828  - R0301  =   R0640            ( )
   A        B    =      C  ]


<< RUNG  139 >>
#####################################################################

Test      Trans.
Reg?=     sDest:
27?       Set Pnt               Dest.
                                = 27.
                                       AQ0561
[ R0828    R0301                       ( )
   A   :     B  ]


<< RUNG  140 >>

Dest.     Dest.
>= 27     = 27                   Dest.
                                 Too
                                 Far
                                       AQ0562
  AQ0560   AQ0561                      ( )
 ]  [  ]/[


<< RUNG  141 >>

Trans.    Test      At
sDest:    Reg?=     Load
Set Pnt   1?        Station        Next
                                   Stop
                                   Load !
                                          Q0872
[ R0301    R0802    Q0257                 ( )
   A   :    B  ]    ]/[


<< RUNG  142 >>

Trans.    Test
sDest:    Reg?=
Set Pnt   0?                        Dest.
                                    = 0 !
                                           AQ0615
[ R0301    R0801                           ( )
   A   :    B  ]
```

###
#   #
This flag will be set if the transporter
destination just loaded is greater than
eighteen (F94) which is the last tank in the
Inspection Etch Line.
###

###
#   #
This flag will be set and hence cause the
sounder to operate to indicate that the
transporter's destination is the load station.
When the transporter arrives at the load
station the sounder will stop automatically.
###

GE SERIES SIX DOCUMENTATION
Plating Line Automation Control Ladder Listing
Instrumentation Systems Limited

```
************************************************************************
<<  RUNG  143 >>

######################################################################
Trans.   Txptr    Reset.                                      Txptr     # This flag will be set when a new destination     #
Dest.    Move     Flags.                                      Go get    # has been loaded into the transporter set         #
Loaded   Cmplete                                              It get    # pointregister (R00301) and will initiate          #
 O0425    O0430   AO0126                                       O0489    # a transporter movement. It also provides the      #
---] [-----] [-----]/[----------------------------------------( )----- # handshake to the find movement request section    #
                                                                        # of the ladder.                                    #
 Txptr                                                                  ######################################################
 Go get                                                                 ######################################################
 It get                                                                 ######################################################
 O0489                                                                  ######################################################
---] [---
************************************************************************
<<  RUNG  144 >>

######################################################################
Hoist    Txptr    Dest =                                      At stat   # This flag will set for one scan to indicate       #
At       Go get   Present                                     Raise     # that the transporter is already at the source     #
Bottom   It get                                               Hoist     # tank and should raise the hoist.                  #
 I0040    O0489    O0482                                       AO0608    ######################################################
---] [-----] [-----] [-----------------------------------------( )----- ######################################################
************************************************************************
<<  RUNG  145 >>

######################################################################
Hoist    Dest =   Txptr                                       Move to   # This flag will set for one scan to indicate       #
At       Present  Go get                                      Stat      # that the transporter should proceed to the        #
Bottom            It get                                                # flightbar source tank with the hoist lowered.     #
 I0040    O0482    O0489                                       AO0607    ######################################################
---] [-----] [-----]/[----------------------------------------( )----- ######################################################
************************************************************************
<<  RUNG  146 >>

######################################################################
At stat  Hoist    Move to                                     Wait      # This flag will be set untill the hoist has        #
Raise    at       Stat                                        Hoist     # raised this will inhibit the transporter          #
Hoist    Top                                                  Raised    # acting on the new destination loaded.             #
AO0608   I0039    AO0607                                       AO0606   ######################################################
---] [-----] [-----] [-----------------------------------------( )----- ######################################################
                                                                        ######################################################
 Wait
 Hoist
 Raised
 AO0606
---] [---
*
```

Plating Line Automatic Control Ladder Listing
GE SERIES SIX DOCUMENTATION
Instrumentation Systems Limited

```
<< RUNG 152 >>

Travers   Diff      Test=
right     travers   Reg=
request   left.     1?

 O0501     R0305    R0802         Destntn
──]/[───────[    A    :    B  ]────────────  -1.Stop
                                              at next
                                              O0502
                                              ─( )──

*************************************************

<< RUNG 153 >>

A01021                                        A01021
──]/[───────────────────────────────────────( )──

*************************************************

<< RUNG 154 >>

A01022                                        A01022
──]/[───────────────────────────────────────( )──

*************************************************

<< RUNG 155 >>

A01023                                        A01023
──]/[───────────────────────────────────────( )──

<< RUNG 156 >>

Travers  Destntn   Hoist   E/Stop  Travers  Hoist   Hoist
left't   -1.Stop   at               right   raise   lower
request  at next   Top              request flag    flag
 O0500    O0502    I0039   I0080    O0501   O0504   O0505    Travers
──]/[──────]/[──────]/[────]/[──────]/[──────]/[──────]/[──   Left
                                                              Demand.
 On      Destntn   Hoist                                      O0002
Station  -1.Stop   At                                        ─( )──
 Prox.   at next   Bottom
 I0038    O0502    I0040
──]/[──────]/[──────]/[──

Manual
TrvLft
Request
 O0341
──]/[──
```

MOMOMOM — Target for MCR WAH
2020202 — Target for MCR TG
1111111 — Target for MCR DAT

##
This rung has the same function as the rung
above, but, is used for traversing right.
##

Plating   GE SERIES SIX DOCUMENTATION
Line Automatic Control Ladder Listing
Instrumentation Systems Limited

```
<< RUNG  157 >>

Travers  Destntn  Hoist   E/Stop   Travers  Hoist   Hoist                            Travers
 right   Hi-Stop   at               left    raise   lower                              Right
request  at next   Top             request  flag    flag                              Demand
 00501    00503   I0039   I0080     00500   00504   00505                              00001
--] [----]/[-----]/[------]/[------]/[------]/[------]/[-----------------------------------( )--
    On    Destntn  Hoist
  Station Hi-Stop   At
   Prox.  at next  Bottom
   I0038    00503  I0040
--] [------]/[-----]/[--
  Manual
  Trv Rht
  Request
   00342
--] [--
```

Plating Line Automatic Control Ladder Listing
GE SERIES SIX DOCUMENTATION
Instrumentation Systems Limited

```
***********************************************************
##########################################################
## The next twenty three (23) rungs form the first started flightbar
## process handler.
##########################################################

<< RUNG 279 >>

 FB1
 Initlsd
 Ready
 00741      Const
--]/[------[ MCR ]-
             003

<< RUNG 280 >>

 FB1
 Data
 Posted
 00724      Const
--] [------[ MCR ]-
             001
 FB1
 Confirm
 Turned
 00854
--] [

<< RUNG 281 >>

 FB1             FB1     FB1            FB1     FB1           FB1
 Tank            Table   Prirty         Table   Prirty        Table
                 Source  High           High    Low           Low
 R00465   MOVE   R00578  R00471         R00546  R00477   MOVE  R00562
-[ A    ]-[ B ]-[ A     MOVE  [ B ]-[ A      [ B ]-[ A          [ B ]-            ( )

<< RUNG 282 >>

 FB1
 Dest
 Station
 R00498   MOVE   R00594
-[ A    ]-[ B ]-

<< RUNG 283 >>

 FB1      Test    FB12          FB1
 Buff1    Reg=    Buff2         Table
 Prirty   Reg?    Destn         Top
 R0500 :  R0802   R00738        R00610
-[ A    ]-[ B ]-[ A    MOVE     [ B ]-            ( )

<< RUNG 284 >>

 FB1
 Initlsd
 Ready
 00741                                            00723
--] [                                            -( )-
```

GE SERIES SIX DOCUMENTATION
Plating Line Automatic Control Ladder Listing
Instrumentation Systems Limited

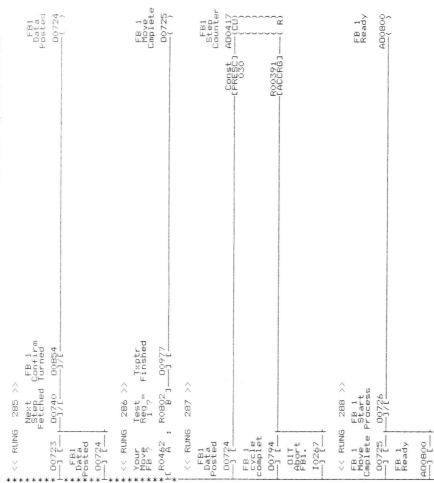

```
<< RUNG 285 >>

       Next      FB 1                                         FB1
       Step      Confirm                                      Data
       Fetched   Turned                                       Posted
       00723     00740     00854                              00724
    ---] [-------] [-------]/[-------------------------------( )

       FB1
       Data
       Posted
       00724
    ---] [

<< RUNG 286 >>

       Your      Test      Txptr                              FB 1
       Move      Reg =     Finshed                            Move
       FB?       1 ?                                          Cmplete
              R0462 :  R0802    00977                         00725
    ---[ A   :   B  ]-----]/[-------------------------------( )

<< RUNG 287 >>

       FB1                                                    FB1
       Data                                                   Step
       Posted                                                 Counter
       00724                                          Const   A00417
    ---] [                                           [PRESC] ---(CU)
                                                      030

       FB 1
       cycle
       complet
       00794
    ---] [                                           R00391
                                                    [ACCRG] ---( R)

       OIT
       Abort
       FB1.
       I0267
    ---] [

<< RUNG 288 >>

       FB 1      FB 1                                         FB 1
       Move      Start                                        Ready
       Cmplete   Process
       00725     00726                                        A00800
    ---] [-------]/[-----------------------------------------( )

       FB 1
       Ready
       A00800
    ---] [
```

****

Plating Line Automatic Control Ladder Listing
GE SERIES SIX DOCUMENTATION
Instrumentation Systems Limited

<< RUNG 289 >>

```
         FB 1        Next           FB 1
         FB1         Step           Start
         Data        Fetched        Process
         Posted
         Ready
A00800   O0724       O0740                            O0726
—] [——————] [————————]/[——————————————————————————————( )—

         FB 1
         Start
         Process
         O0726
——————] [——
```

GE SERIES SIX DOCUMENTATION
Plating Line Automatic Control Ladder Listing
Instrumentation Systems Limited

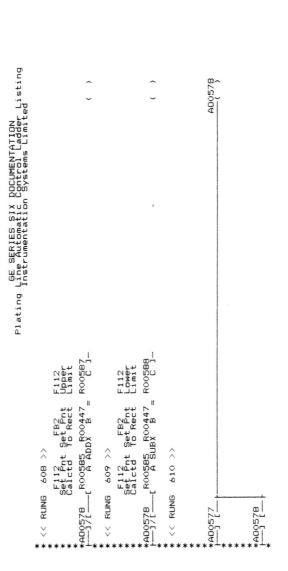

```
*********
<< RUNG  608 >>

        F112      FB2       F112
       Set Pnt  Set Pnt    Upper
       Calctd   To Rect    Limit
*A00578   R00585  R00447 = R00587
|—]/[————[  A ADDX  B  =      C ]—          (   )
*********
<< RUNG  609 >>

        F112      FB2       F112
       Set Pnt  Set Pnt    Lower
       Calctd   To Rect    Limit
*A00578   R00585  R00447 = R00588
|—]/[————[  A SUBX  B  =      C ]—          (   )
*********
<< RUNG  610 >>

                                         A00578
*A00577                                  (  )
|—]  [—————————————————————————|
*********
*A00578
|—]  [—————————————————————————
*—*
```

3 USER PROGRAMMING DOCUMENTS FOR GEC GEM 80

10.10.1 GEM 80 Programming Sheet

This blank form is for GEM 80 rung ladder diagram program development. It allows space for up to 10 series elements across the page and up to two five branch rungs on each page with comments.

10. USER PROGRAM

10.10.2 GEM 80 Data Table Location Usage Record

This blank form is for the programmer to fill in as the GEM 80 rung ladder diagram program is developed to record the data table addresses used, their purpose, and mnemonics.

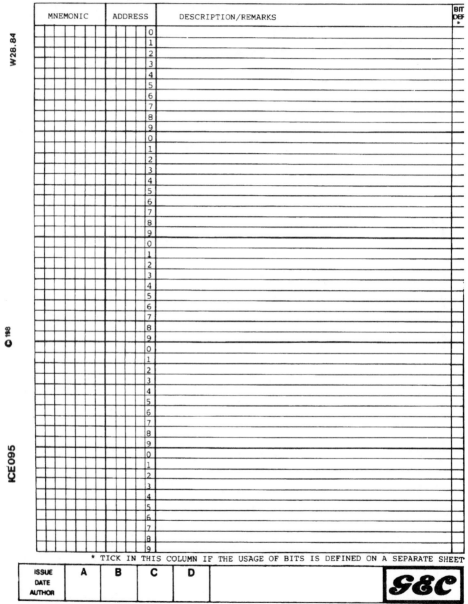

* TICK IN THIS COLUMN IF THE USAGE OF BITS IS DEFINED ON A SEPARATE SHEET

ISSUE	A	B	C	D	
DATE					
AUTHOR					

References and Further Reading

References

CAPIEL, (1982), Draft Standard 65A/WG6 82.09.20, cited by Davie (1985).

Davie, I. D., (1985), 'The PC user's workstation', *PC '85 Conference Proceedings*, Peter Peregrinus Ltd, No. 25.

Hill, R. and Jones, H., (1985), 'The application of programmable controllers to automobile body assembly', *PC '85 Conference Proceedings*, Peter Peregrinus Ltd, No. 25.

Skrokov, M. R., (1980), *Mini and Microcomputer Control in Industrial Processes*, Van Nostrand Reinhold Company.

Yeomans, R. W., Choudry, A., ten Hagen, P. J. W., (1985), *Design Rules For a CIM System*, North Holland.

Yohansson, G., (1985) *Programmable Controllers – an Introduction*, SattControl.

Further reading

Conference Proceedings for PC '85 and PC '86, (1986), Peter Peregrinus Ltd. (Wide range of papers covering P.C. technology and applications.)

Dorf, R. C., (1967), *Modern Control Systems*, Addison–Wesley. (Comprehensive text on control strategies and systems.)

Hollingum, J., (1986), *The MAP Report*, IFS Publications Ltd. (A thorough text that progressively covers all aspects of the 'Manufacturing Automation Protocol', including future developments and product testing.)

Horowitz, P. and Hill, W., (1980), *The Art of Electronics*, Cambridge University Press. (Excellent text and source of information on analog and digital electronics. Highly readable.)

Loxton, R. and Pope, P., (1986), *Instrumentation: a Reader*, Open University Press. (A detailed coverage of several recently developed transducers and their application. Includes useful papers on communications and measurement in process control.)

Morgan, Eric, (1987) *Through MAP to CIM*, Department of Trade and Industry. (A 24-page booklet providing an excellent introduction to computer integrated manufacture (CIM) and the role of MAP in its achievement.)

Programming manuals for GE, GEC, Mitsubishi, SattControl, Square D, etc. (Provide full details of all PC facilities and programming methods for a given model.)

Schmitt, N. M., and Farwell, R. F., (1983), *Understanding Electronic Control Systems*, Texas Instruments Learning Center. (An easily read text covering electronics control, including microprocessor-based systems and PLC applications.)

Tannenbaum, A. S., (1981) *Computer Networks*, Prentice Hall. (A comprehensive text on network communications and open systems (pre MAP).)

Zaks, R. (1980) *Microprocessors – from chips to systems*, Sybex. (Broad introductory textbook.)

Index